Winward Fearon
on
Collateral Warranties

Winward Fearon
on
Collateral Warranties

SECOND EDITION

David L Cornes
BSc(Eng), AKC, FICE, CEng, FCIArb

Solicitor of the Supreme Court
Partner, Winward Fearon

Richard Winward
LLB, FCIArb, F.Inst.CES

Solicitor of the Supreme Court
Partner, Winward Fearon

Blackwell
Science

© David L. Cornes & Richard Winward 2002

Blackwell Science Ltd, a Blackwell Publishing
Company
Editorial Offices:
Osney Mead, Oxford OX2 0EL, UK
    Tel: +44 (0) 1865 206206
Blackwell Science, Inc., 350 Main Street, Malden,
MA 02148-5018, USA
    Tel: +1 781 388 8250
Iowa State Press, a Blackwell Publishing
Company, 2121 State Avenue, Ames, Iowa 50014-
8300, USA
    Tel: +1 515 292 0140
Blackwell Science Asia Pty, 54 University Street,
Carlton, Victoria 3053, Australia
    Tel: +61 (0)3 9347 0300
Blackwell Wissenschafts Verlag,
Kurfürstendamm 57, 10707 Berlin, Germany
    Tel: +49 (0)30 32 79 060

First Edition published 1990
by BSP Professional Books
Second Edition published 2002
by Blackwell Science Ltd

Library of Congress
Cataloging-in-Publication Data
is available

ISBN 0-632-03896-9

A catalogue record for this title is available from
the British Library

Set in 10/12pt Palatino
by DP Photosetting, Aylesbury, Bucks
Printed and bound in Great Britain by
MPG Books Ltd, Bodmin, Cornwall

For further information on
Blackwell Science, visit our web site:
www.blackwell-science.com

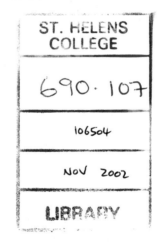

Contents

Preface   viii

1 Principles of Law   **1**
Definition of collateral warranty   1
A comparison of contract and tort   4
Essentials of a contract   5
Form of contract   20
Construing a contract   22
Implied terms   27
Letters of intent   28

2 The Rise of Collateral Warranties   **31**
Negligence   31
1932 to 1988   33
D & F Estates Limited and Others v. *Church Commissioners for*
    England and Others   36
Murphy v. *Brentwood District Council*   38
1990 to 2000   41

3 Contracts (Rights of Third Parties) Act 1999   **47**
The background   47
Contracts (Rights of Third Parties) Act 1999   50
Aspects of the Act   56
Using the Act or not using the Act?   59

4 Assignment and Novation   **64**
Future purchasers and tenants   64
Assignment   64
Prior equities   70
Restrictions on assignment   71
Novation   80

5 Reasonable Skill and Care and Fitness for Purpose   **83**
Reasonable skill and care   83
Fitness for purpose   88
Dwellings   93

6	**Damages and Limitation of Action**	**94**
	Damages	94
	Expectation interest and reliance expenditure	108
	Mitigation and assessment	112
	Assignment	117
	Contribution and apportionment	125
	Limitation of action	129
7	**Developers, Tenants, Purchasers and Funds**	**134**
	The position of a developer	134
	The position of a tenant	137
	The position of a purchaser	143
	The position of the funding institution	145
	Obligations to enter into collateral warranties	147
	JCT enabling clauses	149
8	**Insurance Implications**	**153**
	Principles of professional indemnity insurance	153
	Disclosure of collateral warranties	155
	Particular insurance problems	157
	Other matters of concern to insurers	161
	Policy endorsements for collateral warranties	162
	Problems on changing insurers	166
9	**Typical Terms**	**167**
	General considerations	167
	Typical terms	174
	Contractors and sub-contractors	199
	Guarantees of obligations under warranties	200
10	**Practical Considerations**	**201**
	Does a warranty have to be given?	201
	Commercial balance	203
	Legal costs and consideration	203
	Negotiating and insurance	204
	Warranties must be executed	204
	The givers, receivers and contents of warranties	205
	Standard forms of collateral warranty	207
	Commentaries	211
11	**Other Solutions: Present and Future**	**228**
	Possible solutions – the present	228
	Possible solutions – the future	234
Appendix 1	Housing Grants, Construction and Regeneration Act 1996	237

Appendix 2 The Law Commission Report No. 242: Privity of
Contract: Contracts for the benefit of Third Parties   245

Appendix 3 Contracts (Rights of Third Parties) Act 1999   250

Appendix 4 CoWa/F: The BPF, ACE, RIAS, RIBA, RICS, Form of
Agreement for Collateral Warranty for use where a
warranty is to be given to a company providing
finance for a proposed development (Third Edition,
1992)   256

Appendix 5 CoWa/P&T: The BPF, ACE, RIAS, RIBA, RICS, Form
of Agreement for Collateral Warranty for use where
a warranty is to be given to a purchaser or tenant of
premises in a commercial and/or industrial
development (Second Edition, 1993)   263

Appendix 6 MCWa/F: The JCT Standard Form of Agreement for
Collateral Warranty for use where a warranty is to
be given to a company providing finance (funder)
for the proposed building works by a contractor
(2001 Edition)   269

Appendix 7 MCWa/P&T: The JCT Standard Form of Agreement
for Collateral Warranty for use where a warranty
is to be given to a purchaser or tenant of building
works or part thereof by a main contractor
(2001 Edition)   281

Appendix 8 MCWa/F/Scot(Funder): The Scottish Building
Contract Committee Form of Agreement for
Collateral Warranty for use where a warranty is to
be given by the main contractor to a company
providing finance (funder) for proposed building
works let or to be let under certain SBCC main
contract forms (November 2001)   291

Appendix 9 MCWa/P&T/Scot(Purchaser and Tenant): The
Scottish Building Contract Committee Form of
Agreement for Collateral Warranty for use where
a warranty is to be given by the main contractor to
a purchaser or tenant of the whole of part of the
building(s) comprising the works which have
been practically completed under certain SBCC
main contracts forms (November 2001)   306

Table of Cases   320

Table of Statutes & Statutory Instruments   331

Index   334

Preface

Since the first edition of this book, there have been some immensely important developments in the law relating to collateral warranties and the parties who make use of them. Those developments are in statute, case law and practice.

As to statute, in 1999 Parliament brought into being a radical new Act: The Contracts (Rights of Third Parties) Act. It has created, for the first time in English law, the possibility, in certain circumstances, that parties who are not themselves parties to contracts can enforce those contracts for their own benefit. Prior to that Act, usually, only a party to a contract could sue and be sued on the contract. That ancient rule of privity of contract has, therefore, been very substantially and fundamentally amended. There are some lawyers who are now beginning to think that the rights created by this new Act may enable some collateral warranties to be avoided. These developments are so important that the whole of Chapter 3 is devoted to it in this second edition, including a discussion about the possible practical uses of the Act on development projects.

As to case law, the House of Lords have handed down some very important decisions in recent times, expanding and explaining the extent of 'third party remedies' in the absence of or indeed the replacement of, the contractual solution and clarifying the principles relating to assessment of damages on assignment. These cases are *Linden Gardens Trust* v. *Lenesta Sludge Disposals, St Martins Property Corporation* v. *Sir Robert McAlpine & Sons, Alfred McAlpine Construction Ltd* v. *Panatown, Henderson* v. *Merrett Syndicates* and *White* v. *Jones*. These cases deal with the concepts of recovery of third party loss and a modern statement of the elements of tortious duty: the assumption of responsibility. We have also considered cases such as the House of Lords decision in *Investors Compensation Scheme Limited* v. *West Bromwich Building Society*, which set out the current judicial thinking on the formation and interpretation of contract.

As to practice, people involved in collateral warranties have a much better understanding of the issues. Collateral warranties are being used, not just as contracts that are collateral to other contracts, but to create primary contractual obligations. An example of that approach is the now almost universally used right, created in collateral warranties, for the benefit of a funder of a project to 'step-in' to the project as principal in the event of the developer being in default of the funding agreement. There is also now an increasing trend for the givers of warranties to have their obligations under the warranties guaranteed by their parent companies.

Since the mid 1980s, the use of collateral warranties has been growing in the United Kingdom and that growth rapidly accelerated after the House of Lords decision in *D. & F. Estates Ltd & Others* v. *The Church Commissioners for England & Others* in 1989 and *Murphy* v. *Brentwood District Council* in 1990. Those cases have had the effect of substantially removing the prospects of tenants and purchasers bringing successful actions in tort (and in the absence of contract) in respect of the consequences of defective design and construction on projects.

One of the side effects of the enormous growth in the use of collateral warranties has been the attempt by each of the parties involved to try to have their own interests dominate in the negotiation process. Inevitably, this has produced a plethora of non-standard forms of warranty, some as short as one page and some occasionally reaching 40 pages or more.

Those vested interests have made the production of widely acceptable standard forms of warranty very burdensome for those who have tried. At the time of writing the first edition of this book, there was only one standard form in England and Wales, CoWa/F. That form took three years to agree. The eleven years between the first and second editions of this book have witnessed the publication by representative bodies of the construction industry of new standard form warranties for different procurement routes, the latest being the forms published in the autumn of 2001 by the JCT for main contractors and subcontractors. A detailed commentary on these forms is set out in Chapter 10.

There is one particular issue, which we have not addressed in this book, namely incorporating a power of attorney into a contract to enable the employer to execute a collateral warranty himself where the other party fails to do so. Commercial issues aside, the legality and effect of such powers is a thorny issue, and, rather than attempt to rehearse the arguments here, we felt that our readers would be better served consulting specialist literature on the subject, in particular, Aldridge, *Powers of Attorney* (8th edn, 1991).

Our purpose in writing this second edition, as it was with the first edition, has been to try to explain the law relating to collateral warranties, the developments in the law of privity of contract and the issues that arise in drafting and considering collateral warranties and third party rights, so as to demystify the subject and to clear away some myths. We hope that we have succeeded. Whilst we have sought to explain the law and the issues as we see them, a book is no substitute for good legal advice in the area of collateral warranties and third party rights.

We are grateful to Lucy Welsh and Kay Oliver who have done much of the typing for this second edition, and to all those who encouraged us to embark on this second edition. In particular, we thank Julia Burden, our commissioning editor from Blackwell Publishing, who has been inordinately patient with us during the lengthy preparation of this book. We are also grateful to Emma Le Breton for her input to the research and to Tom Wrzesien, Chloe Shanely and Nicholas Lane of Winward Fearon who also did research. Thanks are also due to our partner, Michael Harlow of Winward Fearon, for his input on landlord and tenant law;

Brandon Nolan of McGrigors, Solicitors, Edinburgh, for his assistance on some aspects of the law of Scotland; Bob Ferguson of AON Limited, insurance brokers, for his input on some of the insurance aspects of collateral warranties; Paul Miller of Morgan Cole for permission to make use of his letter dated 3 December 2001 in relation to Rockwool Limited and their mineral wool product; and Charmian Martin of The Wren Insurance Association Limited for some input on that mutual insurance company's approach to collateral warranties. Notwithstanding all that input, the views expressed in the book are our views.

We are pleased to acknowledge the kind permission of the following, who own the copyright, to reproduce their publications: The British Property Federation, The Association of Consulting Engineers, The Royal Incorporation of Architects in Scotland, The Royal Institute of British Architects and the Royal Institution of Chartered Surveyors to reproduce the Forms of Agreement for Collateral Warranty (CoWa/F and CoWa/P&T); the Joint Contracts Tribunal to reproduce Standard Forms of Agreement for Collateral Warranty (MCWa/F and MCWa/P&T); the Scottish Building Contract Committee to reproduce Standard Forms of Agreement for a Collateral Warranty by a main contractor (MCWa/F/Scot(Funder) and MCWa/P&T/Scot(Purchaser and Tenant)); and Butterworth & Co (Publishers) Limited who kindly gave permission to reproduce some extracts from Volume 22 of *The Encyclopaedia of Forms and Precedents, Landlord and Tenant: Business Tenancies*, which sets out some examples of typical repairing covenants in leases.

We have tried to state the law in England and Wales as at 1 January 2002. We have derived considerable assistance from the cases from Scotland and Northern Ireland that are referred to in the book; we anticipate that readers from Scotland and Northern Ireland, where the practice in relation to collateral warranties is very similar to England and Wales, will find the text very useful.

Winward Fearon   David L. Cornes
35 Bow Street   Richard Winward
London WC2E 7AU   February 2002

Chapter 1
Principles of Law

Definition of collateral warranty

1.1 Construction projects by their very nature and also by their method of procurement create complex legal relationships between the many parties involved in the design and construction process. Further, the finished product has, or should have, a life expectancy calculated not in years but in tens of years giving rise to a class of future owners who had no involvement or control over the original contract works but who may have substantial liabilities if things go wrong. It is because of these characteristics that tort was a useful remedy permitting a much wider forensic examination of liability and a greater potential for remedy and indeed disputes. Tort also permitted a more flexible apportionment of blame amongst the parties involved in the actual construction process, i.e employer, architect, engineer, main contractor, sub-contractor or supplier, sub-sub-contractor and the building control authority. The use of tort as a remedy in construction cases has been substantially eroded for the reasons set out in Chapter 2. In its place there is now widespread use of a contractual remedy in the form of collateral warranties.

1.2 In contract the role of the court is restrictive unlike tort where its role is creative. In contract, subject to statutory exceptions, the basic role of the court is to police a set of rules to ensure that all the component parts of a binding contract have been satisfied; the nature and substance of the bargain between the parties is a matter of private agreement provided of course that the bargain is not illegal or contrary to public policy. In tort the duty and standard of care is created by the courts and is not a matter of private bargain.

1.3 A basic rule of contract is the doctrine of privity of contract. This doctrine states, as a general rule, that only a party to a contract can take the benefits of that contract or is subject to its burdens or obligations. For example, if A promises to B to pay a sum of money to C, as a general rule, C cannot enforce that obligation against A. However, this rule has been fundamentally changed by the Contracts (Rights of Third Parties) Act 1999 (see Chapter 3).

1.4 A collateral warranty has been, and probably remains notwithstanding the Contracts (Rights of Third Parties) Act, one of the ways of overcoming the restriction on remedies created by the doctrine of privity. In law the term collateral warranty has several meanings. It may mean a warranty or representation which is collateral to the main transaction: *De Lassalle* v.

Guildford. L had agreed the terms of a lease with G. However L refused to complete the transaction unless and until he had been given an assurance by G that the drains of the property were in good order. G gave an appropriate verbal assurance and L completed the lease document. After going into possession of the property, L found that the drains were defective. He was not able to bring an action under the terms of the lease; however he claimed damages against G on the basis of a collateral warranty created by G's verbal assurances. L's action succeeded.

1.5   A collateral warranty which is collateral to the main transaction will give an additional cause of action to one of the parties to that transaction but does not introduce a third party. However the term collateral warranty has a further meaning which is more apposite to the subject matter of this book, namely a binding contract entered into between B and C which is collateral to a contract already in existence between A and B whereby B promises to C that he will perform his obligations to A. If B is in breach of those obligations to A then C will have a right of action against B. This collateral warranty or collateral contract may be imputed, that is to say arising impliedly as a matter of fact from the circumstances of a particular case, or by the parties entering into a specific, usually written contract. A further species of collateral warranty is a unilateral undertaking entered into by one party in favour of another, the document being under seal.

1.6   In *Shanklin Pier Ltd* v. *Detel Products,* S who were owners of a pier entered into a contract with G.M. Carter (Erectors) Ltd for the repair and repainting of the pier, the repainting to be carried out with two coats of bitumastic or bituminous paint. Under the terms of their contract with Carter, S reserved the right to vary the specification. D were paint manufacturers who produced a product known as DMU which they represented to S as being suitable for the repainting of the pier in that its surface was impervious to dampness and could prevent corrosion and creeping of rust with a life of 7–10 years. In consideration of this representation S specified to Carter that Carter should use D's paint. The paint proved to be a failure and S sued, not Carter as main contractor, but D as supplier of the paint. D argued that they had not given any such representation or warranty and even if they had it did not give rise to a cause of action because they were not parties to the contract for repainting the pier. The court rejected this argument and held that on the facts there was a warranty and that the consideration for the warranty was that S should cause C to enter into a contract with D for the supply of their paint for repainting the pier. The representations given by D to S were contractually binding in the form of a collateral contract, which arose by imputation.

1.7   In the *Shanklin* case, D was aware that their product was to have a specific use by a third party. This was not the situation in *Wells (Merstham) Ltd* v. *Buckland Sand and Silica Co Ltd*. B were sand merchants and they warranted to W who were chrysanthemum growers that their sand conformed to a certain analysis that would be suitable for the propagation of chrysanthemum cuttings. W ordered a load of sand direct from B and a further two loads via a third party who were builders merchants and not horticultural suppliers, that is to say they supplied sand for various purposes. The third party purchased two loads of sand from B for onward sale and delivery to W and gave no indication to B that the sand was required for W or for horticultural purposes. The sand did not conform with the analysis and as a result W suffered a loss in the propagation of their young chrysanthemums. The court held that B was liable on the basis of a collateral contract and that it was irrelevant that the purchase of the sand had been made through a third party; as between a potential seller and a potential buyer only two ingredients were required to bring about a collateral contract:

(1) a promise or assertion by the seller as to the nature, quality or quantity of the goods which the buyer might regard as being made with a contractual intention; and
(2) the acquisition by the buyer of the goods in reliance on that promise or assertion.

1.8   In contrast to the decisions in *Shanklin* and *Wells*, the court in *Drury* v. *Victor Buckland Limited* rejected the plaintiff's claims against the defendant. In this case D was approached by an agent of V in order to persuade D to purchase an ice cream maker. D had doubts as to whether she could afford the equipment; however she was persuaded to pay a 10 guinea deposit to V and to discharge the balance of the purchase price by way of instalments to a finance company. The full terms of the transaction were set out on an invoice sent to D by V; however, the deal was put through by V selling the machine to a finance company who in turn entered into a hire purchase agreement with D. The machine proved unsatisfactory and D brought proceedings against V on the basis that V was in breach of the implied warranties or conditions as to quality imposed by the Sale of Goods Act 1893. The court held that there was no contractual relationship between D and V. From the case report, it does not appear that D tried to argue that there was a collateral contract. There was evidence before the court that 'in the course of one or other of the interviews in answer to her question "suppose it goes wrong?" he (the defendant's agent) said that if it did she was to ring them up and that she had no need to worry about it as they would service it for 18 months free'. At the very least it would appear arguable to the authors that those words could have constituted a collateral warranty giving rise to an obligation between D and V.

1.9   A helpful description of the nature of a collateral contract was given by Lord Moulton in the case of *Heilbut Symons & Co* v. *Buckleton* when he stated:

'It is evident, both on principle and on authority, that there may be a contract the consideration for which is the making of some other contract. If you will make such and such a contract I will give you £100 is in every sense of the word a complete legal contract. It is collateral to the main contract but each has an independent existence and they do not differ in respect of their possessing to the full the character and status of a contract.'

1.10   Lord Moulton placed emphasis on the word 'contract'. In the cases of *Shanklin* and *Wells* the courts found that collateral contracts arose by implication from the facts of each case. In the construction industry collateral contracts will invariably arise by the acts of the parties entering into a specific agreement. Regardless of whether the contract arises by implication or by specific agreement, a basic understanding of the rules as to the formation and construction of a contract is an essential prerequisite to understanding the problems arising from collateral warranties.

1.11   In the case of *Inntrepeneur Pub Company Ltd* v. *East Crown Ltd* the court held that an entire agreement clause did not merely render evidence of the giving of the collateral warranty inadmissible but it also deprived what would otherwise have been a valid collateral warranty of its legal effect. The entire agreement clause was worded as follows:

> 'Any variation of this agreement which are [sic] agreed in correspondence between the parties shall be incorporated in this agreement where that correspondence makes express reference to this clause and the parties acknowledge that this agreement (with the incorporation of any such variations) constitutes the entire agreement between the parties.'

1.12   The provision of a contractual network of collateral warranties for a typical building project creates a network of considerable complexity. Figure 1.1 sets out in diagrammatical form the contractual arrangements arising from an unmodified JCT 98 Standard Form of Building Contract (without contractor's design). By comparison Fig. 1.2 sets out a typical arrangement after the parties have entered into the collateral warranties that are now commonly sought in respect of a building development.

A comparison of contract and tort

1.13   The classic nineteenth century definition of contract is 'a promise or set of promises which the law will enforce': Pollock, *Principles of Contract* (13th edition). That is to say there is reciprocity of undertaking passing between the promisor and the promisee. Tort is generic in nature and therefore more difficult to define. Basically tort is a collection of civil law remedies entitling a person to recover damages for loss and injury which have been caused by the actions, omissions or statements of another person in such circumstances that the latter was in breach of a duty or obligation imposed at law.

1.14   In contract the rights and obligations are created by the acts of agreement between the parties to the contractual arrangement. In tort the rights and

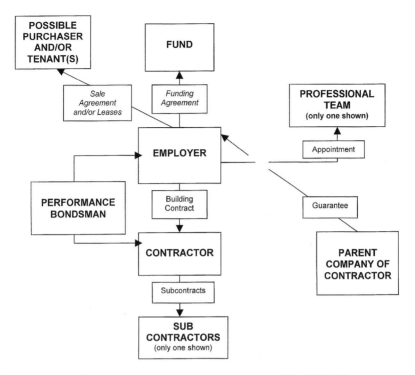

Fig. 1.1 Contractual arrangements arising from an unmodified JCT 98 contract.

obligations are created by the courts applying the common law, which has on the basis of previous authority fallen into three distinct categories: negligence, nuisance and trespass. Historically, actions in contract and in tort derived from the same source – trespass – compared with actions for breach of a deed, which were based upon an action on the covenant. Actions for breach of contract were based on *assumpsit* and actions in tort were *ex delicto*. In the seventeenth century the courts began to draw procedural but not substantive distinctions between *assumpsit* and actions *ex delicto*. These distinctions became substantive differences during the nineteenth century, reflecting the political social and economic philosophy of 'laissez-faire' which emphasised the importance of the legal doctrines of freedom of contract and sanctity of contract.

1.15   The area of tort, which recently has been exhaustively considered by the courts, is the law of negligence. The nature of this tort and the present state of the law, as the authors understand it, are dealt with in Chapter 2.

Essentials of a contract

1.16   Subject to the provisions of the Housing Grants, Construction and Regeneration Act 1996 construction contracts are governed by the

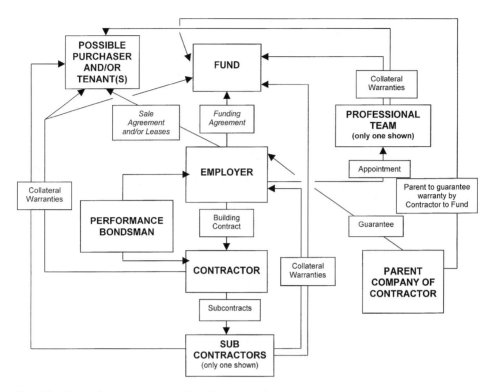

Fig. 1.2   Typical arrangements after the parties have entered into collateral warranties.

ordinary rules of formation and interpretation. What distinguishes construction contracts from other types of contract is their factual complexity and the widespread use of standard form contracts. However these characteristics merely increase the burden of forensic analysis rather than changing the rules of such analysis.

1.17   Ignoring for the moment the effect of the Contracts (Rights of Third Parties) Act 1999, which is dealt with in Chapter 3, there are four essentials of a contract:

(1)   Two or more parties; and
(2)   An intention to create legal relations; and
(3)   An agreement; and
(4)   Consideration

Parties

1.18   There must be two or more parties present to create a contractual obligation. This statement may seem axiomatic. However, the law defines party by reference to legal capacity as well as physical existence. For example, if there is a parent company with subsidiary companies then contracts can be made between the parent and the subsidiary and between the various subsidiaries provided of course they are all registered companies. However, if a company operates by way of a divisional structure then the various divisions do not have a legal capacity to enter into contracts. Similarly in *Henderson* v. *Astwood* the court held that a mortgagee who was selling a property at auction under his power of sale in the mortgage document was not entitled to bid at the auction as otherwise he would be selling to himself.

1.19 If you are drunk, insane, bankrupt, an enemy alien or a minor, then your legal capacity will be impaired. The authors hasten to add that these impairments are stated as alternatives and are not cumulative!

Intention to create legal relations

1.20 Even though the other essentials of a contract are present the courts may consider that a promise is unenforceable if the parties are not *animo contrahendi*, i.e did not intend to create legal relations. There is a presumption in law that domestic or social arrangements are not intended to be legally binding. For example, if A agrees to take B, his son, to the cinema provided that B tidies his room, there is a promisor A and consideration moving from the promisee B; nevertheless the presumption in law will be that no legally binding obligation comes into existence. The presumption is rebuttable depending upon the facts of each particular case.

1.21 Clearly in commercial transactions there is no such presumption although there may be particular circumstances where the parties are not *animo contrahendi*. The House of Lords considered this point in *Independent Broadcasting Authority* v. *EMI Electronics Limited and BICC Construction Ltd*. This case concerned the collapse of a cylindrical steel area mast belonging to the IBA, at Emley Moor in Yorkshire. The mast was nearly a quarter of a mile high with a diameter at its base of only 9 feet. The mast had been designed, supplied and erected by BICC as nominated sub-contractors to EMI who were main contractors to IBA in respect of works, which included the aerial mast. The collapse of the mast was caused by a tension fracture of a flange in a leg of a lattice at 1027 feet. The lattice failure had been caused primarily by vortex shedding, also to a lesser extent by asymmetric ice loading. IBA alleged this was a design fault and proceeded against EMI for damages for breach of contract and negligence and against BICC for damages for negligence, breach of warranty and negligent mis-statement.

1.22 The claim against BICC for the breach of warranty arose from the following circumstances. The Emley Moor mast was one of a series of three cylindrical masts being designed and erected by BICC. Other masts were being erected at Belmont and Winter Hill. Work had started first at the Winter Hill site and on 16 October 1964 when the mast had reached a height of 851 feet it began to oscillate with the result that the labour putting up a building at the foot of the mast left the site and refused to return to it. IBA raised their concerns to BICC by a letter dated 23 October in which IBA expressed concern that very little was known about how such cylindrical structures were likely to behave under certain wind conditions. IBA suggested the matter be investigated fully by taking appropriate readings from instruments attached to the Winter Hill mast in order to confirm the data upon which the design calculations had been based. BICC replied by a letter dated 11 November 1964 when they stated, inter alia, 'I think we will be extremely fortunate if the oscillations at Winter Hill keep within their present limits when the structure is completed. No doubt we shall learn from experience how to overcome these difficulties and I think it should be realised by all concerned that we have achieved something unique in the design of the 1250 ft mast. We expected problems with aerodynamic instability as this phenomenon is well known with cylindrical structures. *However we are well satisfied that the structures will not oscillate dangerously...*' IBA alleged that the words in italics amounted to a contractual warranty. This argument was accepted by the judge at first instance but rejected by the Court of Appeal. Consideration of this particular issue was not necessary for the House of Lords decision; however their Lordships considered the matter and disagreed with the Court of Appeal. The House of Lords rejected the finding that the assurance given in BICC's letter dated 11 November 1964 was a contractual warranty on the basis that there was no evidence that, at the time when it was given, either IBA or BICC intended the letter to create a contractual undertaking. Viscount Dilhorne stated:

> 'In the statement of claim it was alleged that this assurance was a warranty and that in consideration of it IBA forbore from requiring further investigations and from seeking independent advice as to the stability of the mast... In the present case I can find nothing which can by any possibility be taken as evidence that [IBA] when they wrote their letter on 20 October or thereafter had any intention of entering into a contract or that [BICC] when they gave the assurance had any intention of undertaking a contractual obligation.'

On the same issue Lord Fraser of Tullybelton stated:

> 'They [i.e the judge at first instance and the Court of Appeal] treated the letter [dated 11 November 1964] as containing a promise for which they held that ample consideration had been given by IBA in forbearing from insisting on further investigation of the oscillation problem. But I am unable to agree that the letter is capable of being read as containing a promise. The most important sentence in the letter says "We are well satisfied that..." and that expression does not in its terms bear to be a promise or offer. Nor do I think that the circumstances show that [BICC] intended to make a contract. It must be remembered that there

was at that time no direct contract between IBA and BICC to which a warranty could be collateral, so if there was a warranty with contractual force, it can only have been because a separate independent contract on the matter was made and that would require *animus contrahendi* on both sides.'

1.23   It is to be noted that the representation in the IBA case was made after the date of the contract for the construction of the Emley Moor television mast. That is to say the representation was not collateral to the main contract; the contractual result may have been different if the representation had been made before or at the time of the Emley Moor contract and the court may have found that there was a collateral contract.

1.24   By way of contrast to the decision in *IBA* is the case of *Edwards* v. *Skyways Ltd*. E was an aircraft pilot employed by S and a member of S's contributory pension fund which entitled him on leaving the defendants' employment in advance of retirement age to a choice of either withdrawing the sum of his own contributions to the fund or taking the right to a paid up pension payable at retirement age. In January 1962 some 15% of S's pilots, including E, were made redundant. The Pilots' Union BALPA agreed with S (as recorded in the notes of the meeting) that 'pilots declared redundant and leaving [the defendant company] would be given an *ex gratia* payment equivalent to the defendant company's contributions to the Pension Fund'. On the evidence at trial, it appeared that this was an incorrect record of what was agreed as S's representative had said at the meeting that the company would make *ex gratia* payments 'approximating to' S's contribution. E elected to withdraw his contributions to the pension fund and to receive the *ex gratia* payment. S paid the contribution but refused to pay the *ex gratia* payment contending that the recorded agreement was not intended to create legal relations and was too vague and thus was not legally binding. The court held that where there was an agreement and the subject of the agreement related to business affairs, the onus of establishing that the agreement was not intended to create legal relations, (which was on the party setting up that defence) was a heavy onus. S had failed to discharge this burden for the following reasons: the words *'ex gratia'* were used simply to indicate that the party agreeing did not admit any pre-existing liability on S's part and the mere use of the phrase *'ex gratia'* as part of a promise to pay did not show that the promise when accepted should have no binding effect in law. Further the use of the words 'approximating to' did not render the terms of the agreement too vague to be enforceable for at most the phrase would denote on the evidence a rounding off of a few pounds downwards to a round figure.

1.25   This issue was further considered by the Court of Appeal in the case of *Orion Insurance Co plc* v. *Sphere Drake Insurance plc*, the facts of which were concerned with the operation of an insurance pool agreement establishing a no gain no loss underwriting agreement between insurers. Sphere got into financial difficulties and at a meeting, being one of numerous meetings being held between the parties, appeared to agree to a disposal of all liabilities under the pool agreement on the basis that the agreement to the disposal was 'a goodwill agreement and not a legal contract'. Despite the use of those words the presence of a contract was strenuously debated before the Commercial Court, Sphere contending that the parties had reached an agreement which was a final and binding disposal of all liabilities under the pool agreement whilst Orion argued that the agreement should be a goodwill agreement only and not legally binding and that, in consequence, the pool agreements still stood. Hirst J found that on the facts and evidence Orion had succeeded in satisfying, on the balance of probabilities, the heavy burden of proving that the agreement had been made without any intention of creating legal relations and was not therefore binding as a contract. The Court of Appeal upheld the decision of Hirst J. The Court of Appeal also held that it was permissible to adduce parol evidence to establish that there was no intention to create legal relations even if this was not apparent on the face of the written agreement.

Agreement

1.26   The presence or otherwise of an agreement is determined by an objective test and not a subjective test. Objectivity is based upon a reasonable man's understanding of a particular set of circumstances or facts. In its purest form the test excludes that which was actually in the minds of the parties although as a matter of practice when the courts are exploring the issue of consensus they sometimes blur the distinction between objectivity and subjectivity. Of the objective test in relation to an intention to create legal relations, Megaw J in the *Edwards* case stated:

> 'I am not sure that I know what that means in this context. I do however think that there are grave difficulties in trying to apply a test as to the actual intention or understanding or knowledge of the parties especially where the alleged agreement is arrived at between a limited liability company and a trade association; and especially where it is arrived at at a meeting attended by five or six representatives on each side. Whose knowledge, understanding or intention is relevant? But if it be the 'objective' test of the reasonable man, what background knowledge is to be imputed to the reasonable man, when the background knowledge of the ten or twelve persons who took part in arriving at the decision no doubt varied greatly between one another?'

1.27   An exhaustive examination of case law to establish the borderline between objectivity and subjectivity goes well beyond the ambit of this book. Suffice it to say that whilst the courts have well in mind the basic rules as to formation of contract they sometimes adopt a pragmatic solution to the forensic problems created by a strict application of the rules. Since the first edition of this book the courts have moved further down the road of pragmatic and subjective analysis of the issues of contract formation and interpretation as illustrated by the erudite judgment of Lord Hoffman in the case of *Investors Compensation Scheme Ltd* v. *West Bromwich Building Society* dealt with in paragraph 1.66. That being said it is a basic rule that agreement is evinced by offer and acceptance.

The offer

1.28   An offer is a written or oral statement by a person of his willingness to enter into a contract upon terms that are certain or are capable of being made certain. The offeror's intention to enter into a contract may be actual or apparent. An apparent intention is determined by the objective test discussed above. However, in *First Energy* v. *Hungarian Bank*, the Court of Appeal considered that offer letters must be construed against the contextual scene and the court should take into account the surrounding circumstances that reasonable persons in the position of the parties would have had in mind.

1.29   An offer must be distinguished from what is merely a request for information or an invitation to treat. An invitation to treat is a request for an offer. Neither a request for information nor an invitation to treat can be converted into a binding contract by an acceptance. Competitive tendering is a common method of procurement in the construction industry. In the absence of special circumstances the invitation to tender sent out to contractors by the employer or by the consultant on behalf of the employer, is an invitation to treat and not an offer. It is the contractor's submission of tender which constitutes the offer and which in turn must be accepted by the employer to give rise to a formally binding contract. For this reason the employer is not bound by the lowest tender nor as a general rule is the employer responsible for the contractor's costs of preparing the tender.

1.30   An offer can be withdrawn at any time prior to it being accepted. This is so even though the offeror stipulates that the offer shall be kept open for a particular period of time. In *Routledge* v. *Grant*, G offered to buy a house from R stating that R had six weeks in which to decide whether or not to purchase. Before the expiration of the six-week period G withdrew his offer and the court held that he was entitled to do this. In order for the offeror to be bound by a statement that an offer shall be open for acceptance for a particular period, the statement must form part of a contract of option, which is a contract having a separate legal existence to the main transaction.

1.31   The withdrawal of an offer is not effective until the withdrawal has been communicated to the offeree. This communication may be written or oral. If, as in *Routledge* v. *Grant*, the offeror has stated that the offer will be open for a specified time then the offer will lapse at the end of the relevant period even though there has not been any communication of withdrawal between offeror and offeree. It is also arguable that where no time is specified an offer may lapse after a reasonable period of time and no longer be capable of being accepted even though no communication of an intention to withdraw the offer has been made to the offeree. What is a reasonable time will depend upon the particular circumstances of each case.

The acceptance

1.32   An acceptance is a written or verbal expression of assent to the terms of the offer. It must be unequivocal and unconditional. For example, acknowledgement of the *receipt* of an offer is not an unequivocal acceptance nor is an acceptance that fails to distinguish between alternative offers. In *Peter Lind & Co Ltd* v. *Mersey Docks and Harbour Board*, the employer, the Mersey Docks and Harbour Board, invited the building contractor to submit alternative tenders for the construction of a container freight terminal, one tender to be at a fixed price and the other at a price varying with the cost of labour and materials. The employer wrote to the contractor stating that they accepted 'your tender'. The court held that this letter was imprecise because it did not specify which tender was being accepted so that there was no concluded contract at that stage.

1.33   A conditional acceptance is not effective; indeed a conditional acceptance may, on its wording, be a counter offer terminating the original offer so that it is no longer capable of being accepted even though an unconditional acceptance is subsequently sent to the offeror. In *Hyde* v. *Wrench*, W offered to sell his farm to H for a price of £1000. H made a counter offer in the sum of £950, which was rejected. H then purported to accept the original offer; however the court held he was no longer entitled to do so and there was no contract for the sale of the farm.

1.34   There is a general rule that an acceptance must be communicated by the offeree to the offeror to create a binding contract. In the absence of any particular requirements set out in the offer, communication of an acceptance may be oral or in writing. There are certain exceptions to the general rule.

Firstly, posted acceptances are deemed to take effect from the date of posting regardless of whether or not the letter actually reaches the offeror. The same rule applies to telegrams but not to instantaneous communications such as telephone, telex or facsimile transmission and possibly emails. The rationale is that the sender should be aware immediately that the communication has not been received by the offeree and will have a further opportunity to effect the communication of the acceptance. *Quaere* whether the uncertainty relating to emails (e.g. internet service provider receives communication from the sender but unknown to sender fails to transmit to the offeree, or offeree is operating a system which downloads emails at time intervals) takes this type of electronic communication outside the rationale. In *Anson* v. *Trump*, the Court of Appeal held, in respect of service of a defence document in High Court proceedings, that a pleading was served by fax when the complete document was received by the recipient's fax machine even though it might not be printed out until later. Nevertheless, the authors doubt that the postal acceptance rule will be extended to emails. The safe course is therefore to confirm emails by post – a somewhat Luddite solution. See also European Community Directives 00/31 [2000] OJ L175/1 on electronic commerce.

Secondly, communication of acceptance is not necessary for the creation of a unilateral contract as compared to a synallagmatic or bilateral contract. A synallagmatic contract involves the mutual exchange of promises whereas a unilateral contract binds only one party, for example an offer to reward a person who supplies information in relation to a lost article. In *Carlill* v. *The Carbolic Smoke Ball Company*, the defendants who were proprietors of a medical preparation called 'the Carbolic Smoke Ball' issued an advertisement in which they offered to pay £100 to any person who contracted influenza after having used one of their smoke balls in a specified manner for a specified period. C purchased one of the smoke balls on the faith of the advertisement, used it in the specified manner and subsequently contracted influenza. The Carbolic Smoke Ball Company refused to pay over the £100 and upon being sued for breach of contract the court held that C was entitled to recover. Bowen LJ stated:

> 'It is not a contract made with all the world... It is an offer made to all the world; and why should not an offer be made to all the world which is to ripen into a contract with anybody who comes forward and performs the condition.'

Thirdly, the offeree's conduct may preclude him from relying upon a failure of communication of an acceptance. Denning LJ considered this problem in *Entores Limited* v. *Miles Far Eastern Corporation* when he stated:

> 'Now take a case where two people make a contract by telephone. Suppose for instance, that I make an offer to a man by telephone and, in the middle of his reply, the line goes "dead" so that I do not hear his words of acceptance. There is no contract at that moment. The other man may not know the precise moment when the line failed. He will know that the telephone conversation was abruptly broken off, because people usually say something to signify the end of the conversation. If he wishes to make a contract, he must therefore get through again so as to make sure that I heard. Suppose next that the line does not go dead, but it is nevertheless so indistinct that I do not catch what he says and I ask him to repeat it. He repeats it and I hear his acceptance. The contract is made, not on the first time when I do not hear, but only the second time when I do hear. If he does not repeat it, there is no contract. The contract is only complete when I have his answer, accepting the offer.'

1.35   Silence cannot be taken as an acceptance of an offer even though the offer stipulates the same. This is to be contrasted with conduct that may constitute a binding acceptance. Consider for example the situation where a sub-contractor for the supply of ready mixed concrete has forwarded an offer in response to a main contractor's enquiry. The main contractor responds with a purchase order that sets out the main contractor's standard terms of contract, some of which are at variance with the terms of the sub-contractor's offer. The sub-contractor intends to negotiate these matters with the main contractor; however, unbeknown to .he sales department of the sub-contractor the first deliveries of concrete have been sent from the batching plant. The next day the sub-contractor communicates with the main contractor notifying the main contractor that its terms and conditions of contract are not acceptable. In such circumstances the courts have held that the delivery of the concrete by the sub-contractor to the main contractor constitutes an acceptance, by conduct, of a contract based upon the terms of the main contractor's purchase order.

1.36   In the case of G. *Percy Trentham Ltd* v. *Architral Luxfer*, Steyn LJ found that a contract may come into existence even though there was no coincidence of offer and acceptance, during and as a result of the performance of the subject matter of the negotiated contract.

The 'last shot' doctrine

1.37   Commercial negotiations, particularly in the construction industry, can be both protracted and tortuous, offers being met with counter offers and counter offers with counter counter offers, each party endeavouring to impose a standard form contract or their own terms and conditions of trade. As stated above the law of contract requires the offer and the acceptance to be in the same terms, and the courts are often faced with a difficult burden of forensic analysis to decide whether and if so what contract came into existence. The 'last shot' doctrine provides that where there is a series of conflicting documents they shall all be treated as counter offers and the contract, if indeed one comes into existence at all, is to be based upon the last document in time. *Butler Machine Tool Co Ltd.* v. *Ex-Cell-O Corporation (England) Ltd* involved a quotation for a machine tool given by Butler to Ex-Cell-O and stated to be subject to certain terms and conditions (one of which provided for the price to be that prevailing at time of delivery, there being a ten month delivery period) which should prevail over any terms and conditions in the Ex-Cell-O's order. Ex-Cell-O placed an order for the machine, their order form stated to be subject to terms and conditions that were materially different from those of Butler; in particular there was no provision for variation in the price. At the foot of Ex-Cell-O's order form was a tear-off acknowledgement of receipt of order stating 'We accept your order on the Terms and Conditions stated thereon'. Butler completed and signed the acknowledgement slip and returned it to Ex-Cell-O with a letter stating that the order was being entered in accordance with Butler's original quotation. Upon delivery a dispute arose as to price that was greater at time of delivery than that originally quoted. Ex-Cell-O argued that on the facts the contract had been concluded on their terms and conditions and was therefore fixed price. The court rejected this argument and held that Butler's terms and conditions had been stated in their quotation to prevail over any other terms or conditions and the contract should be construed accordingly. Ex-Cell-O's appeal against this decision was allowed by the Court of Appeal who held that Ex-Cell-O's order was a counter-offer, which destroyed the offer in Butler's original quotation. Accordingly when Butler completed and returned to Ex-Cell-O the acknowledgement slip, they accepted Ex-Cell-O's counter-offer. The covering letter with the acknowledgement slip was irrelevant as it merely referred to the price and identity of the machine and did not operate to incorporate Butler's terms back into the contract. Lawton LJ stated:

> 'the modern commercial practice of making quotations and placing orders with conditions attached, usually in small print, is indeed likely, as in this case, to produce a battle of forms. The problem is how should that battle be conducted? The view taken by the judge was that the battle should extend over a wide area and the court should do its best to look into the minds of the parties and make certain assumptions. In my judgment, the battle has to be conducted in accordance with set rules. It is a battle more on classical eighteenth century lines when convention decided who had the right to open fire first rather than in accordance with the modern concept of attrition.'

Lord Denning MR was less emphatic, anticipating the modernist approach of the matrix and recognising that various approaches could be taken to resolve the battle of forms. He stated:

> 'In some cases the battle is won by the man who fires the last shot. He is the man who puts forward the latest terms and conditions ... That may however go too far. In some cases, however, the battle is won by the man who gets the blow in first. If he offers to sell at a named price on the terms and conditions stated on the back and the buyer orders the good purporting to accept the offer on an order form with his own different terms and conditions on the back, then, if the difference is so material that it would affect the price, the buyer ought not to be allowed to take advantage of the difference unless he draws it specifically to the attention of the seller. There are yet other cases where the battle depends upon the shots fired on both sides. There is a concluded contract but the forms vary. The terms and conditions of both parties are to be construed together. If they can be reconciled so as to give a harmonious result, all well and good. If differences are irreconcilable, so that they are mutually contradictory, then the conflicting terms may have to be scrapped and replaced by a reasonable implication.'

1.38 Whilst the last shot doctrine assists in the forensic process, as anticipated by Lord Denning it is not an absolute doctrine and each contract must be considered on its own particular facts. Consider, for example, the facts in *OTM Limited* v. *Hydranautics*. On 8 September 1978, H offered to sell to OTM a device for tensioning chains on a monitoring buoy in the North Sea. The offer incorporated H's terms and conditions including a Californian arbitration clause and a Californian proper law clause, i.e that Californian law should govern the formation, construction and performance of the contract. On 29 September H telexed OTM stating 'it is our intention to place an order for one chain tensioner… A purchase order will be prepared in the near future but you are directed to proceed with the tensioner fabrication on the basis of this telex. The purchase order will be issued subject to our usual terms and conditions.' On 5 October OTM sent to H a purchase order that had the following condition: 'Acceptance of contract. The written acceptance of this contract, the commencement of performance pursuant thereto… by the sellers constitutes an unqualified acceptance by the seller of all the terms and conditions of this contract. This contract… constitutes the entire agreement between the parties either oral or written…'. The purchase order led to an exchange of telexes. However, H made no objection to the above condition although they complained that they had commenced work on the understanding that their offer was acceptable and they were now facing the introduction of new contractual terms. Negotiations continued and an agreement was reached on variations to the terms of the purchase order. On 20 October OTM sent a telex to H agreeing to the one outstanding point and asked H whether in view of the changes to the purchase order H would prefer that OTM re-issued their purchase order. On 20 December H replied to the effect that they saw no need to re-issue the purchase order and enclosed a formal acknowledgement of order which contained the following clause: 'Acceptance of buyers' order is conditional and subject to… the following conditions… Unless buyer shall notify seller in writing to the contrary within five days of receipt of this document the buyer shall be deemed conclusively to have accepted the exact terms and conditions hereof'. A copy of this acknowledgement of order was signed by OTM and returned to H on 3 January. The court held that OTM's telex dated 29 September was not an acceptance of H's offer dated 8 September (it was a letter of intent). Further, OTM's purchase order dated 5 October was a counter offer, which destroyed the original offer in total. The contract was concluded by OTM's telex sent on 20 October. The clause in H's letter dated 20 December was meaningless since there was nothing left to accept: the contract had already been made.

Incomplete agreements

1.39   To create a legally binding contract the process of offer and acceptance must result in the parties being in agreement on all the terms which are essential to their bargain; there must be *consensus ad idem*.

1.40   As a general rule the essential terms of a construction contract are parties, description of the works, price and period for construction. A failure to agree on price or time for performance is not necessarily fatal; in certain circumstances the courts may apply terms as to reasonable price and a reasonable period for performance. Such terms are now implied by the Supply of Goods and Services Act 1982, sections 14 and 15. (Note: sections 2 to 5 of the Act have been amended by Schedule 2 of The Sale and Supply of Goods Act 1994, but not sections 14 or 15.)

1.41   A particular difficulty arises in respect of collateral warranties in so far as the ultimate beneficiary of the warranty might not be known at the time the parties enter into the principal contract. Does this constitute a failure to agree an essential term?

1.42   Lord Buckmaster stated in the case of *May and Butcher Ltd* v. *The King* that:

> 'It has long been a well recognised principle of contract law that an agreement between two parties to enter into an agreement in which some critical part of the contract matter is left undetermined is no contract at all. It is of course perfectly possible for two people to contract that they will sign a document which contains all the relevant terms, but it is not open to them to agree that they will in the future agree upon a matter which is vital to the arrangement between them and has not yet been determined.'

Viscount Dunedin added:

> 'To be a good contract there must be a concluded bargain and a concluded contract is one which settles everything that is necessary to be settled and leaves nothing to be settled by agreement between the parties. *Of course it may leave something which still has to be determined but then that determination must be a determination which does not depend upon the agreement between the parties...* Therefore you may very well agree that a certain part of the contract of sale, such as price may be settled by someone else... as long as you have something certain it does not matter. For instance with regard to price it is a perfectly good contract to say that the price is to be settled by the buyer.'

1.43   It follows that if the contract had been silent it might have been rescued by an implied term, for example under the Supply of Goods and Services Act. In *May and Butcher*, Viscount Dunedin observed:

> 'the simple answer in this case is that the Sale of Goods Act provides for silence on the point and here there is no silence because there is a provision that the two parties are to agree.'

1.44   Paragraphs 7.52 and 7.53 set out examples of precedents that have been used to try and overcome the problem of a non-existent party. The precedent set out in paragraph 7.52 clearly runs a risk of being nothing more than an agreement to agree and therefore being unenforceable in law. However in contrast, whilst the precedent set out in paragraph 7.53 still leaves undecided the identity of the party who is to have the benefit of the collateral warranty, that omission can be dealt with by the *determination* of the developer, that is to say applying the words of Viscount Dunedin: 'A determination which does not depend upon the agreement between the parties'.

1.45　　The case of *Foley* v. *Classique Coaches Ltd* provides a useful illustration of how a contract can be rescued by the use of an implied term. This case concerned an agreement by F to sell to Classique Coaches certain land, which Classique Coaches intended to use for their business of motor coach operators. The sale was subject to a second agreement in which Classique Coaches agreed to purchase from F all the petrol required for the purposes of their business at 'A price to be agreed by the parties in writing and from time to time'. The land was conveyed to Classique Coaches and the petrol agreement operated for three years. Thereafter disputes arose and Classique Coaches purported to repudiate the petrol agreement alleging that it was not binding because there was no agreement as to the price of the petrol. The court rejected this argument holding that a term must be implied in the agreement that the petrol supplied by F should be of a reasonable quantity and sold at a reasonable price and that if any dispute arose as to what was a reasonable price it was to be determined by an arbitration clause in the petrol agreement. The agreement was therefore valid and binding.

1.46　　The issue of incomplete agreements has recently been considered by the Court of Appeal in the case of *Global Container Lines* v. *State Black Sea Shipping*, the facts of which concerned a complex tiered arrangement for the sale of four former USSR military vessels by the Ukranian state shipping company to the USA. The court, referring to the cases of *May and Butcher* and *Foley*, confirmed that the relevant principles to be derived from those cases were that where there was an agreement which provided for the parties to agree something further, then it would be inferred as a matter of construction that the whole contract was conditional upon the further agreement of those matters. However, the court felt that this was not a conclusive presumption because it was always possible to have a contract which did refer to a further agreement in such manner, but which nevertheless constituted a binding agreement. See, for example, *F. & G. Sykes (Wessex) Ltd.* v. *Fine Fare Ltd*. Each case depended upon its own circumstances and had to be construed on the wording used. It was not conclusive that wording which contemplated a further agreement, necessarily meant that there was no complete agreement between the parties sufficient to form a binding agreement, but it was a factor which may lead to that conclusion which indeed was the finding in *May and Butcher*. In the *Global* case the court held that the words 'to be finalised' did not leave anything to be agreed between the parties but instead related to questions of formalisation or performance of documents and transactions. A full reading of the transcript of this case is recommended by the authors, as an illustration of the court's inclination to find in favour of the existence of a contract in the commercial market place.

1.47 Where there is an agreement to agree, which does not constitute a binding contract, the courts have refused to imply a term that the parties must continue to negotiate in good faith: *Walford* v. *Miles*.

Consideration

1.48 Save for contracts made under seal, and subject to the Contracts (Rights of Third Parties) Act, the courts will not enforce gratuitous promises. There must be valuable consideration. Valuable consideration is 'something of value in the eye of the law' (*Thomas* v. *Thomas*). Clearly the payment of money or a promise to pay money is valuable consideration. However 'in the eyes of the law' other acts, however insignificant, may provide valuable consideration. For example a promise to give £50 'if you will come to my house' was held to be valuable consideration in *Gilbert* v. *Ruddeard*. However, as a general rule, a moral obligation does not provide valuable consideration for example a promise made 'in consideration of natural love and affection': *Brett* v. *J.S.* Nor is a pre-existing legal obligation sufficient to provide valuable consideration, often referred to as 'past consideration'. The giving of a collateral warranty by an architect in consideration of terms of appointment that have already been fulfilled or the giving of a guarantee by a construction company in consideration of a payment under the construction contract, are examples of past consideration.

1.49 Valuable consideration need not be adequate consideration in a sense that the courts are not concerned as to the fairness of the bargain between two contracting parties. For this reason payment of nominal consideration is sufficient. Collateral warranty agreements usually provide that consideration for the agreement shall be the payment by the promisee to the promisor of the sum of £1 or £10. Whether it be £1 or £10 this consideration though nominal is valuable consideration for the purposes of enforcing the warranty.

1.50 Valuable consideration must support the promise and not the contract. That is to say there must be some detriment to the promisee (the giving of value) or some benefit to the promisor (receipt of value). In many cases a detriment to the promisee will be the same as a benefit to the promisor, for example a sub-contractor promises to carry out piling works in consideration for which the main contractor promises to pay the price of those works. The carrying out of the piling works and the promise to pay the price of the works are respectively both detriments and benefits. However, whilst consideration must move from the promisee it is not necessary for consideration to move to the promisor. For example, if A promises to B that A will guarantee the repayment of a loan of finance to be made by B to C, whilst A derives no benefit from the transaction if B, in consideration of A's promise, makes the loan to C, then B will suffer a detriment which provides valuable consideration moving from B (the promisee) to A (the promisor) in respect of a contract of guarantee. Similarly in the *Shanklin Pier* case referred to earlier, if Shanklin had not procured that the contractor Carter entered into a contract with Detel for the use of their paints, there would not have been valuable consideration moving from Shanklin, the promisee, to Detel the promisor and Detel's representation as to the quality of their paint would not have been a contractual warranty.

Estoppel contracts

1.51   A discussion of the formation of contract would not be complete without a reference to contracts arising from estoppel by convention. As with the 'matrix', a detailed narrative on this topic is beyond the ambit of this book. Suffice it to state that courts have held that in the absence of the essentials of a contract, referred to in paragraph 1.17 earlier, a contract could still come into existence between the parties if they have so conducted themselves as to a common assumption, in fact or law, that there is a contract and that it would be wrong or unreasonable for the parties thereafter to deny the existence of such contract. Estoppel contracts have received a mixed reception from the courts, finding favour in *G. Percy Trentham* v. *Architral Luxfer* (see paragraph 1.36) and in *Mitsui Babock Energy Ltd.* v. *John Brown Engineering Ltd*, but being rejected in *J. Murphy & Sons Ltd* v. *ABB Daimler Benz*.

Form of contract

1.52   Unless the contract is made under seal there are no special rules governing the form of a construction contract. However, contracts must be made in writing or evidenced in writing if they are to come within the provisions of the Housing Grants, Construction and Regeneration Act (see paragraph 1.72). The contract may be a written contract or an oral contract or partly written and partly oral.

Contracts under seal

1.53   Much of the law relating to the execution of deeds was swept away on 31 July 1990 by the Law of Property (Miscellaneous Provisions) Act 1989 dealing with the execution of deeds by individuals, and by the Companies Act 1989 dealing with the execution of deeds by companies.

1.54   Section 1(3) of the Act provides that a deed must be signed by the person who is to be bound and the signature must be attested by a witness. If this procedure is not followed then the document will not be valid as a deed. Further, section 1(2)(a) provides that a document shall not be a deed unless it is made clear on its face that it is intended to be a deed by the person making it. It follows that the document must be described as a deed and the attestation clause must include a statement that it is being executed as a deed. If such descriptive words are not included then the document will not be valid as a deed. The Act does not abolish the old common law rule that to be effective a deed must be delivered by the person who is to be bound. Delivery does not mean the physical handing over of the document but the evincing of an intention, whether by words or conduct, that the party is to be bound. Such intention was satisfied prior to the Act by the act of signature and it is submitted that signature will continue to satisfy the requirement of delivery.

1.55   Companies may still use their seals to execute a deed provided that this is in accordance with their Articles of Association. Section 130 of the Companies Act 1989 introducing a new section 36(a) to the Companies Act 1985 provides by sub-section 4 that if a document is signed by a director and the secretary of the company or by two directors of the company and is expressed (in whatever form of words) to be executed by the company it will have the same effect as if it had been executed under the common seal of the company, i.e a deed. Further, sub-section 5 provides that a document which makes it clear on its wording that it is intended to be a deed has the effect of a deed even though the specific requirements for execution dealt with in sub-section 4 above are not complied with. For example, a document that is described as a deed and is executed by one director of the company will operate as a deed. Extreme care must be taken to ensure, if it is intended only to create a simple contract, that a company giving a collateral warranty does not accidentally execute the document in such a manner as to give rise to a deed.

Consideration and limitation periods

1.56   There are two important differences between contracts which are not under seal, referred to as simple contracts, and contracts that are under seal. Firstly, simple contracts and contracts under seal have different limitation periods (see paragraph 6.97). Secondly, unlike a simple contract, a contract under seal does not have to be supported by valuable consideration. Thus where the collateral warranty consists of unilateral undertakings by one party, the contract must be a contract under seal if it is to be enforceable. It is important to note that whilst consideration is not necessary for a contract under seal in the absence of valuable consideration and, arguably, in the absence of something more than mere nominal consideration, the remedy of *specific performance* will not be available in respect of the contractual undertakings: *Milroy* v. *Lord*. Specific performance is an equitable remedy requiring the contract breaker to fulfil his contractual obligations, rather than awarding damages for breach. Equity does not assist a volunteer, hence the need for consideration. The authors suggest that a deed, which relies on merely nominal consideration, could be rescued by the addition of a consideration, which consists of 'the mutual undertakings' of the parties to the deed.

Construing a contract

1.57   The same rules of construction apply to both simple contracts and contracts under seal.

1.58   The general rule is that whilst the court strives to give effect to the intention of the parties, it must give effect to that intention as expressed. That is to say it must ascertain the meaning of the words actually used and not try to interpret the motive or state of mind of the parties: *Inland Revenue Commissioners* v. *Raphael* and *Prenn* v. *Simmonds*.

1.59   The courts must have regard to the ordinary meaning of words used unless they are technical or scientific. That being said the courts are concerned to ascertain the intention of the parties and not merely indulge in semantic exercises. In *Lloyd* v. *Lloyd*, Lord Cottenham stated:

> 'If the provisions are clearly expressed, and there is nothing to enable the court to put upon them a construction different from that which the words import, no doubt the words must prevail; but if the provisions and expressions be contradictory and if there be grounds, appearing on the face of the instrument, affording proof of the real intention of the parties, then that intention will prevail against the obvious and ordinary meaning of the words. If the parties have themselves furnished a key to the meaning of the word used, it is not material by what expression they convey their intention.'

1.60   In construing particular terms of a contract the whole of the contract must be considered. Further, if there is no ambiguity or uncertainty then the court must give effect to the intention of the parties, however harsh. In *Trollope and Colls Ltd* v. *North West Metropolitan Regional Hospital Board*, the contract provided for three phases of work, phase 1 to be completed by 30 April 1969 and phase 3 to commence six months after the date of practical completion of phase 1 and to be completed on 30 April 1972, i.e a construction period of 30 months for phase 3. The completion of phase 1 was delayed by a period of 59 weeks, 47 of which were not the fault of the contractor. Practical completion of phase 1 was not achieved until 22 June 1972, leaving a construction period of only 16 months, rather than 30 months, for the completion of phase 3. The court took the view that the intention of the parties was perfectly clear from the wording of the contract and refused to relax the contractors' obligations by implication of the term that the time for completion of phase 3 should be extended by the same period as the extension of time granted in respect of phase 1.

1.61   Where the parties have made manuscript deletions, additions or amendments to a printed form, in the event of any ambiguity in construing the document as a whole, the manuscript will be given greater weight than the printed terms: *Sutro & Co* v. *Heilbut Symons and Co.* Where there is inconsistency between figures and words the court will have regard to the words before the figures: *Saunderson* v. *Pier*.

1.62   If there is inconsistency between different parts of the same contract then, in the absence of an express term resolving the ambiguity (for example condition 2.2 and 2.4 of JCT With Contractor's Design, 1998 Edition) the courts will endeavour to give effect to that part of the contract which expresses the real intention of the parties: *Walker* v. *Giles*.

1.63   If the contract document or documents establish a clear intention, the courts adopt an interventionist approach to resolve any difficulties flowing from the actual words used by the parties: *Gwyn* v. *Neath Canal Company*:

> 'The results of the authorities is that, when a court of law can clearly collect from the language within the four corners of a deed or instrument in writing, the real intention of the parties, they are bound to give effect to it by supplying anything necessary to be inferred from the terms used, and by rejecting as superfluous whatever is repugnant to the intention so discerned.'

1.64   For example in *Mourmand* v. *Le Clair*, the parties expressed the repayment of a debt to be by instalments of 'seven' on a particular day of the month; the court inserted the word 'pounds' after the word 'seven'. In *Simpson* v. *Vaughan* a debtor gave an acknowledgment of debt that was stated to be 'for money borrowed which I promise *never* to pay'; the court struck out the word 'never'.

Parol evidence rule

1.65   Where it is clear that the parties intended the whole of their agreement to be set out in one document or a series of documents, the operation of the parol evidence rule will prevent the parties from relying on extrinsic evidence to try and amend the terms of their agreement. For example, evidence of negotiations that took place prior to the conclusion of the contract will not be admissible.

1.66   The parol evidence rule applies both to oral evidence and to documentary evidence, for example letters and minutes of meetings. It is not an absolute rule however and there are exceptions as the courts adopt a commonsense approach to the question of interpretation of contracts. In particular, if there is some uncertainty or ambiguity as to the intentions of the parties then the courts are prepared to look at what has been described as the *matrix* of a contract, that is to say the surrounding circumstances, background and commercial purpose of the agreement: *Prenn* v. *Simmonds*. A statement of the 'modern approach' to formation of contract and construing the contractual terms was made by Lord Hoffman in the case of *Investors Compensation Scheme Ltd* v. *West Bromwich Building Society*. Lord Hoffman stated five principles:

'(1) Interpretation is the ascertainment of the meaning which the document would convey to a reasonable person having all the background knowledge which would reasonably have been available to the parties in the situation which they were at the time of the contract.

(2) The background is the "matrix of fact" but this phrase is, if anything, an understated description of what the background may include. Subject to the requirement that it should have been reasonably available to the parties and to the exception to be mentioned next, it includes absolutely anything which would have affected the way in which the language of the document would have been understood by a reasonable man ...

(3) The law excludes from the admissible background the previous negotiations of the parties and their declarations of subjective intent ...

(4) The meaning which a document (or any other utterance) would convey to a reasonable man is not the same thing as the meaning of its words. The meaning of words is a matter of dictionaries and grammars; the meaning of the document is what the parties using those words against the relevant background would reasonably have been understood to mean. The background may not merely enable the reasonable man to choose between the possible meaning of words which are ambiguous but even (as occasionally happens in ordinary life) to conclude that the parties must, for whatever reason, have used the wrong words or syntax ...

(5) The "rule" that words should be given their "natural and ordinary meaning" reflects the commonsense proposition that we do not easily accept other people have made linguistic mistakes, particularly in formal documents. On the other hand, if one would nevertheless conclude from the background that some-thing must have gone wrong with the language, the law does not require judges to attribute to the parties an intention which they plainly could not have had.'

On the facts of the *Investors* case the House of Lords held (Lord Lloyd dissenting) that the words in a document stating that 'Any claim (whether sounding in rescission for undue influence or otherwise)' was intended by the parties to mean 'Any claim sounding in rescission (whether for undue influence or otherwise)', i.e. a much more restrictive construction of the word 'claim'. In *Cargill International SA* v. *Bangladesh Sugar & Foods Industries Corp*, the court considered that in general they should look to the intention of the parties rather than the strict letter of the contract's stipulations, and in interpreting their intention, the courts look at the factual matrix of the contract:

'... modern principles of construction require the court to have regard to the commercial background to the context of the contract and the circumstances of the parties and to consider whether against that background and within that context to give the words a particular or restricted meaning would lead to an apparently unreasonable and unfair result.'

Recitals

1.67   Recitals are the introductory statements in a written agreement or deed setting out a précis of the parties' intentions. Recitals usually appear in documents after the words 'whereas' and before the words 'now it is hereby agreed as follows', the latter phrase introducing the operative or main conditions of the agreement. If there is any ambiguity or uncertainty arising from the operative or main conditions of the agreement the courts will look at the recitals in order to establish the intention of the parties to the agreement. It is also important to note that the intention of the parties as evinced by the recitals may be relevant to the court's consideration of whether or not to imply a term into the agreement. Implied terms are dealt with in paragraphs 1.77–1.79.

Ejusdem generis rule

1.68   This is a rule of construction which provides that where a contract condition or clause sets out a list of specific matters so as to create a common category and the specific matters are followed by general words, the courts will construe the general words restrictively so as to confine those words to the common category. For example, the phrase 'and all other deleterious materials' coming after a list of deleterious materials will be construed restrictively.

Contra proferentem

1.69   This is another rule of construction applying to written documents or deeds. The rule provides that if the wording of an agreement is ambiguous or uncertain, *but not otherwise*, the contract should be construed more strongly against the person whose words they are rather than the other party:

> 'We are presented with two alternative readings of this document and the reading which one should adopt is to be determined, amongst other things, by a consideration of the fact that the defendants put forward the document. They have put forward a clause which is by no means free from obscurity and have contended... that it has a remarkably, if

not extravagantly, wide scope and I think that the rule *contra proferentem* should be applied': (*John Lee & Son (Grantham) Ltd* v. *Railway Executive*)

1.70   If a party has incorporated its own standard terms and conditions of trade into an agreement then in the event of ambiguity those terms and conditions will be construed *contra proferentem* that party. Where however the parties execute standard form contracts such as JCT 1998, the *contra proferentem* rule will only operate in respect of amendments or additions to the contract.

1.71   *Contra proferentem* is a particularly important rule of construction in relation to exclusion clauses. Exclusion clauses are dealt with in paragraphs 9.92–9.99.

The Housing Grants, Construction and Regeneration Act 1996 (HGCRA)

1.72   The Housing Grants, Construction and Regeneration Act 1996 applies to contracts entered into after 1 May 1998 provided that the contract is a construction contract and also is a contract in writing or evidenced in writing (section 107). Sections 104 and 105 define a construction contract as an agreement for the carrying out of construction operations whether as main contractor, sub-contractor, or labour only contractor. Construction operations are:

- the construction, alteration, repair, maintenance, extension, demolition or dismantling of buildings or structures (section 105(1)(a)).
- works, e.g. roadworks, powerlines, telecommunication apparatus, aircraft runways, docks and harbours, railways, pipe lines, reservoirs, water mains, sewers (section 105(1)(b)), mechanical and electrical systems (section 105(1)(c)) forming or to form part of the land (whether permanent or not).
- the external or internal cleaning of buildings during the course of their construction, alteration, repair, extension or restoration (section 105(1)(d)) and ancillary operations (section 105(1)(e)).
- painting and decorating (section 105(1)(g)).

1.73   By section 105(2) what could be conveniently categorised as 'strategic operations', e.g. oil and gas extraction, nuclear processing, bulk storage of chemicals, pharmaceuticals, oil, gas, steel, food and drink (sub-sections (a) to (c)), are excluded from the Act. Also contracts for the supply only of plant, equipment or machinery as distinct from such contracts that provide for the supply and installation of plant equipment or machinery (sub-section (d)). Artistic works, e.g. sculptures and murals, are also excluded (sub-section (e)). The detailed wording of Part II of the Act is set out in Appendix 1 of this book.

1.74   Clearly the draftsmen of the statute have endeavoured to produce a detailed definition of a construction contract. Have they succeeded?

Possibly. In *Nottingham Community Housing Association Ltd* v. *Powerminster Ltd*, the court held that a maintenance and repair of a heating system, installed in a building, was a construction operation. However, in *Staveley Industries* v. *Oderbrecht Oil & Gas Services*, the court held that a contract for the design, procurement, supply and installation of fire, gas, electrical and telecommunication equipment in modules to form the living quarters of an offshore gas drilling rig was not a construction operation as the building and structure did not form part of 'the land' within England, Wales or Scotland. In *Gibson Lea Retail Interiors* v. *Macro Self Service Wholesalers*, the court held that a shop-fitting contract which involved the supply of freestanding equipment but also equipment and structures bolted to the floors and walls was not a construction operation as the main purpose of fixing was to achieve stability and not immobility, therefore the fixtures were not a permanent part of the land.

1.75 The HGCRA imposes on the parties to a construction contract, in the absence of express agreement, provisions as to adjudication (section 108), entitlement to stage payments (section 109), dates for payment (section 110), notices for withholding payment (section 111), a right to suspend performance for non-payment (section 112) and prohibition of conditional payment provisions 'pay when paid' (section 113).

1.76 The adjudication scheme, whether statutory or bespoke, has generated a considerable amount of litigation as the construction industry and the courts establish the parameters and efficacy of the process. For those readers who are interested in this subject, a useful analysis of relevant cases can be found on the website www.constructionlawdatabase.com

Implied terms

1.77 The express terms of a contract do not necessarily constitute all the relevant terms of the agreement. In certain circumstances the courts are prepared to imply terms into a contract provided such terms are necessary to give *business efficacy* to the agreement. The leading case on implied terms is *The Moorcock*. In that case Bowen LJ stated:

> 'Now, an implied warranty, or, as it is called, a covenant in law, as distinguished from an express contract or express warranty, really is in all cases founded upon the presumed intention of the parties, and upon reason. The implication which the law draws from what must obviously have been the intention of the parties, the law draws with the object of giving efficacy to the transaction and preventing such a failure of consideration as cannot have been within the contemplation of either side; and I believe if one were to take all the cases, and there are many, of implied warranties or covenants in law, it will be found that in all of them the law is raising an implication from the presumed intention of the parties with the object of giving the transaction such efficacy as both parties must have intended that at all events it should have.'

1.78   Terms may be implied as a matter of law. That is to say they are implied as
        a matter of policy and are of general application to all contracts. Further
        terms may be implied as a matter of fact. That is to say as a matter of
        construction of the presumed intention of the parties to a particular
        contract. The *Foley* case referred to in paragraph 1.45 above is an illus-
        tration of a term being implied as a matter of fact.

1.79   Whilst a term will not be implied unless in the particular circumstances of
        each case it is reasonable to imply such a term, this does not mean that a
        term will be implied merely because it is reasonable. For example, see the
        Trollope & Colls case referred to in paragraph 1.60 earlier where the court
        refused to imply a term to render a harsh contract more reasonable.
        Further, a term will not be implied if it is inconsistent with the express
        terms of the contract. In *Martin Grant & Co Ltd* v. *Sir Lindsay Parkinson & Co
        Ltd*, the Court of Appeal refused to imply a term in a building sub-
        contract:

> 'that (a) the [main contractors] would make sufficient work available to
> the [sub-contractors] to enable them to maintain reasonable progress
> and to execute their work in an efficient and economic manner; and (b)
> the main contractors should not hinder or prevent [the sub-contractors]
> in the execution of the sub-contract works'

where the express conditions of the sub-contract provided for a 'beck and
call' obligation on the sub-contractor; that is to say the sub-contractor was
obliged to carry out his works '... at such time or times and in such
manner as the [main contractor] shall direct or require'.

Letters of intent

1.80   It is the ebb and flow of commercial negotiations that sets the scene for
        considering the contractual effect of letters of intent; for example, the
        effect of a letter of intent in the *OTM* case referred to in paragraph 1.38
        earlier.

1.81   As a general statement the characteristics of a letter of intent are a
        document expressing an intention to enter into a contract at a future date
        but creating no contractual relationship until such future contract has
        been entered into. Such letters are usually sent at a time when it is
        anticipated by the parties that the recipient will be incurring costs and
        overheads in respect of the proposal project. A letter of intent is to be
        distinguished from a provisional contract. This distinction was con-
        sidered in the case of *Hall & Tawse South Ltd* v. *Ivory Gate Ltd*. Judge
        Thornton QC described the comparison in the following terms:

> 'A letter of intent is usually an [sic] unilateral assurance intended to
> have contractual effect if acted upon, whereby reasonable expenditure
> reasonably incurred in reliance upon such a letter will be reimbursed.

Such a letter places no obligation upon the recipient to act upon it and there is usually no obligation to continue with the work or to undertake any defined parcel of work, the recipient being free to stop work at any time. The effect of such a letter is to promise reasonable reimbursement if the recipient does act upon it. However, the letter in question (in the instant case) ... is one which imposes obligations on both parties. It requires the plaintiff to commence the works, being a defined package of work and contract administration. The plaintiff had an option of whether to start or not but, having started, the plaintiff was under an obligation to continue with the works and not to stop, unless the defendant appointed another contractor or gave notice abandoning the work or the contract was superseded by one of the two successor contracts envisaged by the letter. I propose, therefore, to refer to this contract as "the provisional contract".'

1.82   It is important to bear in mind that 'letter of intent' is a term of commercial convenience and not a term having a substantive legal meaning, as for example 'subject to contract'. Each letter of intent must be construed on its own particular meaning. It is suggested that the legal effect of a letter of intent may fall into one or more of the following categories:

(1)   The expression of an intention to enter into a contract at a future date which does not give rise to any legal obligation whether in contract or quasi ex-contractu on a quantum meruit; or

(2)   The expression of an intention to enter into a contract at a future date which does not give rise to any liability in contract but does not exclude or negative a right to recover reasonable expenditure on a quantum meruit; or

(3)   The creation of a conditional or ancillary contractual obligation which may, but not necessarily will, be subsumed by a wider contractual obligation upon formal contracts being exchanged; or

(4)   A legally binding executory contract in that the letter of intent is an offer capable of being accepted or is the acceptance of an offer.

1.83   The essential test is one of substance not form as is illustrated by the following cases. The *OTM* case where the court found that OTM's telex stating 'it is our intention to place an order for one chain tensioner. A purchase order will be prepared in the near future but you are directed to proceed with the tensioner fabrication on the basis of this telex', was a letter of intent and did not constitute an acceptance of an offer. In contrast to *OTM*, the court found there was a binding contract in *Wilson Smithett & Cape (Sugar) Ltd* v. *Bangladesh Sugar & Foods Industries Corporation*. The plaintiffs who were sugar merchants responded by a tender to an invitation sent by the defendants, a nationalised Bangladeshi Corporation. The tender was for the sale of 10,000 tons of sugar cane. The plaintiffs were the second lowest bidder. The Bangladeshi Government decided to import a further 10,000 tons and the defendants were instructed to place an order with the second lowest tenderer, i.e. the plaintiffs, if they were agreeable to matching the successful tenderer's bid. Negotiations were conducted between the plaintiffs and the defendant; eventually the plaintiffs sent to the defendant a letter of offer, which was to remain open until 2 PM local Dacca time on 12 June 1991. The same day the defendants issued a letter of intent, which stated:

> 'We are pleased to issue this letter of intent to you for the supply of the following materials ... all other terms and conditions as per your ... offer dated June 12th ...'

The defendants decided not to proceed with their purchase and they contended that their letter of 12 June was nothing more than an expression of a future intent to enter into a contract. The court rejected this submission and held that the letter, although it used the phrase 'letter of intent', was nevertheless intended to have a contractual significance and effect; accordingly there was a binding contract between the parties. Similarly, in *Turriff Construction Ltd* v. *Regalia Knitting Mills Ltd* the court held that a letter of intent had a contractual effect.

1.84   The concept of 'ancillary contract' or 'if contract' was considered in the case of *British Steel Corporation* v. *Cleveland Bridge & Engineering Company* where the court held that a letter of intent had no contractual effect, as the effect of the material letter was to ask the recipient, British Steel, to proceed immediately with the work pending the preparation and issuing of a form of sub-contract being a document which was still in the state of negotiation not least on the issues of price, delivery dates and applicable terms and conditions. Of an 'if contract' Goff J stated:

> 'As a matter of analysis the contract (if any) which may come into existence following a letter of intent may take one of two forms: either there may be an ordinary executory contract, under which each party assumes reciprocal obligations to the other; or there may be what is sometimes called an "if" contract, i.e. a contract under which A requests B to carry out a certain performance and promises B that, if he does so, he will receive a certain performance in return, usually remuneration for his performance. The latter transaction is really no more than a standing offer which, if acted upon before it lapses or is lawfully withdrawn, will result in a binding contract.'

The possibility of an 'if' contract was considered, but rejected, by the court in *A. Monk Construction Limited* v. *Norwich Union Life Insurance Society*.

Chapter 2
The Rise of Collateral Warranties

2.1   In order to understand why collateral warranties have assumed such importance, it is necessary to look at the tort of negligence in relation to construction problems; this necessarily involves some consideration of the law of negligence and its development immediately prior to the House of Lords decision in 1988 in *D. & F. Estates* v. *The Church Commissioners for England*, the *D & F* case itself and the subsequent decision in the House of Lords in 1990 in *Murphy* v. *Brentwood District Council*, and then looking at cases since *Murphy*, in particular *Henderson and Others* v. *Merrett Syndicates Ltd*.

Negligence

2.2   The tort of negligence is concerned with breach of a duty to take care. In order to succeed in an action for negligence, a plaintiff must prove:

- the defendant owed to the plaintiff a legal duty of care, and,
- the defendant was in breach of that duty, and,
- the plaintiff has suffered damage as a result of that breach.

2.3   The legal duty of care referred to is one that arises independently of a contractual obligation and, indeed, in the absence of contract. Over many years, the courts have produced a long series of decisions to assist in deciding whether or not, on particular facts, a duty of care arises.

2.4   The modern law of negligence really begins in 1932 when the famous decision in *Donoghue* v. *Stevenson* reached the House of Lords. A young lady was bought a bottle of ginger beer by a friend. She had drunk some of the ginger beer, which was in an opaque bottle, before she discovered that there was a decomposing snail in the bottle. It was alleged that she became ill as a result. There was no question in this case of the friend bringing an action in contract under the Sale of Goods Act against the retailer from whom the ginger beer had been purchased because the friend had not suffered any damage. The young lady could not sue the retailer because she had no contract herself with him.

2.5   It was in this way that the House of Lords came to be asked whether the young lady had a cause of action in negligence against the manufacturer. They held by a majority that a manufacturer who sold products in such a form that they were likely to reach the ultimate consumer in the state in which they left the manufacturer with no possibility of intermediate examination, owed a duty to the consumer to take reasonable care to prevent injury. Some understanding of the radical development in English law that this case represented can be gained from the dissenting judgment of Lord Buckmaster, who did not agree with the majority in the House of Lords:

> 'There can be no special duty attaching to the manufacturer of food apart from that implied by contract or imposed by statute. If such a duty exists, it seems to me it must cover the construction of every article, and I cannot see any reason why it should not apply to the construction of a house. If one step, then why not fifty? If a house be, as it sometimes is, negligently built and in consequence of that negligence the ceiling falls and injures the occupier or anyone else, no action against the builder exists according to the English law, although I believe such a right did exist according to the Laws of Babylon.'

2.6   Little did Lord Buckmaster, in his dissenting judgment, appreciate how the *Donoghue* case would be the basis for a rapid expansion of the law of tort in negligence over the following 56 years, along the very lines that he robustly refused to contemplate in his judgment. However, his view did not prevail in 1932; in the same case, Lord Atkin formulated a principle so as to test whether a duty of care exists:

> 'The liability for negligence whether you style it such or treat it, as in other systems, a species of *culpa* is no doubt based upon a general public sentiment of moral wrong doing for which the offender must pay. But acts or omissions which any moral code would sensor cannot in a practical world be treated so as to give a right to every person injured by them to demand relief. In this way, rules of law arise which limit the range of complaints and the extent of their remedy. The rule that you are to love your neighbour becomes in law, you must not injure your neighbour; and the lawyer's question "who then is my neighbour?" receives a restrictive reply. You must take reasonable care to avoid acts or omissions which you can reasonably foresee would be likely to injure your neighbour. Who, in law, is my neighbour? The answer seems to be – persons who are so closely and directly affected by my act that I ought reasonably to have them in contemplation as being so affected when I am directing my mind to the acts or omissions which are called into question.'

2.7   The law in this respect remained fairly static until the early 1960s; through the 1960s and 1970s there was rapid development of the law of negligence, in particular in construction cases.

1932 to 1988

2.8   It took from *Donoghue* in 1932 until 1964 to extend the principle of the *Donoghue* decision to statements that were given negligently: *Hedley Byrne & Co Ltd* v. *Heller & Partners Ltd*. Advertising agents, Hedley Byrne, needed a reference from a banker as to the creditworthiness of a potential customer. They approached their bankers who sought the advice of merchant bankers who in turn reported to Hedley Byrne. The report was headed 'without responsibility' and said that the potential customer was good for ordinary business arrangements. Hedley Byrne proceeded with their contract and by reason of the customer not being good for ordinary business arrangements, lost a considerable sum of money. They sued the merchant bankers. The House of Lords held that a person is liable for statements made negligently in circumstances where he knows that those statements are going to be acted on and they were acted on. However, in this case, the merchant banker escaped liability by reason of having expressed their report to be without responsibility. This case may have assumed new importance since the decision in *Murphy* v. *Brentwood District Council* (see paragraph 2.23).

2.9   The first major extension of the test of Lord Atkin in *Donoghue* in a building case was in 1972 in *Dutton* v. *Bognor Regis UDC and Another* (now overruled by *Murphy*, see paragraph 2.23). A house was built on a rubbish tip and Mrs Dutton was the second owner of the house. The walls and the ceiling cracked, the staircase slipped and the doors and windows would not close; the damage was caused by inadequate foundations. Mrs Dutton sued the builder (with whom she settled before the hearing) and the local authority. The Court of Appeal held that the local authority, through their building inspector, owed a duty of care to Mrs Dutton to ensure that the inspection of the foundations of the house was properly carried out and that the foundations were adequate, and that the local authority were liable to Mrs Dutton for the damage caused by the breach of duty of their building inspector in failing to carry out a proper inspection of the foundations. Lord Denning MR said, applying Lord Atkin's test, 'I should have thought that the inspector ought to have had subsequent purchasers in mind when he was inspecting the foundations – he ought to have realised that, if he was negligent, they might suffer damage.'

2.10   The *Dutton* case was followed on this point in many subsequent and important cases: *Sparham-Souter* v. *Town & Country Developments (Essex) Ltd*; *Sutherland & Sutherland* v. *C. R. Maton & Sons Ltd*; *Anns* v. *Merton London Borough Council*; *Batty and Another* v. *Metropolitan Property Realisations Ltd and Others*. However, as is often the case with changes in orthodoxy in the law, the seeds of destruction of this widening of the law of negligence were sown by a dissenting judgment, that of Stamp LJ in the *Dutton* case in 1972, but it was to take many more years before change came:

'I may be liable to one who purchases in the market a bottle of ginger beer which I have carelessly manufactured and which is dangerous and causes injury to personal property; but it is not the law that I am liable to him for the loss he suffers because what is found inside the bottle and for which he has paid money is not ginger beer but water. I do not warrant, except to an immediate purchaser, and then by contract and not in tort, that the thing I manufacture is reasonably fit for its purpose. The submission is, I think, a formidable one and in my view raises the most difficult point for decision in this case. Nor can I see any valid distinction between the cases of a builder who carelessly builds a house which, although not a source of danger to personal property, nevertheless, owing to a concealed defect in its foundations, starts to settle and crack and becomes valueless, and the case of a manufacturer who carelessly manufactures an article which, though not a source of danger to a subsequent owner or to his other property, nevertheless owing to hidden defect quickly disintegrates. To hold that either the builder or the manufacturer was liable except in contract would be to open up a new field of liability the extent of which could not, I think, be logically controlled and since it is not in my judgment necessary to do so for the purposes of this case, I do not more particularly because of the absence of the builder, express an opinion, whether the builder has a higher or lower duty than the manufacturer.'

2.11   In 1978, the position altered again with *Anns* v. *Merton London Borough Council*. Lessees of flats claimed against a local authority in negligence in relation to the local authority's powers of inspection under the by-laws in that, it was said, they had allowed the contractors to build foundations in breach of the by-laws, with resulting damage to the flats. In a passage in his speech, which later became the excuse for a far-reaching and dramatic expansion of the circumstances in which a duty of care might be held to exist, Lord Wilberforce said in the House of Lords:

'Through the trilogy of cases in this House, *Donoghue* v. *Stevenson*, *Hedley Byrne & Co Ltd* v. *Heller & Partners Ltd* and *Home Office* v. *Dorset Yacht Co Ltd*, the position has now been reached that in order to establish that a duty of care arises in a particular situation it is not necessary to bring the facts of that situation within those of previous situations in which a duty of care has been held to exist. Rather the question has to be approached in two stages. First one has to ask whether as between the alleged wrongdoer and the person who has suffered the damage there is a sufficient relationship of proximity or neighbourhood such that, in the reasonable contemplation of the former, carelessness on his part may be likely to cause damage to the latter in which case a *prima facie* duty of care arises. Secondly, if the first question is answered affirmatively, it is necessary to consider whether there are any considerations which ought to negative, or to reduce or limit the scope of the duty or class of person to whom it is owed, or the damage to which a breach of it may give rise.'

2.12   The development and extension of the law of tort probably reached its climax in the House of Lords in *Junior Books Ltd* v. *Veitchi Co Ltd* in 1983. That case was on appeal to the House of Lords from Scotland. Specialist flooring sub-contractors had laid a floor at the employer's factory and it was said by the employers that the floor was defective. The employers, who were not in contract with the sub-contractor, brought an action in *delict* (which is substantially the same cause of action in Scotland as negligence is in England). Despite the absence of any allegation by the employer that there was a present or imminent danger to the occupier (an essential ingredient of the *Anns* decision), the employers succeeded in their argument that the sub-contractor owed them a duty of care in negligence and that the sub-contractor was in breach of that duty. It was said that there was a close commercial relationship between the employers and the sub-contractors. It is that justification for the decision which had led to enormous difficulties in the minds of those closely involved with the construction industry – after all, what is unusual about an employer engaging a contractor who in turn engages sub-contractors. Notwithstanding these difficulties, Goff LJ in *Muirhead* v. *Industrial Tank Specialists Ltd* felt able to say of the *Junior Books* decision:

> 'Faced with these difficulties it is, I think safest for this court to treat *Junior Books* as a case in which, on its particular facts, there was considered to be such a very close relationship between the parties that the defenders could, if the facts as pleaded were approved, be held liable to the pursuers.'

2.13   By 1983 architects, engineers, contractors and sub-contractors were at risk as to claims in negligence from a fairly wide range of potential plaintiffs. These included not only the people with whom they were also in contract such as the developer, but also subsequent owners and occupiers, including tenants and sub-tenants. The criticism of this state of the law began to mount, particularly from the construction professions.

2.14   The first steps on the road to retrenchment in the law of tort came with *Governors of the Peabody Donation Fund* v. *Sir Lindsay Parkinson & Co Ltd* in 1985. It was alleged in the case that the local authority owed a duty to *Peabody* in relation to drains, which had to be reconstructed after they were found to be unsatisfactory. Lord Keith said:

> 'The true question in each case is whether the particular defendant owed the particular plaintiff a duty of care having the scope which is contended for, and whether he was in breach of that duty with consequent loss to the plaintiffs. A relationship of proximity in Lord Atkin's sense must exist before any duty of care can arise, but the scope of the duty must depend on all the circumstances of the case.'

2.15   In relation to Lord Wilberforce's two-stage test in *Anns* (see paragraph 2.11), Lord Keith also said:

> 'There has been a tendency in some recent cases to treat these passages as being themselves of definitive character. This is a temptation which should be resisted.'

2.16   Of the same passage in *Anns*, Lord Brandon continued the retreat in *Leigh & Sillivan Ltd* v. *Aliakmon Shipping Co Ltd* in 1986:

> 'The first observation which I would make is that the passage does not provide, and cannot in my view have been intended by Lord Wilberforce to provide, a universally applicable test of the existence and scope of a duty of care in the law of negligence.'

2.17   The retreat from *Anns* continued in *Curran and Another* v. *Northern Ireland Co Ownership Housing Association Ltd and Another* in 1987 where the court refused to hold that the Northern Ireland Housing Executive owed a duty of care to future owners of a house to see that an extension had been properly constructed.

2.18   The three cases, *Peabody*, *Aliakmon*, and *Curran* have been referred to as the 'retreat from *Anns*' but in *Yuen Kun Yeu* v. *Attorney General of Hong Kong* in 1987, the two-stage test of Lord Wilberforce in *Anns* was probably put to rest by Lord Keith. Further, in *Simaan General Contracting Co* v. *Pilkington Glass* in 1988, the Court of Appeal held that a supplier of glass units for a new building, who had no contractual relationship with the main contractor and had not assumed responsibility to that contractor, was not liable in tort for foreseeable economic loss caused by defects in the units where there was no physical damage to the units, and the contractor had no proprietary or possessory interest in the property.

2.19   It follows from all this that the extensive duties in tort that had been developed in the 1960s, 1970s and early 1980s were in some disarray by 1988. The whole basis of the decision in *Anns* had received widespread criticism and it was inevitable that sooner or later a challenge was mounted in the House of Lords to their previous decision in *Anns*. The first opportunity was in *D. & F. Estates Ltd and Others* v. *Church Commissioners for England and Others* in 1988.

D. & F. Estates Limited and Others v. Church Commissioners for England and Others

2.20   This case concerned defective plastering carried out by sub-contractors to a main contractor. The non-occupying leaseholder, which was a company, claimed against the main contractor (with whom he did not at any time have a contract) in respect of costs of repair to plastering actually carried out, future repair costs and loss of rent. Lord Bridge delivered the main speech, the remainder of their Lordships agreeing with no substantial dissent.

2.21   The non-occupying leaseholder plaintiff had no option but to bring his case in tort against the main contractors for the simple reason that he had no contract with them. It was a difficult case to frame in the law of tort, if for no other reason, because a contractor has no liability in law for the torts of his independent contractor, namely, the sub-contract plasterers. The plaintiffs therefore put their duty as a duty on the part of the main contractor to adequately supervise the work of the plastering sub-contractors. The judge at first instance found for the plaintiffs but the House of Lords overturned that decision and their reasons are of fundamental importance in the area of negligence liability in the construction industry. Consider these two passages from the speech of Lord Brandon in *D. & F.*:

> 'It is, however, of fundamental importance to observe that the duty of care laid down in *Donoghue* v. *Stevenson* was based on the existence of a danger of physical injury to persons or their property. That this is so, is clear from the observations made by Lord Atkin at pages 581 to 582 with regard to the statements of law of Brett MR in *Heaven* v. *Pender* (1883). It has, further, until the present case, never been doubted so far as I know that the relevant property for the purpose of the wider principle on which the decision in *Donoghue* v. *Stevenson* was based was property other than the very property which gave rise to the danger of physical damage concerned.'

> '... there are two important considerations which ought to limit the scope of the duty of care which it is common ground was owed by the appellants to the respondents on the assumed facts of the present case. The first consideration is that, in *Donoghue* v. *Stevenson* itself and in all the numerous cases in which the principle of that decision has been applied to different but analogous factual situations, it has always been either stated expressly, or taken for granted, that an essential ingredient in the cause of action relied on was the existence of danger or the threat of danger or physical damage to persons or their property, excluding for this purpose the very piece of property from the defective condition of which such danger, or threat of danger arises. To dispense with that essential ingredient in a cause of action of the kind concerned in the present case would, in my view, involve a radical departure from long established authority.'

2.22   The essence of what was being said was that the developments in the law of tort between 1932 and 1988 were tantamount to giving Donoghue, in *Donoghue* v. *Stevenson*, not only damages for her personal injury in being made ill by the decomposed snail in the ginger beer bottle, but also requiring the manufacturer to pay for or provide a new bottle of ginger beer, the thing itself. On this basis, it was easy for the House of Lords in *D. & F.* to come to the view that the plaster, being the damaged thing itself, had not caused damage to persons or property (other than the *de minimis* cleaning of carpets involving an expenditure of about £50) and that the non-occupying lease-holder was not entitled to succeed against the contractor. However, in coming to that decision, the House of Lords had some difficulty in reconciling the *Anns* decision, although they did not overrule it. It follows from the *D. & F.* decision that, for example, tenants, purchasers and funds could not rely in future on the possibility of being able to obtain recompense in tort in respect of defects in design or construction of buildings; hence the immediate and urgent boost in the use of collateral warranties since that decision. The collateral warranty tries to fill the gap in the law of tort by creating a contractual relationship.

Murphy v. *Brentwood District Council*

2.23   The decision in *Murphy* was delivered on 26 July 1990; it was widely known that in argument before the House of Lords, the local authority had asked the House of Lords to depart from their previous decision in *Anns* v. *Merton London Borough Council* – the House of Lords can overrule its previous decisions by reason of the Practice Statement (Judicial Precedent) [1966] 1 WLR 1234. Their Lordships, in some detailed judgments, reviewed the state of the law as it had developed since 1932 in relation to negligence, not only in England and Wales but also in the Commonwealth; they gave consideration to some American tort cases as well as looking at their own previous and recent decision in *D. & F.* Lord Keith in *Murphy*, having expressly approved a passage in a case in the High Court of Australia, *Council of the Shire of Sutherland* v. *Heyman*, which declined to follows *Anns*, said this:

> 'In my opinion, there can be no doubt that *Anns* has for long been widely regarded as an unsatisfactory decision. In relation to the scope of the duty owed by a local authority it proceeded upon what must, with due respect to its source, be regarded as a somewhat superficial examination of principle and there has been extreme difficulty, highlighted most recently by the speeches in *D. & F. Estates*, in ascertaining upon exactly what basis of principle it did proceed. I think it must now be recognised that it did not proceed on any basis of principle at all, but constituted a remarkable example of judicial legislation. It has engendered a vast spate of litigation, and each of the cases in the field which have reached this House has been distinguished. Others have been distinguished in the Court of Appeal. The result has been to keep the effect of the decision within reasonable bounds, but that has been achieved only by applying strictly the words of Lord Wilberforce and by refusing to accept the logical implications of the decision itself. These logical implications show that the case properly considered had potentiality for collision with long established principles regarding liability and the tort of negligence for economic loss. There can be no

doubt that to depart from the decision would re-establish a degree of certainty in this field of law which it has done a remarkable amount to upset.'

2.24   Having then noted that the *Anns* decision had stood for some 13 years and that the House of Lords should be cautious in overruling previous decisions of theirs, he said:

> 'My Lords, I would hold that *Anns* was wrongly decided as regards the scope of any private law duty of care resting upon local authorities in relation to their function of taking steps to secure compliance with building by-laws or regulations and should be departed from. It follows that *Dutton* v. *Bognor Regis UDC* should be overruled, as should all cases subsequent to *Anns* which were decided in reliance on it.'

2.25   The effect of this decision is therefore to substantially remove a cause of action in negligence which had been relied upon by tenants, subsequent owners and occupiers for a considerable period of time to enable them to recover damages in respect of negligent design and construction. Collateral warranties, to create a contractual relationship to fill this gap, are now regarded as being an essential matter as an adjunct to the development of commercial property. However, it may be that the *Murphy* case has put greater importance on to the decision of *Hedley Byrne & Co Ltd* v. *Heller & Partners Limited*; it may also be that *Junior Books Ltd* v. *Veitchi Co Ltd*, which was regarded in *Murphy* as being an application of the *Hedley Byrne* principle, has been given something of a boost, notwithstanding the fact that in a great many recent cases, *Junior Books* has been heavily criticised. That criticism was to be found, for example, in *D. & F.* where Lord Bridge said of *Junior Books*:

> 'The consensus of judicial opinion, with which I concur, seems to be that the decision of the majority is so far dependent upon the unique, albeit non-contractual relationship between the pursuer and the defender in that case and the unique scope of the duty of care owed by the defender to the pursuer arising from that relationship that the decision cannot be regarded as laying down any principle of general application in the law of tort or *delict*.'

2.26   With respect to their Lordships, it is very hard for those involved in the workings of the construction industry to understand how the relationship between an employer and a sub-contractor is 'unique' or that the scope of the duty of care in that case could reasonably have been based on that unique relationship. There really can hardly be a more common relationship in the construction industry than that between an employer and a sub-contractor, nominated or otherwise. However, these issues, and in particular the concept of reliance, require further consideration as a result of the *Murphy* decision.

Hedley Byrne after Murphy

2.27   The approaches adopted by their Lordships in *Hedley Byrne* were some-
       what disparate. However, in so far as it is possible to make one statement
       of principle from their judgments, that statement is set out in the headnote
       at page 575 of the All England Law Report and is as follows:

> 'If, in the ordinary course of business or professional affairs, a person
> seeks information or advice from another, who is not under con-
> tractual or fiduciary obligation to give the information or advice, in
> circumstances in which a reasonable man so asked would know that
> he was being trusted, or that his skill or judgment was being relied on,
> and the person asked chooses to give the information or advice with-
> out clearly so qualifying his answer as to show that he does not accept
> responsibility, then the person replying accepts a legal duty to exercise
> such care as the circumstances require in making his reply; and for a
> failure to exercise that care an action for negligence will lie if damage
> results.'

2.28   In essence therefore *Hedley Byrne* was concerned with negligent mis-
       statements of facts or opinions, in circumstances where it was reasonable
       to expect that the recipient of the information would rely on such infor-
       mation and the recipient did in fact rely upon the information. If these
       factors were present and the recipient suffered financial or economic loss,
       then a claim in negligence could be brought against the person who gave
       the information. Such indeed were the facts of *Hedley Byrne* itself. A useful
       illustration of negligent mis-statement in the construction industry is in
       the *IBA* case, the facts of which are set out in paragraphs 1.21 and 1.22. It
       will be recalled that BICC who had no contractual relationship with IBA
       were considered to be liable in respect of a negligent mis-statement
       arising from their letter to IBA dated 11 November 1964 when they
       negligently mis-stated to IBA that they were 'well satisfied that the
       structures will not oscillate dangerously...'. IBA relied upon that
       assurance.

2.29   Clearly therefore the *Hedley Byrne* principle creates a particular species of
       negligence based upon representation and reliance unconnected with
       physical damage and entitling a party to recover economic loss. Unlike
       Anns the species of negligence established by *Hedley Byrne* survives the
       Murphy decision. However, difficulties arise from the following extracts
       from the speeches of Lord Keith and Lord Bridge:

Per Lord Keith:

> 'It would seem that in a case such as *Pirelli General Cable Works Ltd* v.
> *Oscar Faber & Partners* where the tortious liability arose out of a
> contractual relationship with professional people, the duty extended to
> take reasonable care not to cause economic loss to the client by the

advice given. The plaintiffs built the chimney as they did in reliance on that advice. The case would accordingly fall within the principle of *Hedley Byrne*. I regard *Junior Books Ltd* v. *Veitchi Co Ltd* as being an application of that principle.'

Per Lord Bridge:

'There may of course, be situations where, even in the absence of contract, there is a special relationship of proximity between builder and building owner which is sufficiently akin to contract to introduce the element of reliance so that the scope of the duty of care owed by the builder to the owner is wide enough to embrace purely economic loss. The decision in *Junior Books* can, I believe, only be understood on this basis.'

2.30   It is suggested that the problem arising from the speeches of Lord Keith and Lord Bridge is that in *Junior Books* the relationship between the employer and the sub-contractor is not a 'unique' relationship in the construction industry, indeed it is commonplace. Further there appears to be a complete absence of the representation and reliance that was central to the decision in *Hedley Byrne* and the court's approach in *IBA*. It is submitted that the references to 'reliance' in *Junior Books* were in the context of the test of proximity rather than a *Hedley Byrne* reliance.

1990 to 2000

2.31   In the first edition of this book, the authors expressed the view that their Lordships' attempt to explain the decision on *Junior Books* with reference to the principles in *Hedley Byrne* was in danger of 'locking the front door but leaving open a rear window'. Have the intervening years witnessed an open rear window, partially or fully? The authors think 'that the rear window is partially open'.

2.32   In *Henderson and Others* v. *Merrett Syndicates Ltd and Others*, the House of Lords extracted from the case of *Hedley Byrne* a broader principle than that expressed in paragraph 2.27 above. The facts of the *Henderson* case concerned claims brought by Lloyds names against underwriting agents acting either as members agents or managing agents or both, in respect of huge losses suffered by the names as a result of negligent underwriting. The claims were brought both by names who were in direct contractual relationship with the underwriters and by names with whom the agents had no contractual relationship. The court held, inter alia:

(1)   Where a person assumed responsibility to perform professional or quasi-professional services for another who relied on those services, the relationship between the parties was itself sufficient, without more, to give rise to a duty on the part of the person providing the services to exercise reasonable skill and care in doing so.

(2)   An assumption of responsibility by a person rendering professional or quasi-professional services coupled with a concomitant reliance by the person for whom the services were rendered could give rise to a tortious duty of care irrespective of whether there was a contractual relationship between the parties.

2.33   The leading judgment was given by Lord Goff who looked to the speeches of Lord Morris and Lord Devlin in *Hedley Byrne* for the principle upon which that decision was founded. In so doing, Lord Goff did not feel constrained by such concepts as the 'seeking and giving of information or advice' nor 'the giving of information and advice by a person who is not under contractual or fiduciary obligation' referred to in paragraph 2.27 above. For Lord Goff, and indeed the rest of their Lordships in *Henderson*, the fundamental importance of *Hedley Byrne* was to establish the following principle – an assumption of responsibility coupled with reliance by the claimant, which in all the circumstances made it appropriate that a remedy in law should be available. Furthermore, the assumption of responsibility could arise from a relationship between the parties that was either specific or general to the particular transaction and extended beyond the provision of information and advice to include the performance of other services.

2.34   As already stated, claims for purely economic loss can be pursued under *Hedley Byrne*. Lord Goff added that once a case was identified as falling within the principle of *Hedley Byrne* there was no need to satisfy the 'fair just and reasonable test' (applied to negligence cases generally by the House of Lords in *Caparo Industries plc v. Dickman*).

2.35   Constraints to the broad principle enunciated by Lord Goff can be derived from the references to 'professional' or 'quasi-professional services'. Clearly that definition would include the acts of architects, engineers, surveyors and possibly contractors or sub-contractors undertaking a design function. *Quaere* whether it would extend to contractors or sub-contractors carrying out construction works as distinct from design. If *Junior Books* is followed where the common relationship in the construction industry of employer and nominated sub-contractor was considered to be a special relationship, the 'floodgates' would appear to open, particularly if you consider the contractual structure in *Henderson* – the primary and sub-contractual relationship between the names, the members agents and the managing agent. However, Lord Goff hinted at another potential restraint: the contractual chain. After distinguishing the situation arising in *Henderson* as 'most unusual' he went on to state:

'... that in many cases in which a contractual chain comparable to that in the present case is constructed it may well prove to be inconsistent with an assumption of responsibility which has the effect of, so to speak, short-circuiting the contractual structure so put in place by the parties. It cannot therefore be inferred from the present case that other sub-agents will be held directly liable to the agents principal in tort. Let

me take the analogy of the common case of an ordinary building contract, under which main contractors contract with the building owner
for the construction of the relevant building, and the main contractor
sub-contracts with sub-contractors or suppliers (often nominated by the
building owner) for the performance of work or the supply of materials
in accordance with standards and subject to terms established in the
sub-contract. I put on one side cases in which the sub-contractor causes
physical damage to property of the building owner, where the claim
does not depend upon an assumption of responsibility by the sub-
contractor to the building owner; though the sub-contractor may be
protected from liability by a contractual exemption clause authorised
by the building owner. But if the sub-contracted works or materials do
not in the result conform to the required standard, it will not ordinarily
be open to the building owner to sue the sub-contractor or supplier
direct under the *Hedley Byrne* principle, claiming damages from him on
the basis that he has been negligent in relation to the performance of his
functions. For there is generally no assumption of responsibility by the
sub-contractor or supplier direct to the building owner, the parties
having so structured their relationship that it is inconsistent with any
such assumption of responsibility. This was the conclusion of the Court
of Appeal in *Simaan General Contracting Co* v. *Pilkington Glass Ltd*.'

2.36   Of *Junior Books*, Lord Goff accepted that this case created 'some difficulty'
        to the above analysis; however he left it there, feeling that it was 'unne-
        cessary … to reconsider that decision for the purposes of the present
        appeal'.

2.37   The principle of *Hedley Byrne* was revisited by the House of Lords in the
        case of *White* v. *Jones*, a judgment delivered shortly after *Henderson*.

2.38   The facts of *White* v. *Jones* concerned a claim in negligence against a tes-
        tator's solicitors by intended beneficiaries, the testator's daughters, for the
        solicitors' failure to draw up a will to be executed by the testator
        bequeathing gifts of money to the daughters. The court held that the
        solicitors were liable in tort for the economic loss suffered by the intended
        beneficiaries because:

    (1) (Per Lord Goff and Lord Nolan) The assumption of responsibility by
         a solicitor towards his client should be extended in law to an
         intended beneficiary who was reasonably foreseeably deprived of
         his intended legacy as a result of the solicitor's negligence in cir-
         cumstances in which there was no confidential or fiduciary rela-
         tionship and neither the testator nor his estate had a remedy against
         the solicitor, since otherwise an injustice would occur because of a
         lacuna in the law …

    (2) (Per Lord Browne-Wilkinson and Lord Nolan) Adopting the
         incremental approach by analogy with established categories of
         relationships giving rise to a duty of care, the principle of assump-
         tion of responsibility should be extended to a solicitor who accepted

instructions to draw up a will so that he was held to be in a special relationship with those intended to benefit under it, in consequence of which he owed a duty to the intended beneficiary to act with due expedition and care in relating to carrying out those instructions.

2.39   Lord Goff felt that the broad principle of assumption of responsibility did not apply to the testator's solicitor and an intended beneficiary, although it could apply to the solicitor and the testator. His reasoning was that in the absence of special circumstances, there will have been no reliance by the intended beneficiary on the exercise by the solicitor of due care and skill; the intended beneficiary may not even have been aware that the solicitor was engaged on such a task or that his position might be affected. In the authors' opinion, Lord Goff's position in *White* v. *Jones* is more conservative than the views he expressed in *Henderson* at paragraph 2.33 above. In any event Lord Goff appears in *White* v. *Jones* to be deciding the case on its own particular facts, and retreating from the broad principle stated in *Henderson*. Similarly Lord Browne-Wilkinson was also seen to 'draw some lines in the sand'. He stated:

> 'The law of England does not impose any general duty of care to avoid negligent mis-statements or to avoid causing pure economic loss even if economic damage to the plaintiff was foreseeable. However, such a duty of care will arise if there is a special relationship between the parties. Although the categories of cases in which such a special relationship can be held to exist are not closed, as yet only two categories have been identified, viz (1) where there is a fiduciary relationship and (2) where the defendant has voluntarily answered a question or tenders skilled advice or services in circumstances where he knows or ought to know that an identified plaintiff will rely on his answers or advice. In both these categories the special relationship is created by the defendant voluntarily assuming to act in the matter by involving himself in the plaintiff's affairs or by choosing to speak ... such relationship can arise even though the defendant has acted in the plaintiff's affairs pursuant to a contract with a third party.'

2.40   Lord Browne-Wilkinson did not consider that there was a special relationship within the definition of the two categories set out in paragraph 2.39; however to avoid the situation of there being no remedy in law when justice demanded there be one – the 'legal black hole' mentioned in the *GUS Property Management* case (see paragraph 6.65) – he was content to establish an incremental extension to the categories of special relationship to include the facts of *White* v. *Jones*.

2.41   Lord Mustill delivered a powerful dissenting judgment. He posed the following question:

> 'If A promises B to perform a service for B which B intends, and A knows, will confer a benefit on C if it is performed, does A owe to C in tort a duty to perform that service?'

2.42 To answer this question, Lord Mustill, like Lord Goff in *Henderson*, felt it necessary to determine the principle underlying the decision in *Hedley Byrne*. He detected four themes in the speeches of their Lordships in *Hedley Byrne*: mutuality, special relationship, reliance and undertaking of responsibility. He concluded that, in essence, *Hedley Byrne* was concerned with the undertaking of legal responsibility for careful and diligent performance in the context of a mutual relationship. This was to be contrasted with the principle derived from *Donoghue* v. *Stevenson* where the relationship was imposed externally from the position in which they found themselves. In *Hedley Byrne* the 'liability arose internally from the relationship in which the parties had together chosen to place themselves', i.e. there was mutuality. Lord Mustill concluded, in answer to the question he had posed, 'that to hold that a duty existed, even prima facie, in such a situation, would be to go far beyond anything so far contemplated by the law of negligence'. Lord Mustill had said very little in *Henderson*. Could it be that he was having second thoughts as to the potential breadth of principle inherent in Lord Goff's judgment?

2.43 In *Gable House Estates Ltd* v. *The Halpern Partnership and Another*, Judge Esyr Lewis QC derived the following principles from *Henderson*:

(1) A concurrent duty in tort can exist where the parties are in a contractual relationship if the terms of the contract do not preclude it.

(2) That concurrent duty in tort will arise and enable a plaintiff to recover damages for economic loss where, for example, in a contract for services, certainly where the defendant has special knowledge or skills, the defendant assumes responsibility to the plaintiff. The assumption of responsibility to perform those services with due skill and care will arise from the very nature of the services which the defendant undertakes to provide.

(3) There must, however, be a 'concomitant reliance' by the plaintiff to enable him to establish his claim.

(4) 'Reliance' may take different forms in different circumstances.

(5) The assumption of responsibility and the duty to take care is not confined to cases concerned with erroneous information and advice. For example, a solicitor may be liable if he fails to do something that it was his responsibility to do for his client, which causes loss. In this kind of situation, the fact that the solicitor had been retained and what he failed to do was within the scope of the duties his retainer required of him will be sufficient to establish reliance.

(6) However, where the complaint is that there has been negligent advice or negligent misinformation, the plaintiff must show that he specifically relied on that advice or was actually misled.

2.44 Principles 2 to 5 appear to the authors to be 'propping open the rear window'. Certainly this was the view, albeit *obiter*, of Bedlam LJ in the case of *Barclays Bank plc* v. *Fairclough Building Ltd*, where he stated:

'A skilled contractor undertaking maintenance work to a building assumes a responsibility which invites reliance no less than the financial or other professional adviser does in undertaking his work. The nature of the responsibility is the same though it will differ in extent ... I would hold that [the sub-contractor] in performing the work ... sub-contracted to it owed a concurrent duty in tort to avoid causing economic loss by failing to exercise the care and skill of a competent contractor.'

By way of contrast, in *Plant Construction plc* v. *Clive Adam Associates and JMH Construction Services Ltd* (see paragraph 5.5) the court at first instance held that the facts of that case did not disclose a relationship which would give rise to any duty of care between a third party sub-contractor, JMH, and the employer, Ford. Judge Humphrey Lloyd QC considered that the imposition of a duty of care on the basis of an assumption of responsibility was wholly inconsistent with the contractual structure between employer and main contractor. The court was acting on the constraint enunciated by Lord Goff in *Henderson* (see paragraph 2.35 above). It would appear that the ratio in *Junior Books* is being 'laid to rest' slowly but with dignity!

Chapter 3
Contracts (Rights of Third Parties) Act 1999

The background

3.1   The doctrine of privity of contract has been well established in the law of England and Wales for some time, having developed its modern form in the nineteenth and early twentieth centuries. It was well expressed in 1915 in *Dunlop Pneumatic Tyre* v. *Selfridge* by Viscount Haldane:

> '... in the law of England, certain principles are fundamental. One is that only a person who is a party to a contract can sue on it. Our law knows nothing of a *jus quaesitum tertio* arising by way of contract. Such a right may be conferred by way of property, as, for example, under a trust, but it cannot be conferred on a stranger to a contract as a right to enforce the contract *in personam*. A second principle is that if a person with whom a contract not under seal has been made is to be able to enforce it, consideration must have been given by him to the promisor or to some other person at the promisor's request. These two principles are not recognised in the same fashion by the jurisprudence of certain continental countries or of Scotland, but here they are well established. A third proposition is that a principal not named in the contract may sue upon it if the promisee really contracted as his agent. But again, in order to entitle him so to sue, he must have given consideration either personally or through the promisee, acting as his agent in giving it.'

3.2   More recently, in 1977, Lord Diplock described the doctrine of privity of contract as '... an anachronistic shortcoming that has for many years been regarded as a reproach to English private law' (*The Albazero*).

3.3   Before the Contracts (Rights of Third Parties) Act 1999, it can be seen, therefore, that there are two aspects of privity in English law that are important in the context of collateral warranties. The first is that a contract does not create a right for the benefit of a party who is not a party to the contract. The second is that where a contract is not executed under seal or as a deed, a party cannot enforce the contract unless he has given consideration to the promisor. There has been much debate in the law as to whether these two aspects are in reality separate, or whether they are both just an expression of the rule that consideration must move from the promisee. Collateral warranties are created to deal with the first aspect; it is also primarily the first aspect that is addressed by the Contracts (Rights of Third Parties) Act 1999, although the Act modifies the rules on consideration in a way that is needed to enable the Act to work.

3.4 This rule of privity has been eroded to some extent by the courts in order to remove what might otherwise be seen as an injustice. Examples of this exercise include devices such as collateral contracts (see, for example, *Shanklin Pier Ltd* v. *Detel Products* at paragraph 1.6), agency, assignments of the benefit of contracts (which can be effective without the consent of the promisor under the contract and, of course, the law of tort, which has provided remedies in many cases where actions in contract were prevented by the privity rule. These, necessary, developments in the law were creating illogicalities and making the law difficult to see clearly in every factual situation.

3.5 In construction cases, the difficulties created by the rule of privity of contract have been extensive and led to calls for reform of the law. In *Darlington Borough Council* v. *Wiltshier Northern Limited*, Steyn LJ said:

> 'The case for recognising a contract for the benefit of a third party is simple and straightforward. The autonomy of the will of the parties should be respected. The law of contract should give effect to the reasonable expectations of contracting parties. Principle certainly requires that a burden should not be imposed on a third party without his consent. But there is no doctrinal, logical or policy reason why the law should deny effectiveness to a contract for the benefit of a third party where that is the expressed intention of the parties. Moreover, often, the parties, and particularly third parties, organise their affairs on the faith of the contract. They rely on contract. It is therefore unfair to deny effectiveness to such a contract.'

3.6 Other objections were clear to the continuing existence of the doctrine of privity in the law of England and Wales. The possibility is that a contract-breaker can escape liability if he causes loss to a third party who is not a party to the contract, such as a lessee of a building in which there are defects arising out of the construction of the building under a building contract between the contractor who built the building and the developer/landlord who leases it. In *Linden Gardens Trust v Lenesta Sludge Disposals*, the House of Lords permitted a party to the contract to sue on behalf of a non-party, but the non-party could not require that other party so to sue. This might be said to be a wholly unsatisfactory situation.

3.7 Another objection was that the law of England and Wales was out on a limb compared with most other countries. Many common law jurisdictions had removed and/or modified the doctrine of privity of contract by legislation (for example, Eire, Western Australia, Queensland and New Zealand).

3.8 In Scotland, the rule had never existed in the first place in the form it takes in England and Wales; the Scottish doctrine of *jus quaesitum tertio*, creates the possibility, in theory at any rate, of two contracting parties agreeing in their contract to confer a benefit on a third party, that is enforceable by that party. It may be that the liability of the party liable to make performance to the third party is to be assessed by reference to the consequences for the third party, not the other party to the contract (see the comments of Lord Eassie in *Blyth & Blyth Limited* v. *Carillion Construction Limited*). In any event, notwithstanding the doctrine of *jus quaesitum tertio*, written collateral warranties in the construction field are the norm in Scotland.

3.9 After an attempt by the Law Revision Committee in 1937 to move reform forward, it was not until 1991 that the Law Commission published its Consultation Paper No. 121: *Privity of Contract: Contracts for the Benefit of Third Parties*. The Law Commission examined meticulously, and in considerable detail, the state of the law and what reforms might be considered.

3.10 For example, at paragraph 4.14 of Consultation Paper No. 121, it said:

> 'As a result of cases such as *Murphy* and *D. & F. Estates*, third parties (such as property financiers, purchasers and tenants) frequently seek to protect themselves by means of collateral warranties made with the developer, contractor, sub-contractors and professionals, such as architects, surveyors and structural engineers. In the case of the average shopping centre, one professional may be expected to enter into separate warranty transactions with the financiers, the purchaser and 50 or more tenants. One possible effect of any reform could be to reduce the present complexity by removing the need for so many separate documents. If the contract between developer and contractor were expressed to be for the benefit of financiers, purchasers and tenants alike, it could remove the need for collateral warranties.'

3.11 After the publication of Consultation Paper No. 121, there was considerable consultation which showed a majority in favour of reform, ultimately leading to the publication in 1996 of Law Commission Report 242: *Privity of Contract: Contracts for the benefit of Third Parties*. This report is a very long and thorough examination of the law, following the consultation period. Its conclusion was that legislation was necessary and desirable.

3.12 Law Commission Report 242 set out the arguments for reform:

- The intentions of the original contracting parties are thwarted.
- Injustice to the third party.
- The person who has suffered loss cannot sue, whilst the person who has suffered no loss can sue.
- Even if the promisee can obtain a satisfactory remedy for the third party, the promisee may not be able to, or wish to, sue.
- The development by the courts of exceptions to the doctrine of privity that were not comprehensive.
- Complexity, artificiality and uncertainty.
- Widespread criticism throughout the common law world.
- The fact that most member states of the European Union allow third parties to enforce contracts.

- The difficulties created in commercial life by the doctrine. An extract from the report with specific reference to construction contracts is set out in Appendix 2 of this book. That extract summarises very well the particular difficulties faced on construction projects and how they might be resolved by reform. In short, the Law Commission saw reform as a positive way to assist developers, funders, contractors, sub-contractors and the professional team.

3.13   Some of the important recommendations of Law Commission Report 242 were:

- The rule of English law whereby a third party to a contract may not enforce it, should be reformed so as to enable contracting parties to confer a right to enforce the contract on a third party.
- A right to enforce a contract means (1) a right to all remedies given by the courts for breach of contract (and with the standard rules applicable to those remedies applying by analogy) that would have been available to the third party had he been a party to the contract, including damages, awards of an agreed sum, specific performance and injunctions; and (2) a right to take advantage of a promised exclusion or restriction of the promisor's rights as if the third party were a party to the contract.
- That there should be legislation.
- The legislation should ensure that the rule that consideration must move from the promisee is reformed to the extent necessary to avoid nullifying the proposed reform of the doctrine of privity.

3.14   As part of its Report No. 242, the Law Commission published a draft Contracts (Rights of Third Parties) Bill to be put before parliament. In due course, parliament passed into law the Contracts (Rights of Third Parties) Act 1999.

Contracts (Rights of Third Parties) Act 1999

3.15   The text of the Act is set out in Appendix 3 of this book.
3.16   In the remainder of this chapter, references to 'the Act' mean the Contracts (Rights of Third Parties) Act 1999 and the words 'promisor' and 'promisee' have the same meaning as they do in the Act:

- Promisor means the party to the contract against whom the term is enforceable by the third party, and
- Promisee means the party to the contract by whom the term is enforceable against the promisor.

Commencement and extent

3.17   The Act came into force on 11 November 1999 but, by reason of section 10(2) however, it did not begin to apply to contracts entered into for a moratorium period, which was a period of six months starting on the day the Act came into force, which was 11 May 2000. Section 10(3) provides that the parties to a contract could agree to the application of the Act even if it was entered into during that moratorium period.

3.18   The Act applies in England and Wales and in Northern Ireland (section 10). In the case of Northern Ireland, it applies with some modifications that are set out in section 9 of the Act. Those modifications are related to correctly dealing with affected existing Northern Ireland statutes, so subject to that, the effect in Northern Ireland is much the same as in England and Wales.

3.19   It follows from these provisions of the Act that this new legislation has no application to Scotland.

The right of a third party created by the Act

3.20   Section 1(1) of the Act provides:

> 'Subject to the provisions of this Act, a person who is not a party to a contract (a "third party") may in his own right enforce a term of the contract if –
> (a) the contract expressly provides that he may, or
> (b) subject to subsection (2), the term purports to confer a benefit on him.'

3.21   It is to be noted that the right created is to enforce a *term* of a contract, not the *whole contract* itself. For example, if a building contract contains a term that the contractor is required to use materials of good quality, then that term might be the subject of a third party enforcement right, subject to the other requirements for the existence of the right being met.

3.22   There are, therefore, two categories of potential third parties given this new right in section 1 of the Act to enforce a contract term, although it is not clear from the drafting of the section whether they are intended to be mutually exclusive or, if both have to be considered on particular facts, how they interplay and relate.

3.23   The first category, where the contract expressly confers the right, is probably uncontroversial and certain in its legal effect. It creates by statute a similar effect to that which applies in Scotland by reason of *jus quaesitum tertio* (see paragraph 3.8). It would permit, for example, the express selection of certain terms of the contract and stating them to be enforceable by named third parties The second category is being seen as leading to some uncertainty and requires further consideration.

3.24   The second category is where the term of the contract 'purports to confer a benefit' on the third party (section 1(1)(b)), but is also subject to the provisions of section 1(2) of the Act. Section 1(2) of the Act provides:

> 'Subsection 1(b) does not apply if on a proper construction of the contract it appears that the parties did not intend the term to be enforceable by the third party.'

3.25   The first uncertainty arising on the second category is how the courts will construe the meaning of 'purports to confer a benefit' on the third party. Are contract terms in a building contract that deal with quality and complying with Building Regulations, purporting to confer a benefit on a third party in the absence of any express reference to the types of third parties that there might be? Is there to be a presumption that a development might be sold or leased such that the third party right contemplated by the Act automatically benefits purchasers and tenants by giving them a right against the contractor in the event of breaches of the contract to enforce terms in the contract relating to quality and building regulations? Without very careful drafting of the building contract (by, for example, expressly identifying categories of third parties and those terms of the contract that are to be subject to third party rights), can the tenant or purchaser be sure that he has such rights? Is the promisor under the contract going to find that he has created rights in third parties that he did not intend to create? Is a third party, such as a tenant, able to benefit from a term in a subcontract, when the main contractor is in liquidation?

3.26   A further point is where the third party has a direct contractual relationship with the promisor; for example, an employer has a design and build contract with a contractor. That contractor has employed an architect under an appointment. The architect has given a collateral warranty to the employer. Can it be said that the appointment between the architect and the contractor purports to confer a benefit on the employer? It seems likely that in such circumstances the parties did not intend the relevant term to be enforceable (section 1(2) of the Act) on the basis that the employer has taken the trouble to create a contractual relationship with the third party, which may well be construed as setting out all his rights against the third party. This will, no doubt, be clarified in the courts in due course.

3.27   The second uncertainty is created by the fact that the right under section 1(1)(b) does not apply, if, on a proper construction of the contract, it appears that the parties did not intend the term to be enforceable by the third party (section 1(2)). It is clear that the court can construe *the whole contract*, not just *the term* sought to be enforced by the third party. It also seems clear that there is a presumption of a third party right, rebuttable by the promisor. In other words, the burden of proof to prevent third party rights is on the promisor, not the third party. In seeking to discharge that burden, it seems likely that extrinsic evidence as to the 'factual matrix' surrounding the making of the contract will be admissible (see paragraph 1.66). After all, contracts are not made in a vacuum.

3.28 It is these uncertainties, both for third parties and the contracting parties, that are likely to lead to decided cases in the courts over the next few years. It is the same uncertainties, and the absence of decided cases at present, that are causing the Act not to be used in the way that the Law Commission had hoped (see paragraph 3.10) to prevent the need for collateral warranties in construction. This aspect is developed further in paragraphs 3.49 to 3.60, including the now common use of contract provisions expressly excluding third party rights. This is clearly an option created by the drafting of section 1 of the Act.

3.29 However, by the same route, a developer can benefit from the Act in circumstances where the land is owned by a different company to that entering into the building contract and/or where there may be transfers of the land post-completion of the project between different companies in the same group. As the decided cases show, absent the Act, this can lead to the company that has the right to sue having no loss while the company that suffers the loss has no right to sue. The solution provided by the Act is to name, by express reference in the warranties and the building contract, all the companies in the group of companies to whom third party rights are being given (even generically), including companies that become part of the group after the date of the warranty. In this way, third party rights can be created for every member of a group of companies who may have a legitimate interest in pursuing the contractor or the professional team in the future (see paragraph 3.59).

3.30 In enforcing a term of the contract, the third party has available to him any remedy that would have been available to him in an action for breach of contract as if he had been a party to the contract. It follows that the rules relating to damages, injunctions, specific performance and other relief apply (see section 1(5)).

3.31 The Act also deals with contract terms that seek to exclude or restrict liability and how that is to be applied vis-a-vis the third party. In essence, any term excluding or restricting liability applies to the third party by reason of section 1(6), which states:

> 'Where a term of a contract excludes or limits liability in relation to any matter, references in this Act to a third party enforcing a term shall be construed as references to his availing himself of the exclusion or limitation.'

Who are the relevant third parties?

3.32 The Act sets out at section 1(3) that a third party must be expressly identified:

- in the contract by name
- as a member of a class, or
- as answering a particular description, but need not be in existence when the contract is entered into.

3.33   It is clear that members of a class or answering to a particular description could create very wide categories of third parties in construction. For example, these categories could include purchasers (not limited to a first purchaser) and tenants (not limited to the first tenant or first assignee), even though the names of those parties were not known at the date of contract and even though they were not in existence at the date of contract. This possible wide ranging category of beneficiaries is also a factor in the current climate of exclusion of third party rights in the construction field by an express term of the contract.

Variation and rescission of the contract

3.34   The Act goes some way, once a third party right to enforce a term has arisen, to prevent the contracting parties agreeing to rescind or vary the contract in such a way as to extinguish or alter the rights of a third party, unless the third party consents (section 2), in certain circumstances. Those circumstances are:

- where the third party has communicated his assent to the term to the promisor
- the promisor is aware that the third party has relied on the term, or
- the promisor can reasonably be expected to have foreseen that the third party would rely on the term and the third party has in fact relied on it.

3.35   Clearly, this provision requires careful consideration of the drafting of the contract from the outset. The use of clauses permitting the varying of the contract by the contracting parties without the consent of any third party might be considered but may well fall foul of section 2 of the Act. A safer alternative might be a contract provision that where a third party's consent is required under section 2 of the Act, then such third party shall not unreasonably withhold or delay his consent.

3.36   Another point arises in relation to variations and changes to the work under construction contracts, which is a common provision in construction contracts. Is that kind of provision caught by section 2 of the Act? The answer appears to be that such a variation is a variation in the *work* and not a variation to the *contract*. As such, it would appear that it is not subjected to the rules in section 2 of the Act. Indeed, it seems that the third party's right under the Act would be subject to variations in the work. That analysis is accepted by the Law Commission in its Report No. 242 (see paragraph 9.37 of the report).

Defences available to a promisor

3.37   The Act, subject to any express terms of the contract, reserves to the promisor, in any enforcement of a contract term by a third party, the same defences and/or set-off rights that the promisor would have had available to him if the proceedings had been brought by the promisee (section 3). This provision may come as a surprise to third parties who might otherwise have thought they had a clear run at a third party claim, without the risk of being met by such defences and set-off claims by the promisor. The scope of this statutory provision will be the subject of decided cases in due course; in the meantime, it is possible to see that a compromise in relation to a defects claim between a building owner and a contractor (as promisor) could well be a defence to a claim from a third party promisee.

Protection of promisor from double liability

3.38   At the same time, there is protection of the promisor from double liability. If the promisee has already recovered a sum in respect of the third party's loss in respect of the relevant contract term, then the third party's damages are reduced by an amount that the court or arbitral tribunal think is appropriate (section 5). There is no guidance in the Act as to how a court or arbitration tribunal is to exercise the discretion that is so given. Indeed, the wording appears to contemplate that the promisor's total liability to the third party could be more than the sum that he has already paid to the promisee. It is hard to see how this could arise in practice but it is an outcome that is permitted by the wording of section 5 of the Act.

Dispute resolution and arbitration provisions in contracts

3.39   It is the case that many construction contracts contain arbitration agreements. When the Act was a Bill before Parliament the policy question arose as to how this should be dealt with in considering whether or not the third party should also be bound by the arbitration agreement in the contract when it seeks to enforce its third party rights.

3.40   The Act deals with this at section 8. In essence it is in two parts. The first is to create the possibility of making a third party's right to intervene subject to the application of the arbitration agreement in the contract (section 8(1)). Where that route has been chosen in the drafting of the contract, then the second part is to provide for third parties to be able to insist on arbitration if extra-contractual claims are brought against the third party (section 8(2)). In respect of both of these parts, the arbitration agreement has to be an agreement in writing for the purposes of the Arbitration Act 1996 (sections 8(1)(b) and 8(2)(b)). Otherwise, the remedies must be pursued in court, not in arbitration. This raises interesting questions in relation to arbitration agreements in construction contracts and sub-contracts, particularly name-borrowing arbitrations, which are beyond the scope of this book.

Aspects of the Act

The Housing Grants, Construction and Regeneration Act 1996

3.41   No mention is made in the Act of the Housing Grants, Construction and Regeneration Act 1996, which introduced by statute a right to refer disputes to adjudication in every construction contract. 'Construction contract' is defined in the Housing Grants, Construction and Regeneration Act 1996 and has been the subject of numerous decided cases as to the scope and effect of the provisions in that Act.

3.42   One of the open questions in relation to collateral warranties is whether or not they are 'construction contracts' and this is discussed in paragraphs 9.24 to 9.28. However, what does seem to be clear is that third party rights created by the Act cannot be 'construction contracts' for the purposes of statutory adjudication, either under a contract that has provisions complying with the Housing Grants Act or under the Scheme for Construction Contracts (England and Wales) Regulations 1998, in the absence of an express provision in the contract making third party rights subject to a compliant scheme for adjudication. In such a case, the promisor under a construction contract could find himself subject to adjudication proceedings at the hands of the promisee and, at the same or a different time, court proceedings at the hands of a third party. Adjudication is a much faster process (28 days) than court proceedings. It is likely, therefore, that the promisee will get a much quicker decision than a third party. That means that it is unlikely in practice that adjudication will ever be incorporated in a contract expressly for the benefit of a third party; developers and employers, who usually control the drafting of construction contracts, are likely to want to have the advantage of a quick result in an adjudication and not to extend that to third parties.

Existing cases and the Act

3.43   Quite often in construction contracts, there are provisions against assignment of the benefit of the contract in the absence of the agreement of the other contracting party. A purported assignment in the absence of such consent will be likely to be void and of no effect (see paragraphs 4.38 to 4.49), and, for example, *Linden Gardens Trust Ltd* v. *Lenesta Sludge Disposals Ltd*). Does the Act assist parties who have sold property, after completion of the works carried out under the building contract, with an ineffective assignment to the purchaser of the benefit of the building contract?

3.44   That was the position in *Linden Gardens Trust Ltd* v. *Lenesta Sludge Disposals Ltd*, where a property was transferred from one group company to another, with an assignment of the benefit of the contract that was ineffective because it did not have the consent of the other contracting party, as required by the contract. The House of Lords decided, on the facts of that case, that the employer was entitled to enforce contractual rights against the contractor for the benefit of third parties. It seems that the Act does not change the position in that case and neither does it provide an additional remedy. The Law Commission Report No. 242 at paragraph 5.16 states:

> 'On the facts of *Linden Gardens* itself, there will be no question of a third party having a right of enforcement under our proposed reform. The property in question had been sold to the third party after the contract for the work on the property had been entered into and there was a clause in the works contract barring assignment of the rights under it. The recognition in that case that the promisee could have recovered damages based on the third party's loss will be as important after the implementation of our proposed reform as it is under the present law; and we would not wish our proposed reform to be construed as casting any doubt on the decisions in *The Albazero*, *Linden Gardens* and *Darlington BC* v. *Wiltshier Northern Ltd.*'

3.45   In short, then, the Act leaves to the common law the remedies available to the promisee in a contract otherwise enforceable by a third party in these kinds of circumstances.

3.46   However, the Act clearly offers the possibility of doing away with the need for repeated assignments in circumstances where it is clear that the benefits of building contracts and professional appointments, together with collateral warranties, are intended to be available to future purchasers and tenants. This opportunity is not one that is presently being widely exploited by lawyers and their clients in the property field. It is, perhaps, a reflection of the conservatism of commercial lawyers and developers that collateral warranties, with assignment provisions, continue to be the norm on commercial developments notwithstanding all the difficulties they cause in negotiation and professional indemnity insurance. The reality is that many professional advisers are taking the safer course at present by advising reliance on collateral warranties rather than reliance on an, as yet, untested Act of Parliament (but see paragraphs 7.5 to 7.14 in relation to developers who are not the employer under the building contract).

3.47   In analysing the factual situations in previous cases, deciding whether the Act would have made a difference to the outcome, depends on whether the presumption created by the Act (under section 1(1)(b)) that the third party was intended to have the right to enforce the term of the contract, can be rebutted. One example is *Junior Books* v. *Veitchi*, where there was a chain of contracts from Junior Books, through a main contractor to Veitchi as sub-contractor. The Law Commission, in Report No. 242 (paragraph 7.47), believes that the presumption of an enforceable third party right in this case would be rebutted by reason of that chain of contracts, and Junior Books would have had no rights against Veitchi. However, it has to be said that such an analysis may not be right and there is clearly an argument to be had under the Act as to whether or not the presumption can be rebutted in similar, and very common, factual situations. Most standard form sub-contracts name not only the sub-contractor and main contractor, but also the employer; such sub-contracts describe the work to be carried out under the sub-contract as being part of the work to be carried out under the main contract and do not contain express third party rights to enforce a term of the contract. In short, that is precisely the kind of situation where it might be argued, for example, that a term in the sub-contract purports to confer a benefit on the employer, namely the carrying out of work in accordance with the sub-contract, free of defect. Thus, on the face of it, the hurdle of section 1(1)(b) is overcome by the employer. The real argument will come on section 1(2) as to whether 'on a proper construction of the contract it appears that the parties did not intend the term to be enforceable by the third party'. If it was not so intended, then there can be no third party rights. Is the mere existence of a chain of contracts sufficient to rebut the presumption that third party rights are available? This is a point that is very likely to be litigated, probably in circumstances where there are defects and the chain of contracts is broken, say by main contractor insolvency. The Law Commission's analysis seems to the authors of this book not to give sufficient credit to the contrary arguments that might be made in proceedings on the point.

3.48   A similar analysis in *Woodar Investments Development Ltd* v. *Wimpey Construction UK Ltd*, may well produce a different result. There, the contract between Woodar and Wimpey provided for part of the purchase price of a plot of land to be paid to Transworld Trade Ltd, who was not a party to the contract. The Law Commission in Report No. 242 (paragraph 7.49) takes the view that if the Act had been in force '... Transworld could have brought proceedings for the sum directly. The contract purported to confer a benefit on Transworld, who was expressly identified and, in our view, the purchasers could not have rebutted the presumption that Transworld was intended to have the right to enforce the payment obligation'. That appears to be a correct analysis of the position under the Act.

The response to the Act of the draftsmen of standard forms of contract

3.49   The response of the draftsmen of standard forms of contract in the construction industry has been almost universally to exclude third party rights by new contract terms. This is a reflection of the uncertainties referred to in paragraphs 3.25 to 3.28 earlier and the fact that the construction industry and its professions were very much against the introduction of this legislation in the first place. The opportunity given by the legislation to opt out of third party rights has been the norm in contractual arrangements since the Act came into force, whether on standard forms or bespoke contracts.

3.50   The following organisations have issued amendments excluding the effect of the Act in standard forms: ACE, ICE (5th, 6th and 7th Editions, Design and Construct, Minor Works and Ground Investigation Contract), RIBA and the JCT. This seems to be a reaction to the matters set out in paragraphs 3.25 to 3.28 earlier and may change as the Act is considered by the courts and some of the uncertainties are resolved. None of the standard forms of collateral warranty excluded the operation of the Act vis-a-vis third parties but this has usually been dealt with in commercial situations arising at present by a bespoke amendment to the warranty to exclude third party rights: CoWa/F and CoWa/P&T, both of which were published prior to the Act coming into force. However, the 2001 editions of MCWa/F and MCWa/P&T do expressly exclude the operation of the Act, as do the JCT Subcontractor Warranties published in 2001, SCWa/F and SCWa/P&T. The Scottish Building Contract Committee Standard Forms of Collateral Warranty (MCWa/F/Scot(Funder) and MCWa/P&T/Scot(Purchaser and Tenant) do not need to deal with this because the Act does not apply to Scotland.

Using the Act or not using the Act?

3.51   The Act offers great opportunities to simplify the documentation for projects in relation to collateral warranties. What should be the considerations that determine whether and how this should be done?

3.52   There are three possible approaches, which are considered separately below:

- Exclude all third party rights by an express contractual term
- Let the Act apply by making no reference to it in contracts
- Use the Act to create express rights for expressly stated third parties but excluding other third party rights.

Exclude all third party rights

3.53   The Act leaves open one clear area of uncertainty for those involved with construction contracts. That is under section 1(1)(b), which empowers a third party to enforce a term of a contract where it 'purports to confer a benefit on him' and in relation to ascertaining which third parties are so empowered.

3.54   There does seem to be compelling logic in the construction field for avoiding the uncertain effect of section 1(1)(b) by excluding the whole effect of the Act by a term in the contract. That is the course presently being adopted by most lawyers. This is perhaps taking a sledgehammer to the Act by reason of a disadvantage and, thereby, squashing the possible advantages at the same time. In short, in most projects, the possible benefits of the Act are being lost because of the real uncertainties created by part of the Act.

Let the Act apply

3.55   Another option is just to let the Act apply by not excluding it in contracts. This course will not be appealing for the same reasons that are set out in paragraphs 3.53 to 3.54 above, namely, uncertainty as to the legal effect, particularly of section 1(1)(b).

Use the Act to create express rights

3.56   The logical result of careful consideration of the Act in the construction field is, therefore, either to expressly use the Act or to exclude it. Is it possible to create express third party rights under the contract to a limited, and expressly stated, class of third parties and, at the same time, exclude the potentially uncertain rights that would otherwise be created by section 1(1)(b)?

3.57   The answer appears to be 'yes' because the Act has the word 'or' between sections 1(1)(a) and (b), the former permitting third party rights where the contract expressly creates them and the latter creating a right where any third party can establish that a contract term purports to confer a benefit. However, cautious lawyers will want to see how cases on the Act define and prescribe the scope and effect of the Act before advising that course of action. At present, that judicial guidance is not available.

3.58   However, if that analysis is right, then a contract can create express third party rights for an expressly stated group of third parties, and all other rights could be expressly stated not to apply. It would be necessary to review all the types of clauses that are seen in collateral warranties with a view to checking those now routine matters against the drafting of a suitable principal contract provision.

3.59   Such a drafting approach would then lead to the following considerations:

- The clause should state that the third party rights in relation to the contract are only those expressly created by the contract and that the contract does not create any other third party rights, including those that would, but for the contract, arise under section 1(1)(b) of the Act.
- The third parties in respect of whom the third party rights are to be expressly created will need to be defined. Sometimes they will be able to be named, such as the parent company of a development vehicle, or the company owning the freehold title of the land on which the development is to be constructed, or all the companies in the group of companies of which the employer is part. Such a draft might be as follows where third party rights are generally excluded, but are expressly created for the companies within the group of which the employer is one company:

    '1.1   Save as expressly provided in Clause 1 of this Contract, nothing in this Contract shall, or is intended to, create rights and/or benefits by reason of the Contracts (Rights of Third Parties) Act 1999 (called the "Contracts Act" in this Clause 1) or otherwise in favour of any person who is not a party to this contract (a "third party") and no term or provision of this Contract shall be, or is intended by either party to be, enforceable by any third party by reason of the Contracts Act or otherwise.

1.2 The following parties, whether or not they exist at the date of this Contract and whether or not they have a contractual relationship with the Contractor in relation to the subject matter of this Contract, shall have the right under the Contracts Act to enforce the terms of this Contract against the Contractor as if they had been the Employer under this Contract:

1.2.1 [Group Parent] PLC (Company Number [   ]), and/or

1.2.2 [Developer Subsidiary who is not the Employer] (Company Number [   ]), and/or

1.2.3 Subsidiary companies, holding companies and wholly owned subsidiary companies of the companies at 1.2.1 and 1.2.2 above. In this Clause 1, the terms "subsidiary companies", "holding companies" and "wholly owned subsidiary companies" have the same meaning as in Section 736 of the Companies Act 1985 (as amended) or in any statutory re-enactment or amendment thereof.'

- There will be others who cannot be named but can be described by class or description, to use the terminology of the Act (section 1(3)). These could include purchasers and tenants of the whole and/or any part of the development. This approach could be of great benefit on many projects but particularly on projects where there are a large number of tenants, such as a shopping mall or a large multi-tenanted commercial office building, where there can presently be a multiplicity of warranties.
- The clause should state from when the third party right accrues or crystallises.
- The clause should provide (as in a collateral warranty) what provisions of the contract are to be enforceable by third parties. Usually, these will be those provisions going to quality of materials, workmanship and design but further provisions could be included. For example, third party rights could extend to a term in the contract putting an obligation on the promisor to maintain professional indemnity insurance for a fixed period and not to use certain prohibited materials. Under this kind of drafting approach, those terms that are usually put in the collateral warranty will now be put in the contract itself, otherwise, of course, third party rights cannot be properly created and prescribed.
- Consideration would need to be given to dispute resolution. Is the third party to be required, as a condition of his third party rights, to be bound by a particular method of dispute resolution? What, if any, are the rules to be on joinder of parties? For example, if several tenants and the

landlord want to bring proceedings against the contractor in respect of construction defects, is it sensible to provide for joinder of such actions to the extent that each action deals with the same, or substantially the same, subject matter? This is something that is hard to achieve in collateral warranties. For example, each of the building contract and professional appointments may have separate arbitration provisions with no joinder provisions. Such potential for multiplicity of proceedings cannot be overcome by provisions in collateral warranties and is a problem that is usually ignored in the drafting of warranties because it is so difficult. All of this could be simplified by use of the Act to create and impose joinder provisions, perhaps best by having the court as the venue for all disputes.

- Section 2(3) of the Act permits the parties to a contract to have an express term enabling them to rescind or vary the contract without the consent of a third party. Consideration needs to be given to whether or not a term should be included to affect the result that will otherwise be the case under the Act.

- Are there to be any conditions put on the third party as part of their rights that are in reality obligations (section 3 of the Act)? Section 3 of the Act permits advantage to be taken by the promisor, in proceedings for the enforcement of a term by a third party, of any defence and/or set-off that is created by express contract terms. Consideration should be given to the creation of such terms. For example, although not usually accepted by developers, a net contribution provision (see further at paragraphs 9.87 to 9.91) could be considered in this context. In the same vein, there is nothing to prevent third party rights being expressly limited to the direct cost of remedial work and excluding consequential loss, provided this is clearly stated in the contract.

- Although section 5 of the Act provides some protection of the promisor from double liability at the hands of the promisee, and the third party in relation to the loss in respect of a contract term, there is clearly scope for making the position clearer and more certain in the contract. For example, the manner in which the discretion of a court or arbitral tribunal thinks it appropriate to take into account, in proceedings brought by a third party, a sum paid previously to the promisee could be prescribed and still be within the scope of section 5 of the Act.

- It should be remembered that the approach above can be used to create third party rights for tenants, purchasers and funds. However, the Act cannot be used to create the step-in rights (see paragraphs 9.73 to 9.80) that funds usually require to enable them to take over and complete the project if the developer defaults on the finance agreement or becomes insolvent. It follows that such rights will have to be created by collateral warranties in any event. Such warranties will not only be needed by the fund with the contractor, but also the members of the professional team. This point alone militates against the use of the Act to remove the need for collateral warranties in respect of funders.

3.60   It is too early yet to see the kind of approach set out above being utilised extensively in commercial agreements. The hope expressed by the Law Commission in Report No. 242 that collateral warranties might not be needed once the Act was in place, is unfulfilled. It does, however, seem likely that more use may be made in due course of the possibilities under section 1(1)(a) of the Act to create express rights for enforcement of terms in contracts by a specific and named group of third parties and excluding the rights that might otherwise arise under section 1(1)(b) in respect of a large potential number of beneficiaries, with all the uncertainties that go with that section of the Act. However, this is only likely to happen regularly and routinely when there has been judicial consideration of the Act and in the circumstances that such consideration is favourable for the use of the Act in lieu of collateral warranties.

Chapter 4
Assignment and Novation

Future purchasers and tenants

The problems

4.1 The original developer and often the first purchaser or tenant of a property will have an opportunity to enter into direct contractual arrangements to protect themselves against latent construction defects. It is unlikely that subsequent purchasers or tenants will have such an opportunity, unless they fall within the scope of the Contracts (Rights of Third Parties) Act 1999, and they will be (adopting the terminology of the rule of privity of contract) strangers to the original contractual arrangements with no remedies in contract against the parties responsible for the design and construction of the building.

4.2 The courts have often said in such circumstances that the law of contract provides for a chain of indemnity connecting the ultimate user with the original producer: for example D is the ultimate user or consumer who purchased from C the retailer, C having purchased from B the wholesaler and B having purchased from A the manufacturer. D can sue C for breach of contract but not B or A. However, if C is sued by D, then C will have a right of indemnity against B who in turn has a right of indemnity against A, creating 'the chain of indemnity' that links the manufacturer to the ultimate user. Unfortunately, the strength of a chain of indemnity is only as great as its weakest link. If C the retailer becomes insolvent a critical link in the chain between D and A will have been broken. Further, a purchaser of a freehold building is faced with the difficulty of the principle of *caveat emptor* and the tenant of a leasehold building with the difficulty of full repairing covenants in the lease. *Caveat emptor* and full repairing covenants are dealt with in Chapter 7.

4.3 The future purchaser or tenant must rely on derivative contractual rights. Such rights arise by assignment, which is a unilateral act, or by novation that is synallagmatic (see paragraph 1.34).

Assignment

Assignment of choses in action

4.4 Choses in action are 'all personal rights of property which can only be claimed or enforced by action and not by taking physical possession':

Torkington v. *Magee*. The term includes the benefits arising under a contract and, subject to certain qualifications dealt with in paragraph 4.31 later, rights of action arising by reason of a breach of contract. A chose in action can be a legal chose, for example an interim payment due under a building contract, or an equitable chose such as a legacy under a will or an interest in a partnership. Choses in action can be assigned or transferred unilaterally, for example A the employer enters into a construction contract with B the contractor, requiring B to construct a building to a quality set out in the specification. A can, without the consent of or indeed knowledge of B, transfer the benefit of that contract to a third party C. A is known as the assignor, B the debtor and C the assignee. Only the benefits of a contract can be assigned, not the burdens: *Nokes* v. *Doncaster Amalgamated Collieries*. The burden must be novated (see paragraph 4.50). It is important to understand that the right to assign a chose in action is not derived from contract. It is a statutory right, alternatively a right arising from the rules of equity. Express conditions are commonly found in collateral warranties purporting to grant rights of assignment; these conditions are unnecessary and may have the effect of restricting the rights to assign.

Legal and equitable assignments

Common Law and Equity

4.5 There are four types of assignment:

(1) Statutory or legal assignments of legal choses in action.
(2) Statutory or legal assignments of equitable choses in action.
(3) Equitable assignments of equitable choses in action.
(4) Equitable assignments of legal choses in action.

4.6 A brief knowledge of English legal history is helpful in understanding the dichotomy between legal and equitable assignments. Legal rights derive from the common law of England which was conceived and developed during the period between the Norman Conquest and the fourteenth century. The common law was administered by the King's Justices on circuit through the three common law courts of King's Bench, Common Pleas and Exchequer. There were no courts of equity. However, because of restrictions placed on the continued development of the common law, not least the baronial intimidation of the common law courts and their juries, plaintiffs in search of justice began to petition the King in Council for a resolution of their disputes pursuant to the King's inherent judicial powers. Eventually this practice led to the petitions being referred to the King's Chancellor who initially discharged this function in the name of the King but who subsequently established the Courts of Chancery as an independent tribunal from the King in Council. The jurisdiction of the

Courts of Chancery was based on the cannon law concept of 'conscience' and ultimately developed into the rules of equity. England therefore had two court systems, the Common Law Courts and the Courts of Chancery, each developing their own rules of law. This separation was abolished by statute in 1875, which replaced the old court structure with the present day structure of the Supreme Court of Judicature. Nevertheless the rules of equity remain distinct from the common law.

Legal assignments of choses in action

4.7   The right to make a legal assignment is now governed by statute – section 136 of the Law of Property Act 1925. Sub-section 1 of section 136 provides:

'Any absolute assignment by writing under the hand of the assignor (not purporting to be by way of charge only) of any debt or other legal thing in action, of which express notice in writing has been given to the debtor, trustee or other person from whom the assignor would have been entitled to claim such debt or thing in action, is effectual in law (subject to equities having priority over the right of the assignee) to pass and transfer from the date of such notice:

(a)   the legal right to such debt or thing in action;
(b)   all legal and other remedies for the same; and
(c)   the power to give a good discharge for the same without the concurrence of the assignor.'

4.8   It will be apparent from the wording of the sub-section that certain legal formalities must be complied with if an assignment is to be an effective legal assignment. These formalities are:

(1)   An absolute assignment in writing signed by the assignor;
(2)   A debt or other legal thing in action; and
(3)   Express notice in writing to the debtor.

4.9   An absolute assignment does not include the assignment of part of a debt or thing in action whether or not the part assigned is ascertained or unascertained. In *Walter and Sullivan Ltd* v. *J Murphy & Sons Ltd*, WS were plastering sub-contractors who commenced legal proceedings against M for the sum of £1808 alleged to be due in respect of a sub-contract for plastering works. After the commencement of the proceedings, WS, who were indebted to a third party H & Co, notified M that M were 'to pay to H & Co the sum of £1558 17s 8d from monies owing by you to us... the receipt of H & Co shall be good and sufficient discharge to you in respect of payment made hereunder'. By a second document H & Co agreed with WS that in consideration of the irrevocable authority given by them to M 'we will pay over to you any monies which are paid to us by (the Defendants)... after your debt to us... has been fully repaid...'. The court held that the arrangement between WS and H & Co was an assignment of part of a debt and therefore did not satisfy the requirements of sub-section 1 of section 136 of the Act.

4.10   An assignment that purports to be by way of charge only is not an absolute assignment. This is a complex legal concept outside the ambit of this book. Suffice it to say that the relevant test is to decide whether the assignment merely gives a right to the assignee to payment out of a particular fund by way of security rather than an unconditional transfer of the fund to the assignee. In the *Walter and Sullivan* case, as well as being an assignment of part of a debt the court also held that the assignment purported to be by way of charge. By way of contrast, it was held in *Tancred* v. *Delagoa Bay Company* that an assignment by way of mortgage was absolute because there was a condition for re-assignment on payment of the loan. It is the substance of the transaction and not the titles of documents that determines the nature of the assignment.

4.11   An assignment which is qualified by conditions cannot be a legal assignment. In *Re Williams, Williams* v. *Ball* the assignor purported to transfer the benefit of a life insurance policy but made it conditional upon the assignee surviving the assignor. This was held to be a conditional assignment falling outside section 136 of the Act. The judicial reasoning behind the requirement for an absolute assignment is that the debtor should not be put in doubt or jeopardy by the arrangements between the assignor and the assignee as to whom he is to discharge his obligations. In the cases of *Walter and Sullivan* and *Williams* there were such doubts, but not in the case of *Tancred* where the re-assignment on repayment of the loan would have to be notified to the debtor.

4.12   To create a legal assignment there must be a written document signed by the assignor. Signature by an agent would not appear to be sufficient. Any form of wording may be used provided there is a clear intention to make an absolute assignment. The assignment may be a document passing between the assignor and the assignee, or a written demand from the assignor to the debtor that the debtor pays or discharges his obligations to the assignee. In the latter case, in order to be an effective assignment rather than merely an authority to pay a third party, there must be evidence that the assignee consented to the arrangement between the assignor and the debtor: *Curran* v. *Newpark Cinemas Ltd*. Unlike an assignment, an authority to pay can be revoked prior to the actual payment.

4.13   A debt or other legal thing in action includes both legal choses and equitable choses. The purpose of section 136 of the Act, which replaced but substantially re-enacted section 25, sub-Section 6 of the Judicature Act 1873, was procedural and not intended to create new forms of choses or things in action.

4.14   To create a valid legal assignment, written notice of the assignment must be given to the debtor. No particular form of wording is required; indeed a document can constitute notice even though it was not intended to be a notice. In *Van Lynn Developments Ltd* v. *Pelias Construction Co Ltd*, P's bank overdraft was paid off by Van Lynn in consideration of P assigning the debt to Van Lynn. The assignment was dated 26 June. By a letter dated 27 June, Van Lynn demanded payment from P. In their letter Van Lynn stated, incorrectly, that notice of the assignment had previously been given to P. The court held that a notice of assignment was still good notice to the debtor even though it did not refer to the date of the assignment. Further, as regards Van Lynn's letter dated 27 June, the incorrect statement as to a notice could be ignored as 'an inaccurate surplusage' and it was immaterial that the letter was not written with the intention that it should perform the function of giving notice under the Act. It is not necessary for the notice to the debtor to be given by the assignor or the assignee; it may be given by a third party. In *Bateman* v. *Hunt*, a valid notice was given by the executor of a deceased sub-assignee. In *Herkules Piling Ltd and Another* v. *Tilbury Construction Ltd*, purported notice to the debtor by way of disclosure of documents in legal proceedings in which the debtor was a party, was considered to be insufficient notice of a legal or equitable assignment.

4.15   Once there has been an assignment which complies with the formalities of section 136, there is a transfer to the assignee of the legal right to the chose in action and the assignee can give good discharge upon payment or satisfaction by the debtor. It follows that the assignor has no right to sue in respect of the chose in action unless of course there is a re-assignment to the assignor. The same rules apply to intermediate assignments, thus creating a potential problem where a tenant assigns to a sub-tenant part of the demised property.

4.16   An assignment within the statute does not require consideration, thus voluntary assignments are enforceable between the assignor and the assignee and between the assignee and the debtor.

Equitable assignments

4.17   A failure to comply with the formalities of section 136 of the Act is not necessarily fatal to the transaction; a defective legal assignment may operate as an equitable assignment: *William Brandts Sons & Co* v. *Dunlop Rubber Co*. Indeed a defective legal assignment which takes effect as an equitable assignment may subsequently become a legal assignment if the defect is removed; for example, where an equitable assignee of a defective legal assignment subsequently serves written notice on the debtor to perfect the legal assignment.

4.18   There may be an equitable assignment of an equitable chose or an equitable assignment of a legal chose. No consideration is required for the assignment of an equitable chose provided that the assignor has, at the material time, done all that he can to perfect the gift: *Letts* v. *Inland Revenue Commissioners*. It is suggested that the better view is that the same rule applies to equitable assignments of legal choses although there are judicial dicta to the contrary.

4.19 An equitable assignment may be in writing or oral. Any words will suffice provided they are unambiguous. Referring to the form of an equitable assignment Lord Macnaghten in the *William Brandts* case stated:

> 'It may be addressed to the debtor. It may be couched in the language of commerce. It may be a courteous request. It may assume the form of mere permission. The language is immaterial if the meaning is plain. All that is necessary is that the debtor should be given to understand that the debt has been made over by the creditor to some third person.'

4.20 Lord Macnaghten's judgment in *William Brandt* referred to notice to the debtor. In law there may be a binding equitable assignment between assignor and assignee without notice to the debtor. However, as a matter of practice, notice to the debtor is very important for three reasons. Firstly, in the absence of notice the debtor is entitled to discharge his obligations to the assignor and not to the assignee, whereas if he has notice he does so at his own peril and he may well be required to discharge the obligation a second time to the assignee with no entitlement to recovery from the assignor: *Walter & Sullivan Ltd*. Secondly, the giving of notice to the debtor has an effect on prior equities (see paragraph 4.22 below). Thirdly, the date of notice establishes the order of priority as between successive assignees: *Dearle* v. *Hall*. The notice may be written or oral and the wording of the notice may be informal, although casual conversations may not be sufficient notice: *Re Croggon ex parte Carbis*. Indeed in the case of *Lloyd* v. *Banks*, the court held that a newspaper article was sufficient notice to the debtor.

4.21 An equitable assignment may operate by way of a charge only or be of part of the debt or chose: *Walter & Sullivan Ltd*. Thus, where a developer wishes to dispose of the completed building to more than one purchaser or tenant, it is submitted that he will only be in a position to give each individual purchaser or tenant an equitable assignment of the benefits arising under the principal design and construction contracts. If a legal assignment is required, then the draftsman of the principal contracts should take care to impose an obligation on the designers and contractors to provide a sufficient number of collateral warranties to satisfy the requirements of multi-occupation.

Procedural differences between legal and equitable assignment

4.22 Substantively legal and equitable assignments (provided notice has been given to the debtor) are essentially the same. In the *Herkules Piling* case, it was considered that an arbitration clause in the FCEC form of sub-contract could be assigned by a legal assignment by reason of the wording of section 136 of the Law of Property Act 1925, which stipulated that all attendant remedies were transferred, but not by an equitable assignment as the arbitration clause conferred discrete rights and obligations between the original contracting parties. There are however important procedural differences. As previously stated, a legal assignment within the Act transfers a legal right in the chose to the assignee. Consequently the assignee sues the debtor in his own name. If there is an equitable assignment of an equitable chose in action the assignment being absolute, then again the assignee is entitled to sue in his own name. However, if there is an equitable assignment of a legal chose in action or an equitable chose which is not absolute, for example a part of the debt, the assignor must be joined into the action either as claimant, if he co-operates, or as defendant if he does not. If the assignor is not joined as a party, the assignee's action may well fail although it is important to stress that these requirements are procedural and are not substantive, therefore the courts have a discretion to dispense with joinder of the assignor if they are satisfied that there is no prospect of a further claim by the assignor: *The Aiolos*. Also note that under the Civil Procedure Rules Part 19, the Supreme Court has a wide discretion to order that additional parties should be joined to an action.

Prior equities

4.23   The effect of an assignment, whether it is a legal assignment or an equitable assignment, is to place the assignee in the shoes of the assignor in respect of the benefits (but not the burdens, see paragraph 4.50) arising under the original transaction with the debtor. Consequently the assignee cannot by the assignment obtain a more advantageous position vis-à-vis the original debtor than that which was occupied by the assignor: *Business Computers Ltd* v. *Anglo African Leasing Ltd* where Templeman J stated that:

> 'a debt which accrues due before notice of an assignment is received, whether or not it is payable before that date, or a debt which arises out of the same contract as that which gives rise to the assigned debt, or is closely connected with that contract, may be set off against the assignee.'

4.24   It is important to note that if the set-off arises independently from the original contract between the assignor and the debtor, then it cannot be set off against the assignee if the liability (as distinct from the actual payment) accrued after the date of receipt of a notice of assignment. The giving of notice of assignment is however irrelevant to claims by way of set-off or counterclaim that arise from the original contract or a contract which is closely connected to the original contract. For example, A is the developer, B the architect appointed by A, C the first purchaser of the development from A and D the second purchaser from C. The contract between A and B provides for design works to be carried out by B and payment therefore to be made by A. B also enters into a collateral warranty undertaking to C that he will carry out his design works with reasonable skill and care. C assigns the benefit of the collateral warranty to D. A has not paid all of B's professional fees. In the event that B is in breach of his collateral warranty, if D brings proceedings against B then B will be able to set off the amount of the unpaid fees against D's claims regardless of whether the entitlement to the fees arose after the date of D's notice of assignment to B. This is because the collateral warranty and the original contract between A and B are closely connected contracts. In the above example the same right of set-off arises as between B and C if C were the ultimate purchaser who took an assignment of A's benefits under the original contract with C. In this latter example, the rights of set-off and counterclaim would arise from the same contract.

4.25   A counterclaim for unliquidated damages may be set off by the debtor against any claims brought by the assignee: *Phoenix Assurance Co Ltd* v. *Earls Court Ltd*.

4.26   The debtor's right to counterclaim against the assignee is limited to defending the claims brought by the assignee, the counterclaim being set off in extinction or diminution of the assignee's claims. It does not entitle the debtor to bring positive counterclaims against the assignee, i.e for sums in excess of the assignee's claims. This is because, as stated above, the assignee only takes the benefit and not the burden of the original contract.

Intermediate assignees

4.27   It would appear that where there have been successive assignments the debtor is not entitled to set off against claims brought by the ultimate assignee, counterclaims which the debtor has against intermediate assignees: *The Raven*.

Restrictions on assignment

Statute

4.28   The right to assign may be governed by specific statutory provisions; for example, section 53(1)(c) of the Law of Property Act 1925 provides that equitable assignments of interest in a trust must be in writing. Other examples are bills of lading, copyright, patent rights and life insurance policies. If there is a statutory code this will over-ride the general provisions of section 136 of the Law of Property Act 1925 and also the rules of equity.

Future choses in action

4.29   Future choses in action cannot be assigned either by a legal assignment or by an equitable assignment. The distinction between an existing chose and a future chose is not as obvious as it sounds and can often lead to complex forensic analysis. For example, a right to future payment under an existing contractual right will be an existing chose not a future chose. In comparison, the benefits which are likely to flow from a contract not yet entered into will be a future chose. Further difficulties arise in respect of accrued contractual obligations which may be defeated by a subsequent breach of the contract by the party otherwise entitled to the accrued benefit. In *Hughes* v. *Pump House Hotel Company Ltd*, it was held that a contractor's right to be paid under a building contract was an existing chose even though the right of payment might be defeated by the contractor's subsequent failure to perform his contractual obligations.

4.30   The vital difference between an existing chose and a future chose is that the latter can only be enforced if there is an agreement supported by adequate consideration. In *Re McArdle (deceased), McArdle* v. *McArdle*, M and his wife lived in a dwelling house forming part of the estate of M's father in which M and his brothers and sisters were beneficially interested expectant on the death of the tenant for life. In 1943 and 1944, Mr and Mrs M carried out certain improvements and decorations in and on the house, the cost of which, amounting to £488, was borne by Mrs M. In April 1945, M and his brothers and sister signed a document addressed to Mrs M, which provided:

> 'In consideration of your carrying out certain alterations and improvements to [the dwelling house] at present occupied by you, we the beneficiary under the Will of [the father] hereby agree that the executor... shall repay to you from the said estate when so distributed the sum of £488 in settlement of the amount spent on such improvements.'

In 1948 the tenant for life died and Mrs M claimed payment of the sum of £488. The court held that the consideration for the execution of the document in April 1945 was past consideration and the document could not operate as an equitable assignment for valuable consideration. Further, whilst an equitable assignment could be valid without consideration the document did not constitute such an assignment because, as it contemplated future action by Mrs M to the satisfaction of the signatories, it did not render her title complete and so was not complete and perfect. In other words, the court found that the document purported to be an assignment of a future chose in action, therefore it would not be binding unless it constituted an agreement supported by adequate consideration.

Bare rights of action

4.31   Legal or equitable assignments of bare rights of action, that is to say litigation, are void and unenforceable. This is because the courts consider such transactions to be trading in litigation, or to use the lawyer's terminology 'offending against the rules of maintenance and champerty'. In *Prosser* v. *Edmonds*, Lord Abinger CB stated:

> 'It is a rule not of our law alone, but of that of all countries, that the mere right of purchase shall not give a man a right to legal remedies. The contrary doctrine is nowhere tolerated and is against good policy. All our cases of maintenance and champerty are founded on the principle that no encouragement should be given to litigation by the introduction of parties to enforce those rights which others are not disposed to enforce.'

4.32   Maintenance is an agreement to fund litigation, and champerty (a species of maintenance) is an agreement to divide the proceeds of the litigation. Prior to the Criminal Law Act 1967, maintenance and champerty were both criminal acts and torts. Section 14(2) of the Act abolished the criminal offence and also the tort of maintenance and champerty; however it preserved the rule of law that agreements that tended to either maintenance or champerty would be void and unenforceable.

4.33   Clearly the benefits arising under a contract do not offend against the above rule (see paragraph 4.4). What is the position however if prior to a purported assignment of the benefits of a contract, those benefits have crystallised into rights of action for damages for breach of contract, for example latent construction defects? The courts had to consider a similar situation in *Dawson* v. *Great Northern and City Railway Company*. The Railway Company constructed under statutory powers a tunnel under or near certain houses in which D was interested and in which she carried on business and she claimed to be entitled to compensation on the ground that her interest in the houses had been injuriously affected by structural damage to the houses and by damage to trade stock. The structural damage had occurred prior to the acquisition of her freeholder interest and also the acquisition of her leasehold interest, though both transactions purported to assign rights to recover compensation from the railway company in respect of the structural damage, these rights being statutory rights under the Land Clauses Consolidation Act 1845. The court held that whilst an assignment of a mere right of litigation is bad, an assignment of property is valid even though that property may be incapable of being recovered without litigation. Sterling LJ stated:

> 'Even if the assignment be regarded apart from the conveyance of the lands and buildings... it appears to us that it is good; but we think that great weight must be given to the circumstances that this assignment is incidental and subsidiary to that conveyance and is part of a bona fide transaction the object of which was to transfer to the plaintiff the property of [the vendor] with all the incidents which attach to it in his hands. Such a transaction appears to be very far removed from being a transfer of a mere right of litigation.'

4.34   A similar issue arose in *Ellis* v. *Torrington*. The facts were somewhat
complex but illuminating. The property in question was subject to three
leases: a head lease expiring on 25 December 1917, an underlease expiring
on 18 December 1917 and a sub-underlease expiring on 15 December
1917. All these three leases contained onerous covenants to repair. The
defendant T was the tenant pursuant to the sub-underlease and E the
plaintiff was a sub-tenant of T, although E's covenants to repair were far
less onerous than those imposed on T. On 18 December 1917, E purchased
the freehold interest of the premises which was subsequently conveyed to
him together with the benefits of the covenants to repair in the head lease.
At the expiration of all these leases the premises were substantially out of
repair. T threatened E with proceedings on the basis of E's covenant to
repair in his sub-tenancy from T. Faced with this threat E, who could not
pursue T under the head lease, obtained an assignment of the benefits of
the covenants in T's sub-underlease and then commenced proceedings
against T for breach of covenant as assignee of the sub-under lessor. The
court held that the assignment was free from objection on the ground of
maintenance or champerty, the right of action on the covenants being so
connected with the enjoyment of property as to be more than a bare right
to litigate. Banks LJ stated:

> 'the Respondent is seeking to enforce a right incidental to property, a
> right to a sum of money which theoretically is part of the property he
> has bought.'

4.35   More often than not with construction projects, the assignments of
benefits arising under collateral warranties or collateral contracts even
though they have crystallised into rights of action will be incidental to
property and therefore falling within the principles set out in *Dawson* and
Ellis. In any event, in recent years the courts have become more relaxed
about the issue of maintenance and champerty. In 1968, Lord Denning
stated: 'Much maintenance is considered justifiable today which would in
1914 have been considered obnoxious': *Hill* v. *Archbold*. The leading case is
the House of Lords decision in *Trendtex Trading Corporation and Another* v.
Credit Suisse. Again the facts of this case were somewhat complex involving proceedings in England in respect of an agreement made in Swit-
zerland containing a purported assignment by a Swiss Corporation to a
Swiss Bank of its rights of action against the Central Bank of Nigeria in
respect of a dishonoured letter of credit for the sum of US$14,000,000. The
Bank had assigned to an undisclosed third party all the rights of action
against the Central Bank of Nigeria (CBN) for the sum of US$1,100,000.
Less than five weeks from the date of the assignment the third party had
settled the claim against CBN upon payment by the Bank of US$8,000,000.
One of the issues in the case was the validity of the intermediate
assignment to the Swiss Bank and the subsequent assignment to the third
party. The court held that in determining the validity of an assignment of
a cause of action it was the totality of the transaction that was to be looked

at and *if the assignment was of a property right or interest and the cause of action was ancillary to that right or interest, or if the assignee had a genuine commercial interest in taking the assignment and enforcing it for his own benefit, the assignment would not be struck down as an assignment of a bare cause of action or as savouring of maintenance.*

Accordingly if no parties other than Trendtex and Credit Suisse had been involved the intermediate assignment would have been valid (even though it involved an assignment of Trendtex's residual interest in the CBN litigation) as Credit Suisse had a genuine and substantial interest in the success of that litigation. However, the introduction of the third party rendered the agreement void under English law because the agreement clearly showed on its face that its purpose was to enable the cause of action against CBN to be sold to an anonymous third party with the likelihood of that third party, which had no genuine commercial interest in the claim, making a profit out of the assignment. Thus *Trendtex* introduced a further qualification to the rule against the validity of assignments of bare rights of action, namely where there is a genuine commercial interest in taking the assignment which of course does not necessarily have to be an interest incidental to the use of property. In *Norglen Ltd* v. *Reeds Rains Prudential Ltd,* the Court of Appeal held that there was a genuine commercial interest in:

(1) Shareholders and former directors of a company in liquidation taking an assignment from the liquidator of the company's causes of action against a third party.
(2) A shareholder and director of a company in administrative receivership, who was also a guarantor, taking an assignment from the administrative receiver of the company, of causes of action against a third party.

Nor did the fact that these assignments were made to enable the individuals concerned to make applications for legal aid to pursue the assigned causes of action and also avoid the potential liability for an order to provide security for an opponent's costs, affect the validity of the assignments. The House of Lords subsequently confirmed the Court of Appeal's findings on these points.

Personal contracts

4.36 If the contract is a personal contract then the benefits of that contract cannot be assigned either by a legal assignment or by an equitable assignment. The test, which is an objective test, is whether 'it can make no difference to the person on whom the obligation lies to which of two persons he is to discharge it': *Tolhurst* v. *Association Portland Cement Manufacturers Limited.* It will be noted that the test is not concerned with the personal skill of the debtor. For example, if an architect gives a collateral warranty relating to his design work whilst the design work will involve personal skill on his part and will not be assignable by the architect to a third party, the benefits of the undertakings arising under the collateral warranty given to, say, a tenant will be assignable to a future tenant because, applying the objective test, it can make no difference to the architect whether its obligations lie to the first tenant or to the future tenant. It could however make a difference if a subjective test was the proper test, for example the future tenant could be more litigious than the existing tenant and more likely to bring proceedings against the architect in the event of breach.

4.37   There is a presumption in favour of commercial contracts that the benefits arising under such contracts are assignable. However the presumption is rebuttable. In *Kemp* v. *Baerselman* B contracted with K, a cake manufacturer, to supply him with all the eggs of a specified quality 'that he shall require for manufacturing purposes for one year'. K undertook not to purchase eggs from any other merchant during the year so long as B was ready to supply them. During the relevant year K transferred his business to a company whereupon B claimed to be discharged from his contract and refused to supply any more eggs to K or to the new company. The court held that B's contract was with K personally and that the benefit of the contract was not assignable. The court considered that the personal characteristics of the contract were two-fold: that B's obligations were defined by reference to K's manufacturing purposes and further that K had undertaken not to purchase eggs from any other merchant. In *Tolhurst*, the owner of land had contracted with a company to supply them for 50 years with at least 750 tonnes of chalk per week and so much more as they might require for their manufacture of cement. The original company was a small business, which went into voluntary liquidation and transferred all its assets, including an assignment of the benefit of the supply contract to Associated Portland Cement. Associated Portland Cement was a much larger concern than the original company and carried on business at various places. The court held that the assignment was effective. In *Kemp*, Lord Alverston CJ distinguished *Tolhurst* on the basis that in the latter case the House of Lords had 'treated the contract as a supply to a given cement making place and not as a personal contract'.

Conditions of contract

4.38   Often the contract between the assignor and the original debtor will include an express term purporting to prohibit or restrict the right of assignment (see for example clause 19.1 of the JCT Form of Contract 1998 Edition and clause 12 of the RIBA Form of Collateral Warranty (now treated by the RIBA as being superceded by the BPF Warranty)). Are such conditions effective? Some earlier cases appear to suggest that the right of assignment cannot be prohibited or restricted by the contractual arrangements between the assignor and the original debtor. In *Tom Shaw*

& Co v. *Moss Empires (Ltd) and Bastow*, B was a comedy artist who appeared on stage for a season at the Moss Empires Theatre. His contract with Moss Empires provided by clause 13 that he should not assign his salary which should be paid direct to him and to no other person except in the case of his death. His booking at the Moss Empires Theatre was obtained through his agents Tom Shaw & Co with whom B agreed that 'I hereby agree to pay you or your assigns 10% commission on (my) salary and on all monies which should accrue under the said engagement or a prolongation of the same... and hereby authorise Moss Empires to deduct and pay the said commission from my salary in any manner which you may deem expedient.' Notice of this letter was given by Tom Shaw to Moss Empires. The court held that despite the restriction in B's contract with Moss Empires, B's letter to Tom Shaw & Co constituted a valid equitable assignment. Darling J stated:

> 'The strongest ground for the defence was in the contract ... clause 13. But though Moss Empires might bring an action for breach of that contract, if they could show any damages... it could no more operate to invalidate the assignment than it could to interfere with the laws of gravitation.'

4.39   The strange thing about the Tom Shaw case was that the judge awarded damages and costs against B and not Moss Empires! Also Moss Empires admitted that they were liable to pay the plaintiff.

4.40   In *Spellman* v. *Spellman*, there were conflicting *obiter dicta*. Danckwerts LJ considered that 'the fact that there is a prohibition in the document creating the chose in action against assignment is not necessarily fatal to such claim'. However in the same case, Willmer LJ considered that any prohibitions should be binding, as 'it would be quite impossible for this court to make an order to the contrary because such an order would in effect require the husband to break his contract with the hire purchase corporation'.

4.41   It is suggested that the better view is that expressed in *Helstan Securities Limited* v. *Hertfordshire County Council* where the court held that if the parties to a contract, the subject matter of which was a chose in action, agreed that the chose in action was not to be assigned, any purported assignment was invalid. In this case, Hertfordshire County Council entered into a civil engineering road works contract with a contractor. The contract was the ICE Condition of Contract, Fourth Edition. Condition 3 provided 'the Contractor shall not assign the contract or any part thereof or any benefit or interest therein or thereunder without the written consent of the employer'. The contractor got into financial difficulties and without obtaining the Council's consent assigned to Helstan the amount of £46,437 allegedly owing by the Council to the contractor. The Council refused to discharge the amount to Helstan who brought proceedings against the Council for payment. Croom Johnson J stated:

'If the reported cases are not a sure guide, one is thrown back in this case on the agreement. There are certain kinds of choses in action which, for one reason or another, are not assignable and there is no reason why the parties to an agreement may not contract to give its subject matter the quality of un-assignability.'

4.42   The judge distinguished the *Tom Shaw* case on its own peculiar facts and also on the basis that if that case was good authority for the proposition that a contractual prohibition against an assignment was not effective, the principle was limited to the relationship between assignor and assignee and not between assignee and original debtor.

4.43   Since the first edition of this book, the matter has been clarified by the House of Lords in the case of *Linden Gardens Trust Ltd* v. *Lenesta Sludge Disposals Ltd*. The court held in respect of a prohibition on assignment without consent (under clause 17 of the JCT Standard Form of Building Contract 1963 Edition) that the contractual prohibition effectively prohibited the assignment of any benefit of the contract, including not only the assignment of the right to future performance but also (reversing the Court of Appeal decision) the assignment of accrued rights of action, and an attempted assignment of contractual rights in breach of the contractual prohibition was ineffective to transfer any such contractual rights to the assignee. The court expressed the opinion that the prohibition on the assignment of accrued rights of action was not void as being contrary to public policy since a party to a building contract could have a genuine commercial interest in seeking to ensure that he was in contractual relations only with a person whom he selected as the other party to the contract and there was no public need for the law to support a market in choses in action.

4.44   The decision in *Lenesta Sludge* turned upon the wording of clause 17(1) which provided that 'The Employer shall not without the written consent of the Contractor assign this Contract'. Lord Browne-Wilkinson accepted that at least hypothetically it was possible that there might be a contractual prohibitory term so worded as to 'render invalid the assignment of rights to future performance but not so as to render invalid assignments of the fruits of performance', i.e. accrued causes of action. He stated, 'The question in each case must turn on the terms of the contract in question'.

4.45   The contractual prohibition in the FCEC form of sub-contract is unusual in that it has a proviso to the general prohibition. The relevant clause is clause 2(3), which is worded as follows:

'The Sub-Contractor shall not assign the whole or any part of the benefit of this Sub-Contract ... without the previous written consent of the Contractor provided always that the Sub-Contractor may without such consent assign either absolutely or by way of charge any sum which is or may become due and payable to him under this Sub-Contract.'

4.46 In *Yeandle* v. *Wynn Realisations Ltd (in administration)*, the Court of Appeal held that the proviso in clause 2(3) of the FCEC sub-contract was confined to sums which were or may become due and payable, i.e. liquidated sums, and did not extend to the benefit of the sub-contract, i.e. did not include sums which remained to be ascertained by using the contractual machinery nor the contractual machinery itself. The authors find this decision a little surprising bearing in mind the wording of section 136 sub-section (b) of the Law of Property Act 1925 (set out earlier in paragraph 4.7) providing for the transfer of all ancillary remedies. Nevertheless, *Yeandle* was followed by the Court of Appeal in *David Charles Flood* v. *Shand Construction and Others* where the court held that the proviso only applied to claims which could be expressed simply as a present or future claim for a fixed amount due under the sub-contract. Evans LJ stated:

> 'I would hold that "sum" in the proviso ... means a fixed or liquidated amount. The amount either "is" due and payable at the time of assignment, in which case there is an existing claim in debt, or "may become" due and payable at some future date. In that case, in my judgment, the assignment is of the future anticipated right to claim that amount as a debt, rather than the existing claim or cause of action which may result in the debt becoming due and payable thereafter. If there is a claim for additional remuneration, therefore, as in *Yeandle* that right cannot be assigned, but there could be an assignment of the future right to recover the sum awarded by an arbitrator or a judgment debt.
>
> Similarly, in my view, a claim for damages cannot be assigned until such time as the amount is fixed and there is a finding or an admission that the sum is due. This will probably mean that the claim for damages is replaced by a claim in debt (c.f. a judgment debt) but even if the cause of action continues technically as a claim for damages, I would hold, when liability and amount are both established, that that amount is a "sum due and payable" under the sub-contract ... Again the clause permits a present assignment of the future right to recover that amount.'

4.47 In *Bawejam Ltd* v. *MC Fabrication Ltd* the Court of Appeal held that a prohibition against assignment at law does not necessarily prevent an assignment operating in equity.

4.48 The Court of Appeal had to consider the effect of a qualified contractual prohibition in the case of *Hendry* v. *Chartsearch Ltd*. The relevant clause prohibited the assignment of the agreement in whole or in part without the prior written consent of a party, which consent should not be unreasonably withheld. The court held that in the absence of the prior written consent any purported assignment was invalid and it did not matter that the consent could not have reasonably been withheld. Per Henry LJ: 'prior consent never applied for is never withheld or refused (whether reasonably or otherwise)'. Also note that Evans LJ considered that the fact that the assignee would not be subject to an order for security for costs could be a reasonable ground for withholding consent, though probably not the fact that the assignee could make an application for legal aid. Further, such contractual prohibitions continued after the relevant trading relationships had come to an end. If consent was applied for but refused, the party requiring the consent could then seek a declaration of the court that the consent had been withheld unreasonably (per Henry LJ).

4.49   The emphasis is therefore to be placed on the meaning of the particular contractual prohibition being considered. It also follows that such prohibitions are matters of contract and must satisfy the requirements of an enforceable contract. Accordingly, clauses in collateral contracts which give an unconditional right to a first assignment but then seek to restrain or restrict further assignments, may well fail for lack of consideration. Also, there is the difficulty that the assignee 'steps into the shoes' of the assignor who has, as a consequence of the assignment, an unfettered right to assign.

Novation

4.50   As stated above, it is only the benefits of a contract that can be transferred by way of an assignment. If the parties wish to transfer both the benefit and the burdens then this must be done by a novation agreement. The characteristics of a novation were considered by Staughton LJ in the *Linden Gardens* case.

4.51   A novation occurs when there is a rescission of one contract and the substitution of a fresh contract in which the original contractual obligations are carried out by different parties. For example, A is a developer who enters into a contract with B the contractor. Before practical completion of the construction works and by way of a separate transaction, A agrees a sale of the completed project to C. To give C the benefit of the contractual specification for the building works, A, B and C may enter into a novation agreement that substitutes C for A in the original contract with B. If it is a true novation agreement then its effect will be to release A from all liability in respect of the original contract and C may be sued by B for any breaches of contract by A pre-existing the date of the novation agreement. As a matter of practice however, express terms are often introduced into novation agreements with a view to restricting the retrospective characteristics of novation. Such qualified agreements are really variation agreements rather than novation agreements, and such agreements must be supported by adequate consideration. With a novation agreement the consideration is deemed to be the mutual discharge of the old contract: *Scarf* v. *Jardine*.

4.52   A novation agreement is not possible without consent. It is essential therefore that the principal contracts between developer and consultants and between developer and contractors contain express terms obliging the contractor and the consultant to enter into the novation agreement. To avoid the risk of merely having an agreement to agree which is unenforceable, it is suggested that a specimen form of novation agreement be appended to the original contractual documentation.

4.53   Novation agreements are common where the form of contracting is design and build. Such agreements enable the employer to select his design team and, once the successful contractor has been appointed, to transfer the team to the contractor by way of a novation. It was commonly thought that in such circumstances, the contractor could recover his own losses arising from pre-novation breaches of the designer, e.g. inadequacies in the description of work items in the Employers Requirements. A recent Scottish case has rejected the contractor's right to recover pre-novation losses.

4.54   The meaning and effect of a novation agreement in the context of a building project, in particular the characteristics of claims for losses arising prior to the novation, has now been considered by the Scottish Court of Session in the case of *Blyth & Blyth Ltd* v. *Carillion Construction Ltd*. The facts concerned the novation to the contractor of the terms of appointment between an employer and a consulting engineer for the construction of a leisure complex near Edinburgh. After the novation agreement, disputes arose between the engineer and the contractor, the engineer claiming for unpaid professional fees and the contractor counterclaiming for additional construction costs incurred by them arising from alleged inaccuracies or insufficiencies of information supplied by the engineer, initially to the employer, forming part of the tender documentation provided by the employer to the contractor. The construction contract was the JCT Standard Form of Building Contract with Contractors Design 1981 Edition amended to provide that the contractor should assume responsibility for the design of the works, whether undertaken before or after the execution of the contract. The contractor's primary position taken in their pleadings, but nor pursued at trial, was that the novation agreement was to 're-write' the terms of appointment between the employer and the engineer, substituting 'contractor' for 'employer'.

At trial, the contractor submitted that the novation agreement did not effect a 're-writing' of the terms of appointment; all that occurred on novation was that the contractor became the creditor of the obligation, but the content of the obligation owed by the consultant engineer, to provide advice or services to the employer, remained unaltered. The contractor further submitted that as they had retroactively become a party to a contract as creditors of the obligation, they had contracted with the engineers for the provision of services to a third party, the employer, and if those services were not properly performed, the contractors were entitled to claim their own losses. Lord Eassie described the issue as follows:

'One ought, accordingly, to examine the legally logical position of the defender's analysis to the effect that the Novation Agreement produced an essentially three-sided relationship whereby A (contractor) engages B (consultant) to perform services for, and give advice to, C (employer). Accepting for the moment that analysis to be correct, the questions which arise are whether the sufficiency of the performance by B (consultant) is to be judged by what is required by the destinee of the services C (the employer) and whether in the event of defective performance the losses or costs for which B (consultant) may be liable in damages are those reflecting the need to put C (employer) in the position of having received satisfactory service, or, as the (contractors) contend, those said to have been suffered by A (contractor).'

4.55   Lord Eassie found that on a proper construction of the novation agreement, the contractors could not claim for their own losses caused by breaches by the engineers prior to the date of novation in relation to the duties then owed by the engineer to the employer, the losses sought by the contractors not being losses conceived as having been suffered by the employer.

4.56   The case is, of course, Scottish authority not binding in the English courts. Further the authors understand that the decision is being appealed. However, until this issue is clarified, contractors will have to protect themselves with appropriate contractual conditions where they are taking over the responsibility of the design process by way of a novation. Also, contractors will have to consider entering into appropriately worded collateral warranties with the design teams to protect themselves from pre-novation losses.

Chapter 5
Reasonable Skill and Care and Fitness for Purpose

5.1   One of the common problems associated with collateral warranties is disconformity between the performance obligations set out in the principal contract and the performance obligations in the warranty itself. The former usually provides for a performance obligation of reasonable skill and care and the latter often attempts to impose a performance obligation of fitness for purpose.

Reasonable skill and care

5.2   A contractual obligation to carry out works or services with reasonable skill and care creates a performance obligation which is analogous to the standard of care in negligence. The court considered the standard of care in negligence in *Blyth* v. *Birmingham Waterworks Company* and stated:

> '... negligence is the omission to do something which a reasonable man, guided upon those considerations which ordinarily regulate the conduct of human affairs, would do, or doing something which a prudent and reasonable man would not do.'

5.3   *Blyth* established the appropriate tests for the behaviour of the general public and not for the behaviour of members of a more limited group who have or hold themselves out as having specialist skills such as architects or engineers. In *Bolam* v. *Friern Hospital Management Committee* (approved in *Whitehouse* v. *Jordan*), the court refined the test established in *Blyth* in order to accommodate specialist skills. The court applied the following test:

> '... where you get a situation which involves the use of some specialist skill or competence, then the test of whether there has been negligence or not is not the test of the man on the top of a Clapham omnibus because he has not got this special skill. A man may not possess the highest expert skill at the risk of being found negligent. It is well established law that it is sufficient to be exercising the ordinary skill of an ordinary competent man exercising that particular art.'

5.4   In *J. D. Williams & Co Ltd* v. *Michael Hyde and Associates Ltd*, the Court of
       Appeal set out three qualifications to the test in *Bolam*:

       (1)   In a rare case it may be demonstrated that the opinion alleged to be
             held by a respectable body of the profession cannot in fact withstand
             logical analysis (see also *Bolithio* v. *City & Hackney Health Authority*
             where the court held that the practice relied upon had to be
             respectable, responsible and reasonable, with a logical basis and
             where it involved weighing comparative risks, it had to be shown
             that those advocating the practice had directed their minds to the
             relevant matters and had reached a defensible conclusion).
       (2)   In some cases the evidence given may not establish that the view
             contended for is in fact held by a responsible body of professional
             opinion, but may simply be the personal view of the expert as to
             what he might have done if faced with similar circumstances. This is
             not expert evidence at all, and the judge must discount it and form
             his own view.
       (3)   Where the advice at issue required no special skill, then the *Bolam*
             test was simply irrelevant and should not apply.

5.5   *Plant Construction plc* v. *Clive Adams Associates and JHM Construction Ser-
       vices Ltd* concerned a contract between Ford and Plant to design and build
       two pits for engine mount rigs and a suspension rig at Ford's research and
       engineering centre at Danton. The contract between Ford and Plant pro-
       vided that:

       (1)   Plant was to be responsible for damage to the works caused by its
             own negligence and Ford's negligence.
       (2)   Plant was responsible for all acts and omissions of its sub-
             contractors.
       (3)   Any assistance provided by Ford would not release Plant from
             responsibility for the works.

       Clive Adams were structural engineers engaged by Plant and JHM were
       sub-contractors for the sub-structure work involving shoring excavations
       and roof support. The roof collapsed because of insufficient support. Ford
       sued Plant who settled the claim. Plant then brought claims for breach of
       contract against Clive Adams and JHM. Clive Adams settled. JHM
       defended alleging that they had been following the instructions given to
       them by an engineer employed by Ford in the design and execution of the
       works.

5.6   The Court of Appeal had to consider the principles of the duty to warn
       and the implied term of skill and care in the context of damages known to
       the contractor. The court held:

       (1)   JMH was contractually obliged to carry out the temporary works of
             supporting the roof in the way in which and to the design by which
             they were so instructed by Ford.

(2)   The factual extent of the performance required by the implied term
      that a contractor will perform his contract with the skill and care of
      an ordinarily competent contractor, will depend on all the circum-
      stances.

(3)   Given crucially that the temporary roof support works were
      obviously dangerous and were known to JMH to be dangerous,
      JMH's implied obligation to perform with skill and care carried with
      it an obligation to warn of the dangers which they perceived.

(4)   The facts that the design and details of the temporary works were
      imposed by Ford, that Plant had Clive Adams as their consulting
      engineer, that others were at fault, or that JMH were contractually
      obliged to do what Ford instructed did not negative or reduce the
      extent of performance of the implied terms.

(5)   JMH's duty extended to giving property warnings about risk.

5.7   The facts of *Bolam* were concerned with a medical negligence case; how-
      ever in *Williams*, which approved *Bolam* subject to the three qualifications,
      the problem was a defective gas fired heating system. Clearly, therefore,
      the test set out in *Bolam* is equally applicable to other professional people
      and those exercising specialist skills: see also *Greaves & Co (Contractors)
      Ltd* v. *Baynham Meikle and Partners*.

5.8   An error of judgment or the selection of the wrong method where there is
      a genuine difference of specialists' opinion will not necessarily amount to
      negligence. In *Robinson* v. *The Post Office*, R, a doctor in general practice,
      injected a patient with an anti-tetanus serum without first administering a
      test dose. At the relevant time, medical opinion was moving against the
      use of anti-tetanus serums generally. The Court of Appeal held that since
      R's failure to give a test dose was contrary to accepted procedure, he had
      been negligent, but that no damage had been caused as the result of the
      test would have been negative, and that, since at the time there was still a
      responsible body of medical opinion who favoured anti-tetanus serum, he
      had not been negligent in using it. Similarly, in the case of *Perry* v.
      Tendring District Council and Others which dealt with, inter alia, the failure
      of a consultant engineer to design foundations that would be unaffected
      by long term soil heave, the court considered that the standard of care
      depended on 'what was to be expected of the competent engineer at the
      material date' (i.e time he designed the foundations). This is the 'state of
      the art' defence. There was conflicting expert witness evidence. One
      expert engineer personally knew of heave but was only able to refer to one
      textbook intended for engineers that dealt with it. Another expert had not
      read that textbook and thought that engineers generally would not have
      known of heave although that particular expert had expressed a contrary
      view some 12 years after the material date. Two other engineers stated
      categorically that they had never heard of heave. Judge Newey stated:

      'On the totality of the expert evidence I must, however reluctantly,
      conclude that at the material time a competent engineer would not have
      known of long term heave'.

5.9   A more draconian attitude was adopted by the House of Lords in the *IBA* case, the facts of which are set out in paragraph 1.21. The defendant's submissions that the design and building of the cylindrical mast was work which was 'both at and beyond the frontiers of professional knowledge at that time' was not disputed by their Lordships, nevertheless they held that the designer was negligent. As regards the state of the art submission, Viscount Dilhorne stated:

> 'No doubt all this was true, and bearing in mind the consequences that might ensue if such a mast collapsed – fortunately no-one was killed or injured at Emley Moor, though part of the mast fell across a road and it might have fallen on a farmhouse – it was in my opinion incumbent on [the designer] to exercise a very high degree of care.'

5.10   There are conflicting judicial decisions on the issue of whether a professional man who specialises within his profession has a higher duty than the non-specialist. In *Wimpey Construction UK v. Poole* a consultant held himself out as having especially high skills and was retained on that basis. The court rejected the argument that the test in such circumstances should be that of a main exercising or professing to have especially high professional skills. However in *Ashcroft v. Mersey Regional Health Authority*, the court found that the more skilled a person, the more the care which is to be expected of him, but the test should be applied without gloss either way.

5.11   Competency will invariably be a matter of expert evidence and opinion. However, in the last resort, the courts consider that they have discretion to reject expert evidence as to what is an acceptable practice within a profession. In *Sidaway v. Governors of Bethlem Royal Hospital*, Lord Templeman stated:

> 'Where the practice of the medical profession is divided or does not include express mention, it will be for the court to determine whether the harm suffered is an example of a general danger inherent in the nature of the operation, and if so whether the explanation afforded to the patient was sufficient to alert the patient to the general dangers of which the harm suffered is an example.'

5.12   *Sidaway* was concerned with a surgeon's duty to warn a patient of a potential risk, and on that basis can be distinguished from a designer of a building project. However, it is suggested that the court would have a similar discretion in construction cases. Indeed, the House of Lords adopted a similar position in the *IBA* case.

5.13   The express conditions of the principal contract or the collateral warranty may determine the standard of performance. What, however, is the position if the contract is silent on this particular point? In the *Greaves* case, there was a suggestion that a designer's obligation might extend beyond reasonable skill and care to fitness for purpose. The facts of the

Greaves case concerned G, a building contractor who undertook to design and construct on a package deal basis, a new factory, warehouse and offices for Alexander Duckham Limited. The warehouse was to be used for the storage of barrels of oil. G contracted with B, structural engineers, to design the structure of the warehouse. G informed B that the floors of the warehouse had to take the weight of forklift trucks carrying barrels of oil. After completion and occupation, cracks began to appear in the floors of the warehouse. It was established that the failure of the floors was due to vibrations caused by the use of the forklift trucks. The issues before the court turned upon whether B were in breach of their obligation to carry out their design works with reasonable skill and care, or whether B were in breach of an implied term of the contract between G and B that B's design should be fit for its purpose, namely the movement of loaded forklift trucks. It will be appreciated that a term which is implied as a matter of fact does not have the consequences of a term that is implied as a matter of law in so far as the former only relates to the particular bargain struck between the parties to the contract whereas the latter applies to all bargains unless excluded by the express terms of the contract. In the *Greaves* case, the judgment at first instance appeared to suggest that a fitness for purpose obligation was to be implied as a matter of law. On first reading, this is also the impression given by the judgment of Lord Denning MR in the Court of Appeal. However, on its facts, the *Greaves* case does not create a universal principle of fitness for purpose on the part of designers, in that the court found that whilst there was a contractual term that the designers should design a warehouse that was fit for its purpose, this term was implied as a matter of fact and not law. The court also found that B's design was negligent, that is to say in breach of the obligation to carry out their services with reasonable skill and care.

5.14 Any doubts that lingered from the *Greaves* case were disposed of by the Court of Appeal in *George Hawkins* v. *Chrysler (UK) Limited and Burn Associates*. B were engineers who contracted with C to prepare the design and specification for a shower room at C's factory, which included a new floor and wall coverings. G, the plaintiff, was an employee of C and he slipped on a puddle of water in the shower room after having used the shower. G sued C and C in turn brought proceedings against B. C settled G's claim but continued the third party proceedings against B. The main issues in the case were:

(1) was there an implied term of the contract that B would use reasonable skill and care in selecting the material to be used for the floor of the shower room?

(2) was there an implied warranty or term that the material used for the floor would be fit for use in a wet shower room?

The judge at first instance found against C in respect of the first issue. However, on the second issue the judge found that B was in breach of an implied warranty that they would provide 'as safe a floor as was practicable in the expertise of the profession to provide a safe floor for these men in these conditions'. On appeal, the Court of Appeal held, inter alia, that although a party contracting for both the design and supply of a product will usually be under an implied contractual duty to ensure that it is reasonably fit for the purpose for which it is intended, where the contracting party is a professional man providing advice or designs alone (i.e. without supplying any product), no warranty will normally be implied beyond a term that reasonable skill and care will be taken in giving the advice or preparing the design. There was nothing in the present case to require the implication of any term other than a duty to take reasonable care and skill in preparing the design.

5.15   If the party being called upon to enter into a collateral warranty is an architect or engineer with a design function, and his principal contract expressly provides for a performance obligation of reasonable skill and care or is silent on this matter, for that party to enter into a collateral warranty with a fitness for purpose obligation will be increasing his liabilities. The important distinction between reasonable skill and care and fitness for purpose is that fitness for purpose is an absolute obligation and provided the obligation is clearly established or defined by the contract document, the party in breach will not be able to plead as a defence that he has discharged his services with reasonable skill and care. The facts in *Samuels* v. *Davis* provide a useful illustration of the dichotomy between reasonable skill and care and fitness for purpose. In *Samuels*, the Court of Appeal held that where a dentist undertakes for reward to make a denture for a patient, it is an implied term of the contract that the denture will be reasonably fit for its intended purpose. Du Parcq LJ stated:

> '... if someone goes to a professional man ... and says: "Will you make me something which will fit a particular part of my body? ..." and the professional gentleman says "Yes", without qualification, he is then warranting that when he has made the article, it will fit the part of the body in question ... If a dentist takes out a tooth or a surgeon removes an appendix, he is bound to take reasonable care and to show such skill as may be expected from a qualified practitioner. The case is entirely different where a chattel is ultimately to be delivered.'

5.16   It is important to note that if an architect or engineer extends his potential liability by entering into a collateral warranty providing for a fitness for purpose obligation, there may well be serious repercussions in respect of his professional indemnity policy (see paragraph 8.32 to 8.34).

Fitness for purpose

5.17   The first section of this chapter has dealt with the legal obligation of a professional man. What is the position of the contractor or sub-contractor, in particular the design and build contractor?

5.18   Clearly, as stated above, if the contract expressly deals with the standard of the contractor's performance then, in the absence of ambiguity, the express terms will determine the extent of the contractor's or sub-contractor's legal obligation. However, if the contract is silent on these matters it has long been held that a contractor, or sub-contractor, who agrees to carry out construction works impliedly warrants (that is to say there is a term implied by law) that he will carry out his works with reasonable skill and care (often referred to as the obligation to carry out the works in a good and workmanlike manner). The standard of performance is the same as reasonable skill and care in negligence. The contractor or sub-contractor also warrants that the materials he supplies for the purposes of works will be of a merchantable quality, that is to say good of their kind. This warranty is an absolute warranty and extends to latent defects and it will not help the contractor to show that he has exercised reasonable skill and care in the selection of those materials: *Young & Marten Limited* v. *McManus Childs Limited*. M were developers of a residential housing estate and Y were a firm of roofing sub-contractors. Y provided an estimate for the supply and laying of certain roof tiles subsequent to which M specified that Y should use a particular roof tile called 'Somerset 13'. These tiles were supplied by only one manufacturer, J. Beale & Co. The tiles supplied by Beale appeared to be sound; however, 12 months after completion of the roofs a large number of tiles began to disintegrate, a consequence of a latent defect. M was sued by the purchasers of the houses and M sought indemnity against Y. At first instance, the court rejected M's submission that there was an implied term that the Somerset 13 tiles should be reasonably fit for their purpose and should be of merchantable quality. On appeal the Court of Appeal held:

(1)   Unless the circumstances of a particular case suffice to exclude then there will be implied into a contract for the supply of work and materials a term that the materials used will be of merchantable quality and a further term that the materials used will be reasonably fit for the purpose for which they are used; and

(2)   In this particular case the circumstances sufficed to exclude the term that the tiles would be reasonably fit for the purpose for which they were required; and

(3)   In this particular case the circumstances were not sufficient to exclude the term that the tiles were merchantable. The fact that these tiles were obtainable from only one manufacturer was not a circumstance which excluded the implication but, per Lord Reid, if the tiles had been made by only one manufacturer who was willing to sell only on terms which excluded or limited the ordinary liability (under statute) and if that fact was known to the employer and to the contractor when they made the contract, then it would be unreasonable to place upon the contractor a liability for latent defects; and

(4)   Y supplied and fixed tiles which were latently defective and thereby breached the implied term (of merchantable quality).

5.19   In *Gloucestershire County Council* v. *Richardson,* the House of Lords found
        that the particular circumstances of the case excluded both the implied
        warranty of suitability and the implied warranty of merchantable quality.
        In that case R entered into a contract with G for the construction of an
        extension to a technical college. The contract was in the RIBA Form 1939
        Edition, 1957 Revision. The bills of quantities provided for a PC sum for
        concrete columns to be supplied by a nominated supplier. R contracted to
        erect the columns. Clause 22 of the conditions of contract dealing with
        nominated suppliers, unlike clause 21 which dealt with nominated sub-
        contractors, did not entitle R to make reasonable objection to a proposed
        supplier, nor to object on the ground that the supplier would not
        indemnify him in respect of his main contractor's obligation. G's architect
        instructed R to accept a quotation given by C W & Co for the supply of the
        concrete columns. CW's standard conditions of trade restricted their lia-
        bility in respect of good supply by them. The columns supplied by CW
        had latent defects because of faulty manufacture and after erection cracks
        appeared in them; the columns were unsuitable for use as structural
        members of the extension. The House of Lords considered that the cir-
        cumstances set out above indicated an intention on the part of G and R to
        exclude from the main contract any implied terms that the concrete
        columns should be of good quality and fit for their required purpose.

5.20   The Court of Appeal in *Rotherham Metropolitan Borough Council* v. *Frank
        Haslam Milan & Co Ltd* and *M. J. Gleeson (Northern) Ltd* v. *Taylor Woodrow
        Construction* has given extremely useful guidance to the operation of the
        obligations of fitness for purpose and merchantable quality (now satis-
        factory quality) in the context of a construction project. The case con-
        cerned the use of steel slag as a fill material, which although suitable for
        some fill purposes, was not fit for the particular purpose of fill in a con-
        fined area where the fill had to be inert. Steel slag was not inert. The
        emphasis of the court was that the matter should be approached not from
        the test as to whether there should be an implication of an obligation of
        fitness for purpose, but whether in all the circumstances and the 'matrix'
        there was or was not reliance on the contractor's or supplier's skill and
        judgment. The judgments identified the types of circumstances that
        would be relevant to the test. The case also provides a useful comparison
        of the obligations of fitness for purpose and merchantability.

**Supply of Goods and Services Act 1982 as amended by the Sale and Supply of
Goods Act 1994**

5.21   The common law rules have now in part been replaced (as to goods) and
        in part supplemented (as to services) by a statutory code, the Supply of
        Goods and Services Act 1982 (as amended by the Sale and Supply of
        Goods Act 1994), which governs all contracts made after 4 January 1983.
        Section 4 of the Act, as amended, sets out the code for quality and fitness.
        Section 4(2) of the Act provides that goods shall be of a satisfactory

quality, i.e. they should meet the standard that a reasonable person would regard as satisfactory taking account of any description, price and all other relevant circumstances. By section 4(3) the following matters were excluded from the statutory warranty:

(1) Matters which are specifically drawn to the transferee's attention before the contract is made; or
(2) Where the transferee examines the goods before the contract is made, matters which that examination ought to reveal; or
(3) Where the property is transferred by reference to a sample, matters which would have been apparent on reasonable examination of the sample.

Sections 4(4) and 4(5) provide that where the purpose for which the goods are being acquired is made known to the party contracting to buy the same, whether expressly or by implication, there is an implied condition that the goods should be reasonably fit for their purpose, whether or not that is a purpose for which such goods are commonly supplied. Section 4(6) creates a proviso whereby the fitness for purpose condition does not apply if the other party does not rely, or it was unreasonable for him to rely, on the skill or judgment of the party providing the goods.

5.22   Section 1(3) of the Act provides that a contract is a contract for the transfer of goods even though services are to be provided under the same contract. It follows that section 4 applies to the materials part of a contract for the supply of work and materials. A building contract is a contract for the supply of work and materials.

5.23   Section 13 of the Act deals with a contract for services and provides that there shall be an implied term that the contractor will carry out the services with reasonable care and skill. Unlike the statutory warranties as to quality and fitness of goods, section 13 does not exclude the common law rules in so far as those rules impose a stricter duty than the Act. It follows that the decisions in *Young & Marten* and *Gloucestershire* are still of relevance.

Design and build contractors

5.24   What is the standard of performance of the contractor who undertakes a design obligation? Is it the same as the professional man, that is to say reasonable skill and care, or is it the higher duty of fitness for purpose? This issue was considered in the *Greaves* case. Lord Denning stated:

> '... now, as between the building owners and the contractors, it is plain that the owners made known to the contractors the purpose for which the building was required, so as to show that they relied on the contractors' skill and judgment. It was, therefore, the duty of the contractors to see that the finished work was reasonably fit for the purpose for which they knew it was required. It was not merely an

obligation to use reasonable care. The contractors were obliged to ensure that the finished work was reasonably fit for the purpose.'

5.25   In *Viking Grain Storage Limited* v. *T. H. White Installations Limited*, W were package deal contractors for the design and construction of a grain drying and storage installation. The installation was not fit for its purpose and V contended that there were implied terms of the contract that W would use materials of good quality and reasonably fit for their purpose and that the completed works would be reasonably fit for their purpose, namely that of a grain drying and storage installation. The court held that there were no terms of the contract or any other relevant circumstances which were inconsistent with the implied terms of quality and fitness for purpose and further that there was no reason to differentiate between W's obligation in relation to the quality of materials and their obligation as to design. V had relied upon W in all aspects, including design, and on the skill and judgment of W, and in the circumstances, the terms contended for should be implied. By way of contrast the Supreme Court of Ireland in the case of *Norta Wallpapers (Ireland)* v. *Sisk & Sons (Dublin)*, held that where a roof structure, which had been supplied and erected by a specialist sub-contractor, subsequently leaked and was unsuitable for its purpose, the fact that the main contractor was given no choice but to use the specialist sub-contractor, his design and price constituted circumstances which meant that there was no reliance by the employer on the main contractor and, accordingly, there was no fitness for purpose obligation on the main contractor in respect of the specialist sub-contractor's failure.

5.26   In the (unreported) case of *Trolex Products Limited* v. *Merrol Fire Protection Engineers Limited* there was an interesting issue as to whether a design obligation was created by the bringing together of what otherwise would have been standard components. T were the sub-contractors for the supply of an electronic control system which was incorporated into M's own works comprising the installation of a fire protection system in the Ras Abufontas power and water station in Quatar. In response to the submission by T that there was minimal design obligation in the sub-contract, Potter J stated:

'I should perhaps add that at one stage I had evidence from T which minimised the work of design carried out, suggesting that it was no more than, in effect, a matching of pieces of standard equipment to make up a package to do the job; or as Mr B put it "logic design work created from standard equipment". Even if that was so in fact, I am satisfied from the answers of Mr B that there was a conscious realisation that design work was involved and that T were consulted as experts in their field. Further, it is clear that substantial time was spent on this work. Again, whether or not that was so, it is not suggested that anything was said by T to delimit or belittle the design work involved and the construction of the written contract is clear in my view, namely as one for work of design as well as the supply of goods.'

Dwellings

5.27   Where a contractor is involved in the construction of a residential dwelling there is an implied term, implied by law, that the contractor will carry out his work in a good and workmanlike manner, that he will supply good and proper materials and that the dwelling will be reasonably fit for human habitation: *Hancock* v. *B. W. Brazier (Anerley) Limited*. This common law obligation has now been supplemented by a statutory code set out in the Defective Premises Act 1972, which came into force on 1 January 1974. Section 1(1) of the Act provides:

> 'A person taking on work for or in connection with the provision of a dwelling (whether the dwelling is provided by the erection or by the conversion or enlargement of the building) owes a duty,
>
> (a)   if the dwelling is provided to the order of any person, to that person;
>
> and
>
> (b)   without prejudice to paragraph (a) above to every person who acquires an interest (whether legal or equitable) in the dwelling;
>
> to see that the work which he takes on is done in a workmanlike or, as the case may be, professional manner with proper materials and so that as regards that work the dwelling will be fit for habitation when completed.'

5.28   Unlike the Supply of Goods & Services Act discussed above, the Defective Premises Act does not operate via the parties' contract but in fact creates a statutory duty. Accordingly, it has a much wider effect and is akin to tortious liability which is not dependent upon a contract. It follows that future owners are entitled to sue the builder if he is in breach of the statutory duty. The Act is, however, restricted to the provision of dwellings and does not apply to commercial developments although the term dwellings includes dwellings that are created by conversion or enlargement. There is also a statutory exception: Section 2(1) provides that where the construction of the dwellings is subject to 'an approved scheme' the Act does not apply. The National House Building Council operates a warranty scheme for dwellings. The NHBC's schemes dated 1973, 1975, 1977 and 1979 are approved schemes under the Act (see The House Building Standards (Approved Scheme) Order 1979 Statutory Instrument 1979/381). As far as the authors are aware, the most recent NHBC scheme is not yet an approved scheme under the Act.

5.29   The Act prohibits any attempt to exclude its operation by section 6(3), which makes void any term in a contract that purports to exclude or restrict the operation of the Act.

Chapter 6
Damages and Limitation of Action

Damages

Nature of damages

6.1   Damages for breach of contract are intended to be compensatory, that is to say so far as they can, they are intended to place the innocent party in the same position that such party would have been in had the other party performed his contractual promises: *British Westinghouse Electric Company Ltd* v. *Underground Electric Railways*. In contrast, an award of damages in respect of tortious liability is intended to place the innocent party back in the position he was before the breach of the duty. Awards of damages for breach of contract may therefore be greater than awards of damages in tort. In *Muirhead* v. *Industrial Tank Specialists Ltd*, M entered into a contract with a third party for the supply of pumps to be used at M's lobster farm. The electric motors for the pumps were supplied by ITS. The motors failed causing the pumps to fail, which resulted in the death of M's lobsters. There was no contract between M and ITS and M brought proceedings in negligence. M claimed three heads of damage: the value of the dead lobsters; the loss of profits on those lobsters; the loss of profits that would have been earned by M's business had the pumps functioned properly. It was held that M was entitled to recover the first two heads of damage but not the third head of damage. The court considered that the third head of damage might have been recoverable had there been a contract between M and ITS.

6.2   Damages may be general damages, which can be assessed in an approximate figure, or special damages, which represent past pecuniary loss and have to be calculated and pleaded with sufficient particularity so as not to take the defendant by surprise at trial. Further, damages may include future damages, that is to say those which it is anticipated are likely to flow from the breach of contract. Future damages are known as prospective loss and must be claimed at the same time as past or present loss in respect of the same breach of contract: *Conquer* v. *Boot*. This rule does not apply to breaches of different promises within the same contract or breaches of recurring obligations; such breaches are considered to give rise to separate causes of action. Damages for breach of contract (which term includes liquidated and ascertained damages) must be distinguished from claims arising under the terms of the contract, for example a claim for direct loss and expense under clause 26 of JCT 1998.

Such claims are not subject to the general principles on damages set out below. They are however subject to the rules of construction of a contract dealt with in paragraphs 1.57 to 1.71.

6.3   The parties to a contract may agree on the amount of damages to be awarded in the event of a breach of contract. Such damages are known as liquidated and ascertained damages and are commonplace in construction contracts, usually related to the completion obligations of the contractor. Provided the agreed sum is a genuine pre-estimate of loss and not a penalty, it will be enforced by the courts: *Dunlop Ltd* v. *New Garage Co Ltd*. Damages may also include consequential losses. Consequential losses, often referred to as 'economic loss', are losses such as loss of profit, which are not directly related to physical damage: *Muirhead*. (See also paragraphs 9.92 to 9.99.)

Causation

6.4   There must be a causal link between the breach of contract and the damage suffered by the innocent party. In *Quinn* v. *Burch Brothers (Builders) Ltd*, Q was a sub-contractor to B in respect of certain building works. There was an implied term of the sub-contract that B would supply, within a reasonable time, any plant or equipment reasonably necessary for carrying out the sub-contract works. B was in breach of that term in failing to supply a stepladder requested by Q. To prevent any delay, Q used a trestle, which he knew to be unsuitable unless it was footed by another workman. The trestle was not footed and it moved causing injury to Q. It was held that B's breach of contract provided the occasion for Q to injure himself but was not the cause of his injury; the injury was caused by his own voluntary act in using the trestle. Accordingly B was not liable to pay damages for Q's injury, for his damage was not a natural and probable consequence of the breach of contract even if, as was in fact doubtful, it was a foreseeable consequence of that breach.

6.5   Damages for breach of contract may only be awarded for the breach itself and not for any loss caused by the manner of the breach: *Malik* v. *Bank of Credit & Commerce International*. Further, the breach of contract must be the effective or dominant cause of the loss: *Leyland Shipping* v. *Norwich Union*. But it need not be the sole cause: *Galoo* v. *Bright Grahame Murray*.

6.6   The causal link may be broken by the intervening acts of the plaintiff, as in *Quinn's* case, or by the intervening acts of a third party. It must be noted however that the courts are wary of laying down any general principles and many cases turn upon their own particular facts. For example, in the case of *Weld-Blundell* v. *Stephens*, S negligently and in breach of contract permitted a libellous letter written by W to fall into the hands of a third party who communicated the contents of the letter to the person who had been libelled. That person sued W, who in turn sued S. The court held that the acts of the third party broke the causal link between S's initial breach and W's damage.

6.7   A more recent example arose out of the conflict between Iran and Iraq in the 1980s in the case of *Bank of Nova Scotia* v. *Hellenic Mutual War Risks Association (Bermuda) Ltd (The Good Luck)*. The claimant bank had financed the purchase by a Greek shipping group (called Good Faith) of various vessels including the 'Good Luck'. The finance had been secured by mortgages, which required the taking out of contracts for marine insurance which extended to war risk cover. The insurance was placed with the defendants and was conditional upon the insurers' right to specify 'additional premium areas' and 'prohibited zones'. At the time, the Persian Gulf was designated both an additional premium area and, at the northern end of the Gulf, a prohibited zone. The bank took out mortgagee interest insurance with the defendant who, by letter of undertaking, agreed to advise the bank promptly if they ceased to insure.

6.8   The shipowners chartered their vessels, including the 'Good Luck', to Iranian charterers. The defendants were aware that the shipowners were deliberately allowing their ships to be sent to both additional premium areas and prohibited zones in the Persian Gulf and that their vessels were uninsured as a consequence. The defendants took no steps to inform the bank. In April 1982, the shipowners commenced negotiations with the bank for a rescheduling of their loan by way of an increased facility. On 6 June 1982, the 'Good Luck' was hit by Iraqi missiles whilst proceeding up the Khor Musa Channel to Bandar Khomeini, both an additional premium area and a prohibited zone. She was badly damaged and ultimately declared a constructive total loss. The shipowners made a fraudulent claim on the defendants in respect of this loss. The bank at this time was still in the process of completing the negotiations with the shipowners for the refinancing and although the bank was aware of the loss of the 'Good Luck', did no more than cursorily investigate with the managing agents of the defendants, the circumstances of the loss. The bank's refinancing was capped at 67% of the security value of the shipowners' vessels, a value of US$50,225,000 (including the insurance value of the 'Good Luck' of US$4,800,000), i.e. US$33,650,750. The refinancing was completed in July 1982 when the shipowners drew down the total loan of US$33,650,750, of which US$30,971,630 was used to pay off the existing loans, discharging the existing mortgages and the mortgage interest insurance. The balance of US$2,679,120 was used by the shipowners as additional working capital. On 4 August 1982, the defendants rejected the shipowners' insurance claim for the loss of the 'Good Luck'. The bank called in the loans and as a consequence of inadequate security suffered a loss.

6.9   The case was a rich tapestry indeed for arguments as to causation; there was the fraudulent claim by the shipowners, the ineptitude of the bank, and the loss of the mortgagee interest insurance caused by the bank accepting a repayment of the original loans to the shipowner. Hobhouse J gave judgment for the bank for the amount of the additional sum that the bank had allowed the shipowners to borrow on the basis that the insured value of the 'Good Luck' was included within the calculation of the security value. The judge found that if the bank had been notified in writing that the ship was in a particular zone and therefore not covered by war risks, then the bank would have adopted a different approach to the refinancing – the insured value of the 'Good Luck' would have been excluded from the calculation of security value reducing it to US$30,434,750. As a consequence the existing mortgage on the 'Good Luck' would not have been discharged and the position under the mortgagee interest would have been preserved. The judge further found that the strong probability was that the whole matter of the 'Good Luck' and any value to be attached to its hull or its insurances, would have been put on one side until the situation had been clarified and that would have meant that no draw-down in respect of working capital would have been permitted in July and August 1982.

In so finding, Hobhouse J rejected the defendant's submissions that the bank's own negligence was the sole cause of its loss. The judge apportioned blameworthiness in the proportion of two-thirds to the defendants and one-third to the bank. However, as the claim was in contract independent of negligence, he held that the bank's claim should not be reduced on the basis of contributory negligence. The Court of Appeal, reversing the decision on liability for breach, but not on the judge's findings as to causation, expressed agreement with Hobhouse J that if the defendant was in breach of the letter of undertaking, such breach would have been at least a cause of the bank's loss. The Court of Appeal further stated that it would be clearly wrong in all the circumstances to hold that the bank's action amounted to a *novus actus interveniens*, breaking the chain of causation between the defendant's breach and the bank's loss. Finally, the Court of Appeal agreed with the judge's finding on the issue of contributory negligence. The House of Lords, whilst reversing the Court of Appeal's decision on liability, concurred with both the judge's and the Court of Appeal's findings in the issues as to causation. Lord Goff stated:

'The club sought to draw a distinction between the cause of the advance being made by the bank to Good Faith and the cause of the bank's inability to recover the additional loan when it eventually sought to realise its securities. The latter, it was submitted, was the true proximate cause of any loss. With this submission I am unable to agree. On the findings made by the judge, the failure of the club to comply with its obligations left to the bank making an advance which was inadequately secured and which it would not otherwise have made, and the bank's loss was accordingly caused by the breach of the club's letter of undertaking. The club further submitted that what caused the advance was the bank's improvident decision to grant the further advance to Good Faith and/or the false assurances given by Good Faith to the bank, each of which was a *novus actus interveniens*. Again I do not agree with this submission, which is inconsistent with the concurrent findings of the judge and the Court of Appeal – findings with which I find myself to be in complete agreement.'

6.10   In *Brown* v. *KMR Services Ltd*, Mr Brown, an underwriting name at Lloyds, brought claims against his member's agent for the agent's failure, on breach of contract and negligently, to warn Mr Brown of the dangers of excess of loss reinsurance, or advise on a proper spread of risk. The Court of Appeal considered the question of causation in relation to the giving of professional advice. The court held that where a member's agent was in breach of its duty to provide appropriate information and advice to a Lloyd's name, the question of causation was to be approached by identifying first what specific advice the name ought to have received and then what the name could prove, on the balance of probabilities, would have been the consequences of his receipt of such information and advice.

6.11   Where there are intervening *events,* as distinct from intervening acts, of the claimant or a third party, such events will not break the causal link between breach and damage if they were foreseeable. In *Monarch Steamship Co Ltd* v. *Karlshamms Hjesabrikes (A/B)*, the defendant ship owners chartered a ship to the plaintiffs for the carriage of a cargo from Manchuria to Sweden on terms that the voyage would be completed by July 1939. In breach of the charter party the ship was unseaworthy, resulting in a delay to the voyage, and in September 1939, on the outbreak of World War II, the British Admiralty diverted the ship to Glasgow resulting in transhipment of the cargo to Sweden on neutral vessels involving additional cost. The court held that in July 1939 the parties to the charter party should have had in mind the possibility of the outbreak of war and should have foreseen the events that subsequently happened. Accordingly, the defendant was entitled to recover damages for breach of contract.

Measure of damages

6.12   The term 'measure of damages' can be used either in a wider sense to include both categories of damage and the calculation or quantification of damage, or in a more restricted sense, namely the principles of law which define the categories or heads of damage which will be recoverable when there has been a breach of contract, commonly referred to by lawyers as the 'rules of remoteness of damage'. It is the latter restricted meaning which is adopted for the purposes of this sub-section.

6.13   The basic rule as to measure of damages is often referred to as the rule in *Hadley* v. *Baxendale*. This was the name of a case heard in 1854 involving a claim for breach of contract by a mill owner against a carrier and arising from the carrier's failure to deliver a crankshaft within the time specified by the contract of carriage. Unbeknown to the carrier the crankshaft was critical to the whole of the output of the mill. The plaintiffs brought a claim against the defendants claiming a loss of profit for the whole of their production between the dates when the crankshaft should have been delivered and the date when it was actually delivered. In rejecting their claim for loss of profits, Alderson B stated:

'Where two parties have made a contract which one of them has broken, the damages which the other party ought to receive in respect of such breach of contract should be such as may fairly and reasonably be considered either as arising naturally, i.e according to the usual course of things, from such breach of contract itself, or such as may reasonably be supposed to have been in the contemplation of both parties, at the time they made the contract, as the probable result of the breach of it.'

6.14 The judgment in *Hadley* v. *Baxendale* was explained and indeed developed in two leading cases in the twentieth century: *Victoria Laundry (Windsor) Ltd* v. *Newman Industries Ltd* and *Koufos* v. *Czarnikow Ltd (The Heron II)*. In the *Victoria Laundry* case, the rule was explained with reference to three main propositions:

'(1) In cases of breach of contract the aggrieved party is only entitled to recover such part of the loss actually resulting as was at the time of the contract reasonably foreseeable as liable to result from the breach.

(2) What was at that time reasonably foreseeable depends on the knowledge then possessed by the parties or, at all events, by the party who later commits the breach.

(3) For this purpose knowledge "possessed" is of two kinds; one imputed, the other actual. Everyone, as a reasonable person, is taken to know the "ordinary course of things" and consequently what loss is liable to result from a breach of contract in the ordinary course... But to this knowledge, which a contract breaker is assumed to possess whether he actually possesses it or not, there may have to be added in a particular case knowledge which he actually possesses, of special circumstances outside the "ordinary course of things" of such a kind that breach in those special circumstances would be liable to cause more loss.'

6.15 The wording of the judgment in *Hadley* v. *Baxendale* caused much lively debate on the issue of whether there were two branches to the rule, the first branch being damages arising naturally and the second branch being damages in the reasonable contemplation of the parties. This issue was not purely semantic for it will be appreciated that if the first branch of the rule was unqualified by the parties' reasonable contemplation, or, to put it another way, by the particular bargain struck between the parties, then the first branch of the rule would tend towards the reasonable foreseeability test applicable to the measure of damages in tort, that is to say damage which should have been foreseen by a reasonable man as being something of which there was a real risk: *The Wagon Mound (No. 2)*. However, the judgments in *Victoria Laundry* and *Koufos* support the view that there is really only one rule in *Hadley and Baxendale*, and that damages which may reasonably be supposed to have been contemplated by the contracting parties are damages which arise naturally from a breach of contract. What was in the reasonable contemplation of the parties is decided on both an objective basis and a subjective basis. The objective test turns upon the contemplation of a reasonable person – that is to say, it is imputed knowledge – whereas the subjective test turns upon the actual knowledge of the parties, or the particular party, who is in breach of contract.

6.16   The application of the rule in *Hadley* v. *Baxendale* can be usefully illustrated by reference to the facts of the *Victoria Laundry* case and the *Koufos* case.

Victoria Laundry

6.17   V entered into a contract to purchase from N, an engineering firm, a boiler, which was installed on N's premises. V, who were a firm of launderers and dyers, required the boiler to extend their business to increase their general turnover and also they had in mind the prospect of certain profitable dyeing contracts being obtained from the Government. V and N agreed that the boiler should be delivered to V's premises on 5 June 1946. The boiler was damaged whilst it was being dismantled on N's premises and delivery to V was delayed until 8 November 1946. It was found as a matter of evidence that N were aware of both the nature of V's business and that V intended to put the boiler into use as quickly as possible. V sued N for breach of contract, and their claim for damages included their loss of business profits. At trial the judge disallowed the claim for loss of profits on the ground that it was based upon special knowledge, which had not been drawn to the attention of N. V appealed and the Court of Appeal, reversing the trial judge's decision, held that N, an engineering company with knowledge of the nature of the plaintiff's business, having promised delivery by a particular date of a large and expensive plant, could not reasonably contend that they could not foresee that loss of business profit would be liable to result to the purchaser from a long delay in delivery; and that although N had no knowledge of the dyeing contracts which V had in prospect, it did not follow that V was precluded from recovering some general and perhaps conjectural sum for loss of business in respect of the contracts to be reasonably expected. The Court of Appeal was applying the objective test of reasonable contemplations as regards the general business profits of the launderers. However the specific profits, which would have been earned from the prospective dyeing contracts with the Government, were not claimable as N did not have actual knowledge of those contracts.

Koufos

6.18.   C was the owner of the *SS Heron II*. C entered into a charterparty with K for the consignment of sugar from Constanza to Basrah. At the time of making the contract, the ship was docked in Piraeus and the shipowners anticipated that it would be ready to load in Constanza by about 25 to 27 October 1960 after which date it would proceed at all convenient speed to Basrah. The ship arrived at Constanza on 27 October, was loaded with sugar and departed from that port on 1 November. A reasonably accurate prediction of the length of the voyage between Constanza and Basrah was 20 days. C knew that K were sugar merchants and that there was a sugar market in Basrah but C had no actual knowledge that K intended to sell the sugar promptly after its arrival. In breach of the charterparty contract, the *SS Heron II* deviated from the voyage by calling at Berbera, Bahrain and Abadan, delaying the voyage by some ten days the ship arriving at Basrah on 2 December and not 22 November. The prices on the sugar market at Basrah tended to fall in October and November to a low point in December. Between 22 and 28 November the price of sugar in the Basrah market was £32.10s per tonne and the price between 2 and 4 December was £31.2s 9d per tonne. K brought a claim against C for breach of contract claiming damages based upon a loss of profit by reason of the price differential. The House of Lords held that, since prices in a commodity market were liable to fluctuate, shipowners should reasonably contemplate that it was not unlikely that, if their ships delayed their voyage, the value of marketable goods on board their ships would decline, and that therefore, where there was wrongful delay in the delivery of marketable goods under a contract of carriage by sea, the measure of damages was a difference between the price of the goods at their destination when they should have been delivered and the price when they were in fact delivered. Again the court was applying the objective test and did not feel it necessary to consider the actual knowledge of the contract breaker. See the decision of the Court of Appeal in *Hotel Services Ltd* v. *Hilton International Hotels (UK) Ltd* for a recent rationale of *Hadley* v. *Baxendale* and the dichotomy between direct and consequential damages.

6.19 In *Brown* v. *KMR Services* the Court of Appeal reaffirmed the test of remoteness of damage stated by the Master of the Rolls, Sir Thomas Bingham, in *Banque Bruxelles Lambert SA* v. *Eagle Star Insurance Co Ltd*:

> 'The test is whether, at the date of the contract or tort, damage of the kind for which the plaintiff claims compensation was a reasonably foreseeable consequence of the breach of contract or tortious conduct of which the plaintiff complains. If the kind of damage was reasonably foreseeable it is immaterial that the extent of the damage was not.'

6.20 On the facts of *Brown*, the Court of Appeal rejected the defendant's submissions in respect of excess of loss catastrophe reinsurance that a reasonable person in the position of the contracting parties would not have predicted that in the years 1988, 1989 and 1990 there would have been such a concentration of major catastrophes, and that the recoverable damages should be limited to such losses as could reasonably have been foreseen – that only one or two catastrophes would have occurred in any of those years. (See also *Bank of Nova Scotia*.)

Recovery of third party losses

6.21   Construction projects are characterised by a class of future owners out-side the ambit of the originating contractual relationships, with potential for a dichotomy between the legal right to recover and the incurring of loss, the 'legal black hole' referred to by Lord Keith in the *GUS Property Management* case (see below at 6.65). The problem flows from the general principle that a claimant may only recover damages for a loss which he has himself suffered: *British Westinghouse*. These issues have now been considered by the House of Lords in *Linden Gardens* and *St Martins Property Corporation Ltd* v. *Sir Robert McAlpine & Sons Ltd* (both appeals being heard jointly by the House of Lords and the Court of Appeal – see below) and in *Alfred McAlpine Construction Ltd* v. *Panatown Ltd.*

Linden Gardens and St Martins Property Corporation

6.22   In *Linden Gardens* the lessee of the premises, Stock Conversions Ltd, entered into three contracts for the removal of asbestos:

(1)   In June 1979 with Lenesta Sludge (first defendant) as prospective sub-contractor
(2)   In July 1979 with McLaughlin & Harvey (second defendant) as main contractor
(3)   In February 1985 with Ashwell Construction (third defendant)

Completion of the works under contracts (1) and (2) took place in March 1980 and under contract (3) on 16 August 1985. On 3 July Stock Conver-sions commenced legal proceedings against Lenesta for damages for breach of contract. Stock Conversions made the following assignments, presumably for market value, to Linden Gardens Trust Ltd:

(1)   On 1 August 1985 of its interest in the 3rd, 5th and 6th floors,
(2)   On 2 December 1986 of its interest in the 4th floor,
(3)   On 14 January 1987 of the rights of action under the contracts with McLaughlin and Ashwell.

Proceedings commenced by Stock Convertions in July 1985 against Lenesta Sludge were amended in 1987, Linden Gardens becoming the claimant, and McLaughlin and Ashwell the second and third defendants respectively.

6.23   In *St Martins Property Corporation Ltd and Another* v. *Sir Robert McAlpine & Sons Ltd*, St Martins Property Corporation entered into a contract with McAlpine for a mixed commercial and residential development. The contract was dated 29 October 1974. On 25 March 1976, St Martins Corporation transferred the property and assigned the benefit of the contract to St Martins Property Investments Ltd for its market value.

Subsequent to practical completion of the podium deck on 1 November 1979, substantial defects were discovered in the structure requiring remedial costs in excess of £800,000. Both Corporation and Investments brought proceedings against McAlpine to recover the cost of the remedial works.

6.24   The court held in *Linden Gardens* that the claimant was entitled to recover the cost of remedial works carried out by Stock Conversions in 1985 (£22,205) and also the works carried out by Linden Gardens (£236,000) after the date of the assignment of the rights of action in January 1987. In *St Martins* the court held, inter alia, that as Corporation had sold the property for market value to Investments, Corporation could only recover nominal damages. In both *Linden Gardens* and *St Martins Property Corporation* the purported assignments of the contractual rights were held to be invalid for reasons set out in paragraph 4.43, therefore the courts were looking at the right to recover damages by the original contracting party after a disposal of the relevant property at market value and where the remedial costs had been incurred, in part or in whole, by a third party. The House of Lords confirming the Court of Appeal's decision (although reversing the Court of Appeal on the meaning of the contractual prohibition) held that a party to a building contract was entitled to recover substantial damages from the contractor in breach even though that party had disposed of the damaged property and had not incurred the remedial costs, since the parties were to be treated as having entered into the contract on the footing that the original contracting party would be entitled to enforce contractual rights for the benefit of those who suffered from defective performance but who, under the terms of the contract, could not acquire any right to hold the defendants liable for breach. The majority of their Lordships felts that this decision was one which did not create new principle, but fell within the exceptions to the rule that a party can only recover its own losses established by the case of *Dunlop* v. *Lambert* as explained in *Albacruz (cargo owners)* v. *Albazero (owners)* ('The Albazero'). However, Lord Griffiths was in favour of new principle finding that with contracts for the supply of work, labour and the supply of materials, the recovery of damages for breach of contract was not dependent or conditional on the plaintiff having a proprietary interest in the subject matter of the contract at the date of breach. Further it was irrelevant who actually paid for the repairs and where a tortfeasor's liability was temporarily discharged by payment by a third party, on the plaintiff's behalf, the plaintiff ought not to be prevented from suing the tortfeasor for damages. In his judgment, Lord Griffiths added a constraint which was thought previously not to be material: 'The court will of course wish to be satisfied that the repairs have been or are likely to be carried out ...'. Contrast this with the House of Lords decision in *Ruxley Electronics* v. *Forsyth* where it was accepted that 'the courts are not normally concerned with what a plaintiff does with his damages' (see also the Court of Appeal in *Dean* v. *Ainley*).

Panatown

6.25   Panatown Ltd entered into a building contract with Alfred McAlpine
        Construction Ltd for the design and construction of an office block and
        car park on land owned by Unex Investment Properties Ltd (UIPL).
        Panatown and UIPL were both parts of the Unex group of companies.
        McAlpine also entered into a duty of care agreement (DCD) with UIPL
        whereby McAlpine agreed with UIPL that they would carry out their
        obligations under the building contract with Panatown with reasonable
        skill and care. Disputes arose between Panatown and McAlpine in respect
        of building defects and Panatown commenced arbitration proceedings
        against McAlpine claiming damages for breach of contract. McAlpine
        defended the proceedings on the ground, inter alia, that Panatown had no
        proprietary interest in the site and therefore had suffered no loss. This
        defence was taken as a preliminary issue in the arbitration and was
        rejected by the arbitrator. On appeal, the Court of Appeal confirmed the
        arbitrator's decision.

6.26   Before considering the House of Lords decision in *Panatown* it is helpful to
        look at the judgments of the House of Lords in respect of *St Martins
        Property Corporation* and a subsequent decision of the Court of Appeal,
        Darlington Borough Council v. Wiltshier Northern Ltd.

6.27   In *St Martins Property Corporation*, as previously stated, the assignment
        from Corporation to Investment was invalid, therefore the court had to
        consider Corporation's right to recover damages even though Corpora-
        tion had no proprietary interest and the remedial costs had been expen-
        ded by Investments. The majority of the House of Lords held that
        Corporation was entitled to recover substantial damages on the basis that
        the facts fell (per Lord Browne-Wilkinson) 'within the rationale of the
        exceptions to the general rule that a plaintiff can only recover damages for
        his own loss – "*the narrow ground*"'. The reasoning of the court was (per
        Lord Browne-Wilkinson):

        'The contract was for a large development of property which, to the
        knowledge of both Corporation and McAlpines was going to be occu-
        pied, and possibly purchased, by third parties and not by Corporation
        itself. Therefore, it could be foreseen that damage caused by a breach
        would cause loss to a later owner and not merely to the original con-
        tracting party, Corporation. As in contracts for the carriage of goods by
        land, there would be no automatic vesting in the occupier or owners of
        the property for the time being who sustained the loss of any right of
        suit against McAlpines. On the contrary, McAlpines had specifically
        contracted that the rights of action under the building contract could *not*
        without McAlpines' consent be transferred to third parties who became
        owners or occupiers and might suffer loss. In such a case, it seems to me
        proper ... to treat the parties as having entered the contract on the
        footing that Corporation would be entitled to enforce contractual rights
        for the benefit of those who suffered from defective performance but

who, under the terms of the contract, could not acquire any right to hold McAlpines liable for breach. It is truly a case in which the rule provides "a remedy where no other would be available to a person sustaining loss which under a rational legal system ought to be compensated by the person who caused it".'

6.28   Lord Griffiths came to the same result, but on the 'broader ground' that there was a principle that the original contracting party had suffered a loss because he had not received the bargain he had contracted for.

6.29   In *Darlington Borough Council* v. *Wiltshier Northern Ltd*, Wiltshier, a construction company, entered into two contracts with Morgan Grenfell (Local Authority Services) Ltd for the construction of a recreational centre for and on a site owned by Darlington Borough Council. As pre-agreed, Morgan Grenfell assigned to Darlington all its contractual rights and causes of action against Wiltshier. Darlington claimed damages against Wiltshier for breach of contract relating to building defects. The court held that Darlington, as assignee, was not entitled to claim damages other than nominal damages. The Court of Appeal held that as the building contracts, to the knowledge of both parties, were entered into for the benefit of Darlington and it was foreseeable that damage caused by breach of those contracts would cause loss to Darlington, Darlington as assignee of Morgan Grenfell's rights against Wiltshier, could claim substantial damages for loss caused by Wiltshier's breaches of contract and the damages should be assessed on the normal basis as if Darlington had been a party to the original contracts.

6.30   It will be noted from the above facts that the assignor, Morgan Grenfell, unlike Corporation in the *St Martins Property Corporation* case, had never acquired or transmitted to Darlington any proprietary interest in the development. The Court of Appeal did not consider this to be relevant to their decision that the case fell to be decided 'by direction application of the rule in *Dunlop* v. *Lambert* as recognised in a building contract context in Lord Browne-Wilkinson's speech in the McAlpine case'. Dillon LJ stated:

'[The respondent] also sought to distinguish the decision in the McAlpine case on the ground that in the present case Morgan Grenfell never acquired or transmitted to the council any proprietary interest in [the development]. I do not see that that matters as the council had the ownership of the site of [the development] all along. It was plainly obvious to Wiltshier throughout that [the development] was being constructed for the benefits of the council on the council's land.'

6.31   The decision in the *Darlington* case was based upon the 'narrow ground'. However, Steyn LJ, whilst accepting that the decision in *Linden Gardens* could with 'only a very conservative and limited extension' apply by analogy to the facts in *Darlington* and did so apply, nevertheless agreed with the 'broader ground' set out in the judgment of Lord Griffiths. Steyn LJ stated:

'The rationale of Lord Griffiths' wider principle is essentially that, if a party engages a builder to perform specified work and the building fails to render the contractual service, the employer suffers a loss. He suffers a loss of bargain or of expectation interest. And that loss can be recovered on the basis of what it would cost to put right the defects ... subject to one qualification, it will be clear from what I said earlier that I am in respectful agreement with the wider principle. It seems to me that Lord Griffiths based his principle on classic contractual theory.'

6.32   The qualification went to the requirement mentioned by Lord Griffiths for the court to be satisfied that the relevant repairs had been or were likely to be carried out. Steyn LJ preferred the line of authority in *Dean* and *Ruxley* that the courts were not concerned with what the plaintiff proposed to do with the damages, or that the plaintiff intended to undertake any repairs.

6.33   In *Darlington*, Dillon LJ and Waite LJ also decided, *obiter*, that if Morgan Grenfell had commenced proceedings, prior to the assignment, for damages for breach of contract, it would have done so as constructive trustee for Darlington and would have been accountable in equity to Darlington for any damages recovered. On this reasoning it follows that on an assignment and the assignee stepping into the shoes of the assignor, there is an available cause of action carrying a right to substantial damages. The equitable remedy was considered appropriate, albeit *obiter*, by Judge Thornton in *John Harris Partnership (a firm)* v. *Groveworld Limited* which concerned proceedings for breach of contract against a firm of architects, by a developer who had a 45% exposure to the loss by reason of its joint venture partnership. The judge's principal finding was that Groveworld was entitled to recover the whole of the loss and it was irrelevant that the plaintiff intended to pass on 55% of the loss recovered from the defendants, to its joint venture partner. The relevant loss was the loss, which was immediate to the party claiming the loss. As to the equitable remedy, the judge stated the components to be:

(1)   the nature or terms of the relevant contract being such that it was contemplated by the plaintiff and the defendant that the third party's loss would be recoverable at the suit of the plaintiff.

(2)   the plaintiff had an equitable obligation to pursue the defendant for the loss caused to the third party by the defendant and to account to the third party for any recovery.

6.34   The judge also considered that what he described as the *Linden Gardens* exception was not confined to building contracts but could be extended to contracts involving the provision of professional services.

6.35   Back now to *Panatown*. A reading of the full judgments is recommended. Essentially however, the House of Lords confirmed the 'narrow ground' basis of recovery of third party losses and approved the 'broader ground' principles enunciated by Lord Griffiths and accepted by Steyn LJ. There was however a 'sting in the tail' for Panatown. Their Lordships held by a majority (Lord Goff and Lord Millett dissenting) that as regards both the 'narrow ground' and the 'broad ground', Panatown's claims failed because UIPL had a direct cause of action against McAlpine because of the DCD, even though the DCD provided for different and less onerous obligations than the principal contract and only gave subsidiary remedies to UIPL.

6.36 Lord Clyde considered that 'there was a plain and deliberate course adopted whereby the company with the potential risk of loss was given a distinct entitlement directly to sue the contractor and the professional advisers' and therefore he 'did not consider that an exception can be admitted to the general rule that substantial damages can only be claimed by a party who has suffered substantial loss'. Lord Clyde was not impressed with the 'broader ground' and did not feel it necessary to express a view on the effect of the DCD on that principle of recovery.

6.37 Lord Jauncey stated as regards the 'narrow ground':

> 'What is important, as I see it, is that the third party should as a result of the main contract have the right to recover substantial damages for breach under his contract even if those damages may not be identical to those which would have been recovered under the main contract in the same circumstances. In such a situation the need for an exception to the general rule ceases to apply.'

Of the 'broader ground' Lord Jauncey stated:

> 'I consider Panatown is not entitled to recover under [the broader ground] . . . because UIPL have a direct right of action against McAlpine under the DCD.'

6.38 Lord Browne-Wilkinson stated as regards 'the narrow ground':

> '. . . the whole contractual matrix relating to this development envisaged that McAlpine's obligations under the building contract were to be enforceable against McAlpine not only by Panatown but also to a very substantial extent by UIPL and its successors in title under the DCD. It was suggested in argument that the purpose of the DCD was to give purchasers of the site from the Unex Group undoubted causes of action for breach of a tortious duty of care. Even if that is so, it does not alter the fact that under the DCD, UIPL itself has the right to claim substantial damages for any negligent performance of the building contract, a right which will cover most of the claims arising under the building contract. In my judgment the direct cause of action which UIPL has under the DCD is fatal to any claim to substantial damages made by Panatown against McAlpine based on the narrower ground.'

As regards the 'broader ground' Lord Browne-Wilkinson stated:

'I will assume that the broader ground is sound in law and that in the ordinary case where the third party (C) has no direct cause of action against the building contractor (B) A can recover damages from B on the broader ground. Even on that assumption, in my judgment, Panatown has no right to substantial damages in this case because UIPL (the owner of the land) has a direct cause of action under the DCD.'

6.39   Collateral warranties and other species of direct agreements could, after *Panatown*, have the effect of excluding the remedy of assignment of original contractual rights to the successive owners of property. It follows that care must now be taken to ensure that collateral warranties give a much broader class of rights than that previously thought necessary, for example, fitness for purpose obligations and not simply reasonable skill and care, also remedies in respect of contractual delays and consequential losses.

6.40   Potential, unanswered, questions arising from the judgments in *Panatown* on the 'broader ground' are:

(1)   to what extent is the carrying out of the repairs a precondition to recovery?
(2)   are consequential losses recoverable?

6.41   Lord Jauncey and Lord Clyde appeared to be saying that carrying out repairs was a precondition to recovery and that consequential losses were not recoverable unless they came within the expenditure of the repair costs. Lord Goff could see no reason why consequential losses should not be recovered, for example delay damages. As to the first question he felt it appropriate to have regard to such matter when the reasonableness of the plaintiff's claim to damages was under consideration (per Lord Lloyd in *Ruxley*). Lord Millet considered it to be a question of reasonableness.

Expectation interest and reliance expenditure

Expectation interest

6.42   The rules as to measure of damages give rise to two broad categories of damage: expectation interest and reliance expenditure. A possible third category was suggested in *Ruxley* in respect of defective building works: compensation for loss of expectation performance or amenity. This category would appear to be confined to 'consumer' type actions and is probably better dealt with as a fact of the expectation interest category rather than separate category of loss. Expectation interest is the primary category of damage and represents the difference between the value to the promisee of a promise, which has been performed satisfactorily, and the value to the promisee of a promise, which has been performed defectively or incompletely. The loss of profits on the sale of the sugar in the *Koufos* case is an example of expectation interest. Similarly where there are building defects the expectation interest will be the diminution in value, that is to say, the difference between the market value of the property without defects and the market value of the property with the defects.

Reliance expenditure

6.43   Save for claims against building surveyors where the measure of damage is diminution in value (see *Perry* v. *Sidney Phillips & Son*), the measure of damage in respect of building defects will often be based on reliance expenditure and not expectation interest. A reliance expenditure award of damages is payment of compensation for wasted expenditure incurred by the promisee in reliance on the promisor's promise to perform. There are four broad categories of reliance expenditure. First, there is expenditure incurred by the promisee in order to perform his part of the contract, for example the cost of labour or materials. Secondly, expenditure incurred by the promisee prior to entering into the contract but which will be wasted if there is not a satisfactory performance of the contract, for example the taking on of additional laundry staff in the *Victoria Laundry* case. Thirdly, expenditure incurred prior to the breach of contract not related to the promisor's performance but which will be wasted if the promisee fails to fulfil his contractual obligations; for example in the *British Westinghouse* case the plaintiffs incurred the costs of extra coal used by defective generators. Fourthly, expenditure incurred after the breach of contract, for example the costs of repairs paid to a third party.

6.44   It is the fourth category of expenditure that is most commonly awarded in the construction industry as the measure of damage for defective buildings and is often referred to by lawyers as 'substituted performance'. In *East Ham Corporation* v. *Bernard Sunley & Sons Ltd*, B constructed a school for E under the then current RIBA Form of Contract. Several years after completion of the work stone panels fixed to the exterior walls fell off owing to defective fixing. The court held that the proper measure of damages was the cost of replacing the stone panels.

6.45   As a general rule the person entitled to damages for breach of contract may elect to recover expectation interest or reliance expenditure. However, where the damages arise from a breach of a construction contract, because of defective building works or incomplete building works, the courts appear to require the plaintiff to demonstrate an intention to carry out the works or an intention to continue to occupy the relevant building or premises. The proper measure of damage is more than an academic debate. It can have a profound effect on the damages recovered particularly when the costs of carrying out the remedial works or completing outstanding works can be substantially greater than the diminution in the market value of the property. Such were the circumstances in *Radford* v. *De Froberville and Lange*. R was the owner of a large house in Holland Park, London, which was let into flats. The house had a large garden and R decided to sell for building purposes part of the garden which fronted on the highway. R obtained planning permission to build a new house on the proposed plot and he agreed to sell the undeveloped site to D. In the document of transfer D agreed to build a new house in accordance with the planning permission and also to construct a wall in accordance with the detailed specification along the boundary of R's retained land and the new plot, the wall to be situated on the new plot. These obligatio.is had to be carried out within a certain time, which was subsequently extended. However D eventually sold the plot to L and the proposed house and the dividing wall remained unbuilt. R sued D for breach of contract because of D's failure to construct the dividing wall. D admitted liability but denied that R had suffered any loss on the basis that the proper measure of damage was diminution in value and there was no diminution as the absence of a physical barrier on the boundary of the property was unlikely to bring about any significant diminution in letting value. The cost of erecting the wall was £3400. The court held that as R intended to make good D's breach by building a wall himself on his own property the proper measure of damages was the cost of carrying out the work on his own land.

6.46   In the *Radford* case the defendant had failed to carry out or complete the works. In *Harbutts Plasticine Ltd* v. *Wayne Tank & Pump Co Ltd* (over-ruled by *Photo Productions* v. *Securicor Transport* on a different point), the defendants W were responsible for defective work. W entered into a contract with H to design and install equipment for storing and dispensing stearine in a molten state (at temperatures of between 120°F and 160°F) at H's factory, which was an old building. W specified Durapipe, a form of plastic pipe, which was to be heated by electrical tapes wound round the pipe controlled by a thermostat. In fact, Durapipe was wholly unsuitable for this purpose because it was liable to distort at temperatures of about 187°F and had a low thermal conductivity. The installation was completed on 5 February 1963 and both parties intended to test it the next day. As it was very cold, to ensure that the stearine would be molten for the test, an employee of W switched on the heating tapes on the night of 5 February and the installation was left unattended during that night. In the early hours of 6 February there was a fire, which destroyed the factory. H rebuilt the factory but because of planning restrictions they had to replace their old five-storey mill with a new two-storey factory. The cost of building the new factory was £67,973 compared with the diminution in value of the old mill before and after the fire of £42,538. The court held that the proper measure of damage was the cost of reinstating the factory not the difference in its value before and after the fire. Widgery LJ stated:

'The distinction between those cases in which the measure of damage is the cost of repair of the damaged article, and those in which it is the diminution in value of the article, is not clearly defined. In my opinion each case depends on its own facts... If the article damaged is a motorcar of popular make, the plaintiff cannot charge the defendant with the cost of repair when it is cheaper to buy a similar car on the market. On the other hand, if no substitute for the damaged article is available and no reasonable alternative can be provided, the plaintiff should be entitled to the costs of repair. It was clear in the present case that it was reasonable for the plaintiffs to rebuild their factory, because there was no other way in which they could carry on their business and retain their labour force.'

6.47   The question of the correct measure of damages was considered by the House of Lords in *Ruxley Electronics and Construction Ltd* v. *Forsyth*. The defendant had contracted for an enclosed swimming pool to be con-structed in his garden. The contract expressly provided that the max-imum depth of the pool should be 7 ft 6 in. After completion, the defendant discovered that the maximum depth was only 6 ft 9 in and only 6 ft at the point where people would dive into the pool. The defendant refused to pay the balance of the price due under the contract and counterclaimed for damages for breach of contract. The breach was admitted; the damages claimed as a consequence of the breach were disputed. The trial judge awarded the defendant £2500 on his counter-claim on the basis that the shortfall in depth had not decreased the value of the pool. The £2500 was awarded for general damage for loss of amenity. The defendant had claimed £21,500, the cost of reconstructing the pool to the contractual specification. The Court of Appeal allowed the defendant's appeal finding that it was not unreasonable to award as damages the cost of replacing the swimming pool in order to make good the breach of contract, even though the shortfall in the depth of the pool had not decreased its value. The House of Lords reversed the decision and held:

(1)   In assessing damages for breach of contract for defective building works, if the court took the view that it would be unreasonable for the plaintiff to insist on reinstatement because the expense of the work involved would be out of all proportion to the benefit to be obtained, then the plaintiff was confined to the difference in value.

(2)   The plaintiff's intention, or lack of it, to reinstate was relevant to reasonableness and hence to the extent of the loss which was sus-tained, since, if the plaintiff did not intend to rebuild he had lost nothing except the difference in value, if any.

(3)   Where the diminution in value caused by the breach was nil, it was not correct to award the cost of reinstatement as an alternative to the difference in value.

(4)   The cost of reinstatement and diminution in value were not the only available measures of recovery for breach of contract for defective building works and the court was not confined to option for one or the other. Where there had been a breach of performance resulting in loss of expectation of performance, satisfaction of a personal preference or a pleasurable amenity but there had been no

diminution in value, the court would award modest damages to compensate the plaintiff.

6.48   In *Tito and Others* v. *Waddell and Others No. 2*, (*The Ocean Island* case) the court held that a plaintiff can establish that his loss consisted of or included the costs of doing the work if he could show that he had done the work, or intended to do it, even though there was no certainty that he would.

Claims for both expectation interest and reliance expenditure

6.49   To what extent can a party claim both expectation interest and wasted reliance expenditure? The Court of Appeal has held that a party cannot claim both heads of damage; he must seek either expectation interest or wasted reliance expenditure: *Cullinane* v. *British 'Rema' Manufacturing Co Ltd*. It may be however that *Cullinane* can be distinguished on the basis that in that case the plaintiff claimed an expectation interest based on his gross profits and not net profits; if the claim is limited to net profits there is no duplication of damages. Similarly where there is a claim for damages for building defects it would appear arguable that a claim for the costs of repairs together with a claim that the property had suffered a diminution in its market value because of the fact that it is a repaired structure, are not overlapping claims for damages and both heads of damage should be recoverable. Indeed both costs of repairs and diminution in market value were recovered in the case of *Thomas and Others* v. *T. A. Phillips (Builders) Ltd and Taff Ely Borough Council*. This case involved negligence and not breach of contract. However the decision on measure of damages should also be relevant to a breach of contract case.

Mitigation and assessment

6.50   Lawyers often refer to the claimant's duty to mitigate his loss. To talk about duty is probably adopting too high a standard of conduct; it is probably more helpful to consider mitigation in terms of reasonableness.

> 'A plaintiff was under no duty to mitigate his loss, despite the habitual use by lawyers of the phrase "duty to mitigate". He was completely free to act as he judged to be in his best interest. On the other hand, a defendant was not liable for all the loss suffered by the plaintiff in consequence of his so acting. A defendant was only liable for such part of the plaintiff's loss as was properly to be regarded as caused by the defendant's breach of duty': *Sotiros Shipping Inc. and Another* v. *Sameiet Solholt*

Essentially therefore a claimant will not be allowed to recover damage which could be avoided had the claimant acted reasonably. The burden of proof rests on the defendants to show that the claimant behaved unreasonably. The level of behaviour is one to be decided on the facts of each particular case although as a general rule the courts tend to favour the claimant and are often unimpressed with defendants' attempts to demonstrate that, with the benefit of hindsight, the claimant's behaviour was unreasonable. For example, the courts do not expect a claimant to do anything other than that which is in the ordinary course of a business: *Dunkirk Colliery Co* v. *Lever*.

6.51 If a claimant's reasonable attempts to mitigate the loss fail and result in additional loss or damage, such losses or damage may be recoverable from the defendant: *Banco de Portugal* v. *Waterlow & Sons Ltd*. However, if the claimant takes greater steps than he need have done and these result in a reduction of the loss and damage, then the defendant is entitled to the benefit of that reduction.

6.52 Mitigation is often described as the mirror image of the rules of remoteness discussed above and also the rules of assessment discussed below in paragraph 6.55. That is to say the courts often disregard strict application of the rules and are more concerned to answer what has been described as the real question, namely what is the loss to the claimant. 'In the end the question seems to me to come down to a very short point. The cost is a loss if it is shown to be a loss', per Megarry VC in *The Ocean Island* case.

Betterment

6.53 Betterment is a topic of particular relevance to defective building works and involves consideration of both measure of damage and mitigation. In the *Harbutts Plasticine* case betterment was considered under the heading of measure of damage. The court held that the plaintiffs were not required to give credit under the heading of betterment merely because they had replaced the old building with a new one of modern design. Widgery LJ stated:

> 'The plaintiffs rebuilt their factory to a substantially different design, and if this had involved expenditure beyond the costs of replacing the old, the difference might not have been recoverable, but there is no suggestion of this here. Nor do I accept that the plaintiffs must give credit under the heading of "betterment" for the fact that their new factory is modern in design and material. To do so would be the equivalent of forcing the plaintiffs to invest their money in the modernising of their plant which might be highly inconvenient for them.'

6.54 In *Governors of the Hospital for Sick Children and Another* v. *McLaughlin and Harvey Plc and Other* the issue of unnecessary expenditure was approached from the point of view of mitigation. The defendants endeavoured to argue that despite the plaintiffs' reliance on expert opinion the plaintiffs had not acted reasonably in selecting their repair scheme. The court rejected the defendants' argument. (See also *Skandia Property (UK) and Another* v. *Thames Water Utilities Ltd.*)

Assessment

6.55   The appropriate date for assessment of damages is often considered adjunctively with mitigation. The general rule is that damages shall be assessed at the date of breach of contract: *Miliangos* v. *George Frank (Textiles) Ltd.* However as a matter of practice the court's approach is invariably based on a finding as to the reasonableness of the claimant's conduct, particularly where the claim is for wasted reliance expenditure, for example, the cost of building repairs. In the *East Ham Corporation* case, the court held that the cost of repairing the stone panels should be assessed at the time when the defects were discovered and put right. In that case the breaches of contract occurred in May 1954 and the remedials were carried out in 1960. The case of *Radford* (discussed in paragraph 6.45) adopted the test that the costs of repairs should be assessed at the date when, in all the circumstances, it was reasonable for the plaintiffs to commence the repairs. In *Dodd Properties Ltd* v. *Canterbury City Council* (a case involving the tort of nuisance but of guidance to the issue of the date of assessment of the costs of building repair works) the court considered it reasonable for a plaintiff to delay carrying out repairs until such time as a judgment had been awarded against the defendant. These cases would appear to support the argument that impecuniosity or lack of credit may be one of the factors which can be taken into account in deciding whether or not the claimant has behaved reasonably in postponing the carrying out of repairs. It is a question of whether, in all the circumstances, it was 'a matter of commercial good sense' to delay the carrying out of the remedial works. This is to be contrasted with the situation where impecuniosity, or lack of funds, is the sole reason for not carrying out the remedial works; in such circumstances the increased costs of the remedial works may well be irrecoverable: *Liesbosch Dredger* v. *Edison SS*.

Difficulties of ascertainment

6.56   Provided the head of damage satisfies the test of remoteness the courts will award damages for breach of contract even though the precise quantification of the loss is not possible. The courts will make approximate assessments of damages. What is the position however if that assessment is dependent upon a contingency? For example, an undertaking given by a developer to a consultant in a collateral warranty agreement that the developer will obtain collateral warranties in like terms from all the other consultants involved in the development. If the developer is in breach of his undertaking will damages be recoverable by the consultant? This situation has often been described as the loss of a chance or opportunity.

6.57 The question of whether damages could be claimed in respect of a loss of chance was considered in *Chaplin* v. *Hicks*. H, a well known actor and theatrical manager, published a letter in a London daily newspaper stating that with a view to dealing once and for all with the numerous applications he received from young ladies desirous of obtaining engagements as actresses, he was willing that the readers of the newspaper should by their vote select 12 ladies to whom he would then give engagements. The ladies were to apply by way of a photograph endorsing their name and address on the rear face of the photographs. H received a tumultuous response, some 6000 ladies, so that he decided to pre-select some 50 ladies on a regional basis from which the readers of the newspaper would then select the final 12. C was one of the ladies who went forward to the pre-selection stage; however by reason of some confusion about her address she was not notified in time to attend the pre-selection interview and her photograph was not put forward to the final selection by the readers of the newspaper. C sued H for breach of contract and it was argued on behalf of H that even if there had been a breach, C's claim for damages was speculative as there was no certainty that she would be one of the 12 ladies selected by the readers of the newspaper. This argument was rejected by the court and it was held that where, by a contract, a person has a right to belong to a limited class of competitors for a prize, a breach of that contract by reason of which that person is prevented from continuing as a member of the class and is thereby deprived of all chance of obtaining the prize, is a breach in respect of which the person may be entitled to recover substantial and not merely nominal damages. *The existence of a contingency, which is dependent on the volition of a third person, does not necessarily render the damages for a breach of contract incapable of assessment.*

In *Cook* v. *Swinfen* (not followed in *Midland Bank* v. *Hett Stubbs & Kemp* but not on this point), S was a solicitor who was acting for C the wife in divorce proceedings. The husband had brought divorce proceedings against C, and S was negligent in not defending the husband's petition nor cross petitioning the divorce on the basis of the husband's adultery. The husband's petition was unopposed and C brought proceedings against S claiming damages for a loss of a chance to obtain a divorce against her husband and maintenance for herself and her child. The court held that C was entitled to recover such damages. The court recognised that in so far as it was anticipating the outcome of hypothetical proceedings brought by C against her former husband, to that extent the court's assessment was speculative; nevertheless the court considered that it was in a position to make an assessment based on the balance of probability. It follows, in the authors' submission, that a court would be faced with similar speculation if in the example set out in paragraph 6.56 above the developer failed to obtain collateral warranties from other consultants; the court would have to consider whether on the balance of probabilities there was a likelihood of those other consultants being responsible for defective design which would have given rise to a right of contribution between all the consultants. It is presumed that in the light of the decisions in *Chaplin* and *Cook* the courts would be prepared to award damages in such circumstances.

6.58   The loss of a chance was considered by the Court of Appeal in *Allied Maples Group Ltd* v. *Simmons & Simmons*. The facts concerned the negligent advice given by solicitors in respect of the acquisition of a certain business and properties from a third party; because of restraints on alteration and difficulties with planning consents, it was agreed that a subsidiary company rather than the properties would be acquired by Allied Maples. This involved ensuring that the subsidiary company was a 'clean company', and the giving of certain indemnities. In the event the conditions of sale did not properly protect Allied Maples against contingent liabilities on assigned lessees. Allied Maples suffered a substantial loss. The court held, on the issue of causation, that if the defendants had given proper advice as to the contingent liabilities, the plaintiffs would have taken steps to protect themselves from the contingent liabilities. Further, that on the balance of probability, the third party would have offered some form of protection against these liabilities if asked and that had the properties subject to the liabilities not been included in the sale, the deal would not have gone ahead. The defendants appealed.

6.59   The Court of Appeal held:

(1)   Where the plaintiff's loss resulting from the defendant's negligence depended on the hypothetical action of a third party, either in addition to action by the plaintiff or independently of it, the issue fell within the sphere of quantification of damages dependent on the evaluation of the chance that the third party would have taken the action which would have enabled the loss to be avoided.

(2)   It was not a question of causation where the plaintiff could only succeed if he showed on the balance of probability that the party would have taken that action.

(3)   Once the plaintiff had proved on the balance of probability as a matter of causation that he would have taken action to obtain a benefit or avoid a risk, he did not have to go on to prove on the balance of probability that the third party would have acted so as to confer the benefit or avoid the risk to the plaintiff.

(4)   The plaintiff was entitled to succeed provided he showed that there was a substantial, and not merely a speculative, chance that the third party would have taken the action to confer the benefit or avoid the risk to the plaintiff.

(5)   The evaluation of a substantial chance was a question of quantification of damages, the range lying somewhere between something that just qualified as real or substantive on the one hand and near certainty on the other.

Assignment

6.60   The topic of damages merits individual consideration in respect of assignment of choses in action. There would appear to be two important questions in relation to collateral warranties, the answers to which have been troubling the construction industry:

(1)   Does the assignee have any right to recover damages against the original debtor if the assignor has not incurred any expenditure on repair or rebuilding costs or has sold his property for its full market value; and

(2)   Does the assignee have any right to recover against the original debtor greater damages than would have been recoverable by the assignor?

Question 1

6.61   In the *Ocean Islands* case, Megarry VC stated:

'If the plaintiff has suffered little or no monetary loss in the reduction of value of his land, and he has no intention of applying any damages towards carrying out the work contracted for, or its equivalent, I cannot see why he should recover the costs of doing work which will never be done. It would be a mere pretence to say that this cost was a loss and should be recoverable as damages.'

6.62   In *Perry* v. *Tendring* (on its facts based upon a claim in negligence but of some relevance on the issue of damage), the court expressly doubted whether an assignor who had sold for full market value had suffered any loss. Judge Newey stated, *obiter*:

'... I am also uncertain as to what damages the assignee could recover, since the assignor would not have expended money on the remedying of undiscovered defects and would presumably have obtained market price for his property.'

6.63   What is the position with building defects? If the defects are patent at the time of sale and assignment of benefits, the purchase price received by the assignor will invariably reflect the diminution in market value, and the assignment of benefits will be qualified accordingly so that it is the assignor who sues those responsible for the defective construction works. If the defects are latent then more than likely the property will have been transferred for its full market price. As the assignor has received full market price and will not in fact be expending any money on carrying out necessary repairs, has he suffered any loss which can be recovered by the assignee? These questions were considered by the Court of Appeal in the case of *Dawson* v. *Great Northern and City Railway Co* (referred to in paragraph 4.33) in 1904 and some 78 years later by the House of Lords in *GUS Property Management Ltd* v. *Littlewoods Mail Order Stores Ltd* and, more recently, in *Linden Gardens*.

Dawson

6.64   As stated, Dawson's case was concerned with a claim of a right to statutory compensation. The head note to the case states that it concerns a claim of a right to compensation and not a claim for damages for a wrongful act. This headnote is somewhat misleading. The Court of Appeal was troubled by the principle of law that a bare cause of action was not assignable and it is submitted that this was the reason for the court's emphasis on the right of statutory compensation and that their comments on the issue of damages are still helpful to formulating an answer to question 1 in paragraph 6.60 above. In *Dawson* both the freehold and leasehold interests were sold to D for market value and the vendor did not appear to have suffered any loss. However, the court held that the assignments of the benefit of the right of compensation which were made at the same time as the conveyance of the freehold and the transfer of the leasehold, entitled D the assignee to recover in respect of the assignor's rights of compensation against the original debtor, the railway company. Sterling LJ stated:

> 'It appears to us that the intention of this deed was to place the plaintiff precisely in the same position as regards the defendants with respect to the lands conveyed as was previously occupied by [the assignor] and in particular to transfer to the plaintiff the compensation for structural damage to the conveyed property.'

GUS Property Management

6.65   The facts of the *GUS Property Management* case concerned a building in Queen Street Glasgow owned by Rest Property Co Ltd which was damaged in late 1970/early 1971 by piling works carried out on neighbouring property for and on behalf of Littlewoods Mail Order Stores Ltd. Serious structural damage was caused to the building owned by Rest. Rest was a wholly owned subsidiary of a holding company which in April 1972 adopted a policy of rationalising its property portfolio, involving the transfer to a newly created wholly owned subsidiary company, GUS Property Management Ltd, the plaintiffs, of various properties including the Queen Street building. Accordingly in March 1975 Rest conveyed its Queen Street building to GUS for a figure of £259,618, representing its book value. In June 1976, Rest assigned to GUS all of its claims arising out of the negligent building operations carried out on behalf of Littlewoods. Following the assignment, GUS brought proceedings against Littlewoods claiming damages for delict (negligence) against Littlewoods, the consulting structural engineers (subsequently abandoned), the main contractors and the specialist piling sub-contractors. GUS claimed alternative heads of damage. Their first head of claim was for the sum of £350,000 representing the diminution in value of the building based on the difference between the respective values of the building in a damaged and undamaged state at the time of the building operations. The second head of claim was for the sum of £55,450 in respect of the costs of repairing the damage to the building, which costs had been incurred by GUS not Rest, after the date of the conveyance of the building but before the date of the assignment. The main contractors and the sub-contractors contended that GUS's claims should be dismissed for the following reasons:

(1) GUS were really seeking to pursue a claim which Rest itself could have pursued at the date of the assignment; and

(2) the only relevant loss which GUS could claim title to recover was loss suffered by Rest and recoverable by Rest at the date of the assignment; and

(3) accordingly, the sums spent on repairs by GUS themselves were irrecoverable; and

(4) the alternative claim for diminution in the building's value was irrelevant since the costs of the repairs, being all that was necessary to achieve compensation, represented the proper measure of loss; and

(5) in any event, since the property had been transferred at book value without any regard for the fact that the building had been damaged, Rest had suffered no loss and accordingly had no claim to assign to GUS.

6.66   In the Court of First Instance in Scotland, the Lord Ordinary rejected the defendants' submissions and the defendants appealed to the First Division, the Appellate court in the Scottish system. The First Division reversed the decision of the Lord Ordinary and dismissed the claims brought by GUS against the main contractor and the piling sub-contractor. GUS appealed to the House of Lords.

6.67   It will be apparent that the central issues in this case concerned the precise nature of an assignee's claim for damages against the original debtor. The First Division approached the issue from the correct starting point, namely that GUS, as assignees, could only sue in respect of claims which were vested in Rest at the time of the assignment; that is to say the assignee stands in the shoes of the assignor. However, the First Division went on to find that the assignor Rest, and consequently the assignee GUS, could not have brought a claim at the date of the assignment for the costs of repair works as Rest did not carry out those works nor had it incurred any expenditure nor any obligation in respect of those works. As for the alternative claim, the First Division felt that it did not have to deal with this claim as its quantification substantially exceeded the claim for remedial works. Nevertheless, the First Division commented that Rest would have had no claim in respect of the diminution in value because they had suffered no loss; the price, which Rest received, was the book value and would have been exactly the same even if the building had been in an undamaged state. The court considered that the position would be the same if a building with latent defects was sold at its market value in an undamaged state. The House of Lords, reversing the decision of the First Division, held:

'(1)   that the best measure of the loss sustained was likely to, but need not necessarily, be the difference between the price obtained in the sale of the property in its damaged condition and the price it would have fetched in an undamaged state; and

(2)   that in this case the figure of price was of no relevance in estimating the plaintiffs' loss; and

(3)   that the depreciation in value and the costs of reinstatement of the building were alternative approaches to estimating the damages, the appropriate measure emerging after the leading of evidence; and

(4)   that the facts as to the costs of remedial works carried out by the plaintiffs' themselves might have evidential value for the purpose of arriving at an estimate of the loss suffered by the assignor Rest.'

6.68   Lord Keith delivering the leading judgment in the House of Lords stated as follows:

'Where the property is disposed of in an arm's length transaction for the price which it is fairly worth in its damaged condition, the difference between that price and the price which it would have fetched in an undamaged condition is likely to be the best measure of the loss and damage suffered. It may happen that the owner of the property disposes of it otherwise than by such a transaction. He may, for example, alienate it gratuitously... It is absurd to suggest that in such circumstances the claim to damages would disappear... into some legal black hole so that the wrongdoer escaped Scot-free. There would be no agreed market price available to form an element in the computation of the loss and so some other means of measuring it would have to be applied, such as an estimate in the depreciation in value or the cost of repair... It is undeniable on the pleadings that Rest suffered some loss through the defendant's operations. Its building was seriously damaged. How is the loss to be measured in money terms? One approach is to consider the extent to which the value of the building was depreciated as a result of the damage to it. Another is to assess the cost of repairs necessary to restore the building to the condition it was in before the defendant's operation. Both these approaches involve a process of estimating, an exercise familiar to courts of law... There is no doubt that the plaintiffs' pleadings were not drawn with that degree of accuracy which Counsel might normally hope to achieve. The drafts-man does not appear to have had in the forefront of his mind a sound grasp of the true legal position, namely that the plaintiffs are suing not for their own loss, but for that suffered by Rest. It would, however, not be right or just to dismiss the action by reason of this formal pleading defect, which is capable of being put right by a similar amendment without any prejudice being suffered by the defendants. The plaintiffs' averments about their own expenditure on repairs to the building are not open to any objections so far as they are averments of facts. They have relevance, in my opinion, as indicating the scale of expenditure in which it is likely that Rest would have required to incur if they had continued to own the building. The facts averred may thus have evi-dential value for the purpose of arriving at an estimate of the loss suffered by Rest which is what the plaintiffs, as assignees of the claim, are in substance seeking to recover.'

6.69 The claims in this case were claims for damages arising from negligence. It is submitted however that the House of Lords decision in so far as it relates to the measure of damages of an assignee is equally applicable to a claim for damages for breach of contract. It follows that the answer to question 1 in paragraph 6.60 must be that the sale by the assignor for full market value or the fact that the assignor does not carry out the remedial works, go to matters of *measure of damages* and do not prevent an assignee from recovering its cost of remedial works or diminution in market value. Some commentators have expressed the view that it is essential to the recovery of damage by an assignee that the assignment of the benefits of a collateral warranty is contemporaneous with the conveyance or transfer of the property, and that the assignment should, as well as the property, be expressly stated to be part of the transaction for which the purchase price is being paid (see for example Cartwright: 'The Assignment of Collateral Warranties' *Construction Law Journal* 1990 vol. 6 no. 1). The authors agree that such suggestions are very sensible practicable steps; however they are not necessary as a matter of law in the light of *GUS Property Management*.

6.70 In the *Linden Gardens* case, the Court of Appeal had to consider this question in relation to the assignment of causes of action for damages for breach of contract. Staughton LJ stated:

'But it is said that in such a case the assignee can recover no more as damages than the assignor could have recorded.

That proposition seems to me well founded. It stems from the prin-ciple already discussed, that the debtor is not to be put in any worse position by reason of the assignment. And it is established by [*Dawson*] see also [*GUS Property Management*] ... But in a case such as the present, one must elucidate the proposition slightly; *the assignee can recover no more damages than the assignor could have received if there had been no*

assignment, and if the building had not been transferred to the assignee.'
[italics added]

6.71   The assumption which forms the root of the statement, i.e. 'no assignment' and 'no transfer', confirms the approach adopted in *Dawson* and
GUS Property Management. Staughton LJ went on to state:

> 'Stock Conversion had an accrued cause of action against McLaughlin
> & Harvey by 25 March 1980, the date of practical completion of that
> company's works. They subsequently incurred the cost of remedial
> works in the sum of £22,205.02. That claim was validly assigned to
> Linden Gardens Trust, who can recover in respect of it. Further brea-
> ches of contract were later discovered, giving rise to remedial works in
> 1987 and 1988, and to expense and loss in the sum of £236,000. Stock
> Conversions acquired a cause of action in respect of some of these
> defects against McLaughlin & Harvey, dating from 1980 and in respect
> of others against Ashwell Construction for defective work in 1985. It is
> immaterial that during the Ashwell Construction works Stock Con-
> version disposed of part of their interest in the building. In my opinion,
> Stock Conversion acquired the right to substantial damages for those
> breaches of contract, and did not lose it when they disposed of the rest
> of their interest in the building for its sound market value, on 12
> December 1986. After the assignment on 14 January 1987, Linden
> Gardens Trust as assignees became entitled to enforce Stock Conver-
> sions' claim. It is immaterial that Linden Gardens Trust subsequently
> incurred the expense of remedial work and suffered loss of rent while it
> was carried out, although the cost and loss may assist them in estab-
> lishing the damage which would, but for the assignment and transfer of
> property, have been recoverable by Stock Conversion.'

6.72   Sir Michael Kerr stated:

> 'The next point is then that an assignee can recover damages from the
> debtor to the same extent as his assignor could have done, but he cannot
> enforce any claims, let alone under new heads of damage, which would
> not have been available to his assignor. However, his right to damages,
> limited to that extent, is enforceable by him even though he is not
> accountable to his assignor. These principles are illustrated by [*Dawson*]
> and [*GUS Property Management*].
>   There is no problem about Linden Gardens' right to recover the
> £22,205.02 which Stock Conversions had spent in remedying the ori-
> ginal breach by McLaughlin & Harvey. But in my view Linden Gardens
> are also prima facie entitled to recover the £236,000 odd which they
> themselves expended in remedying the other breaches by these con-
> tractors and the further breaches by Ashwell, all of which had been
> committed prior to Stock Conversions' assignment to Linden Gardens.
> Both are claims for damages which had vested in Stock Conversions

and which were validly assigned to Linden Gardens. The fact that, at the time of the assignment, Stock Conversions were aware of the full extent of the breaches, and therefore of the extent of their claims for damages, does not appear to me to make any difference to the validity and effect of the assignment to Linden Gardens. The only limitation upon Linden Gardens' right of recovery is to the extent to which the defendants may be able to show that Linden Gardens' claims exceed what would have been recoverable by Stock Conversions if there had been no assignment. But that is merely a question which goes to quantum, like the discussion in the speech of Lord Keith in *GUS* as to what would be the appropriate measure of damages in the circumstances.'

6.73   Of the suggestion that Lord Keith's reasoning in *GUS Property Management* can be distinguished on the grounds that in that case the price for the transferred property was not market value but was '. . . fixed in an internal group transaction and for accounting purposes only, without any reference to the true value of the building', Staughton LJ stated:

'It may well be said that this passage appears to accept that, if Rest had sold the building for its sound market value, they would have suffered no loss, and, consequently the pursuers could have recovered only nominal damages. The claim would then indeed have disappeared into a legal black hole. However, I do not consider that Lord Keith expressed an opinion to that effect, even obiter and sub silentio. There was no issue as to what would have happened if Rest had received the full market value of the building in sound condition.'

6.74   In the *St Martins Property Corporation* appeal, Staughton LJ considered the position of the assignee where there was no breach of contract prior to the assignment. His Lordship stated:

'In my judgment St Martins Corporation by the assignment transferred to St Martins Investments the contractual right to have the building properly constructed. St Martins Investment can sue for the breach of obligation, and recover substantial damages – which they in fact incurred. The damages must be no more than St Martins Corporation would have suffered if there had been no assignment and no transfer of property, since McAlpines must not be put in any worse position by reason of assignment. That is established by Dawson's case. But it seems unlikely that the cost of remedial work in the present case would have been any different if there had been no assignment.

In the course of the hearing I did not understand it to be disputed that an assignee of the benefit of a contractual right could, in principle, recover damages if it is not performed after [sic] the date of the assignment. That appears to follow from the decision of the House of

Lords in [*Tolhurst* v. *The Associated Portland Cement Manufacturers Ltd*].'

6.75   Although the main decision by the Court of Appeal on the nature of contractual prohibitions on assignments was subsequently overturned by the House of Lords, the above mentioned passages on the assessment of an assignee's damages were not considered by and, it is submitted, survived the House of Lords decision. Per Lord Browne-Wilkinson:

'(5) What is the measure of damages recoverable by the assignee?
    In view of my decision on the earlier issues, this issue does not arise for determination. I mention it only to explain that the Court of Appeal considered that the assignee was entitled to recover what the assignor could have recovered had there been no assignment.'

Question 2

6.76   The answer to question 2 in paragraph 6.60 is concerned with the consequential losses. If A enters into a collateral warranty with B, the first tenant who then assigns his lease to C together with the benefit of the collateral warranty in the event that remedial works are necessary to the building causing consequential losses, for example loss of business profits, is the original debtor A liable for B's consequential losses or C's consequential losses? It may well be of course that, because of the nature of C's business, those losses are much greater than the losses that would have been incurred by B and vice versa. In the *GUS Property Management* case the plaintiffs' claim included a claim for consequential losses in respect of the loss of rental income for the period of the carrying out of the remedial work. The House of Lords did not specifically deal with this point; however it is suggested that in the light of the wording of the judgment of Lord Keith a claim for consequential losses suffered by the assignor (B in our example) would be recoverable, but not the consequential losses of the assignee (C in our example).

6.77   Consequential losses were specifically dealt with in the *Dawson* case. The plaintiff's claim in respect of her freehold interest consisted of:

(1)   £666 13s 4d in respect of structural damage, that is to say the amount which would have to be spent on the property in order to re-instate it in the condition in which it was before the defendants' works were executed; and

(2)   £700 damage to trade stock which was an estimate of the sum which would be sufficient to recoup the plaintiff's loss occasioned by her by disturbance of her drapery business carried on at the property and by damage caused, or likely to be caused, to stock during the period occupied in the re-instatement of the building.

6.78   As regards claim 1, the court held that this claim was recoverable stating:

> '[this sum] we understand to be the amount which would have to be spent on the property in order to reinstate it in the condition in which it was before the defendant's works were executed; and we are unable to see that there ought to be any difference in this amount whether the property was in the occupation of [the assignor] or of the plaintiff, or whether the proceedings were taken in [the assignor's name] or the plaintiff's... We think, therefore, that so far as this item is concerned, the defendants have not had any greater burden imposed on them than they would have had to bear if the proceedings had actually been taken in [the assignor's] name.'

6.79   However, the court rejected claim 2, stating:

> 'The amount has been arrived at on the assumption that the Plaintiff was the person in occupation of the property, and it is contended that it ought to have been ascertained on the basis that [the assignor] was the occupier. In our opinion the plaintiff cannot... recover a greater amount of compensation than [the assignor] could have got... It seems to us that, in these circumstances, [the assignor] could not recover any damage under the head of "damage to trade stock", and, for the reasons already given, neither can the Plaintiff.'

6.80   See also the judgments of Staughton LJ and Sir Michael Kerr in *Linden Gardens* at paragraphs 6.70 to 6.72.

6.81   It is suggested that the answer to question 2 is that the assignee will not be entitled to recover his own consequential losses but should, subject to questions of proof, be entitled to recover the consequential losses of the class or category, which could have been suffered by the assignor had there not been the assignment or transfer of property. It may well be, however, that the courts will use the rules as to measure of damages to restrict what otherwise might be unreasonable awards of damages to the assignee.

Contribution and apportionment

6.82   Where, by reason of a breach of contract or tort, one or more parties are liable to the claimant for the same *damage*, the claimant is entitled to recover the whole of his loss against each defendant: *Cassell & Co v. Broome*. Whilst the courts will make apportionments of loss as between the defendants (see paragraph 6.84 below) the courts will not apportion as between the defendants and the claimant. This rule is important in respect of collateral warranties; for example, if a building defect is caused by the negligent design of two consultants and only one of those consultants has given a collateral warranty, then that consultant may be liable for the whole of the plaintiff's loss.

6.83   If, however the parties' breaches have caused *different damage* then the defendant's liability will be restricted to the damage for which he was responsible: *Baker* v. *Willoughby*.

6.84   The Law Reform (Contributory Negligence) Act 1945 empowers a court to apportion damages as between the defendant and the claimant where the claimant's fault has contributed to the damage, for example where defective building works have been negligently overlooked by the architect and negligently overlooked by the employer's Clerk of Works. It would appear that save for one exception, contributory negligence is confined to damages for tortious liability and not damages for breach of contract: *Basildon District Council* v. *J. E. Lesser (Properties) Ltd and Others*; see also, *Bank of Nova Scotia* at paragraph 6.7 to 6.9. The exception is confined to cases where the breach of contract is co-extensive with a tortious liability for negligence, the latter liability existing independently from the contract: *Forsikringsaktieselskapet Vesta* v. *Butcher and Others*. See also *Bank of Nova Scotia* (paragraph 6.7).

Contribution

6.85   As between the defendants, the court is empowered to apportion blame by virtue of contribution awards under the Civil Liability (Contribution) Act 1978. Section 1(1) of the Act provides that 'any person liable in respect of any damage suffered by another person may recover contribution from any other person liable in respect of the same damage (whether jointly with him or otherwise)'. It follows that the courts can apportion blame between several defendants regardless of whether or not the defendants' liability arises from different contracts or from a tortious liability: *Birse Construction Ltd* v. *Hastie*. In deciding the issue of liability, section 1(4) provides that a bona fide compromise entered into by the party from whom contribution is sought, shall be conclusive evidence of liability provided that the factual basis for the claim can be established. Section 1(5) of the Act provides that any judgment of a court shall be conclusive evidence of liability.

6.86   In *Jameson and Another* v. *Central Electricity Generating Board (Babcock Energy Ltd,* third party) the Court of Appeal also distinguished the meaning of the word 'damage' as used in the Act and 'damages' in its normal sense of compensation. Auld LJ stated:

> 'In the Act of 1978 the word 'damage' is not defined, but, in my view, its meaning is plain in the various contexts in which it appears. It is the wrong causing injury ... the scheme of the Act is to provide contribution in respect of "compensation" for "damage".'

6.87   The words "the same damage" in section 1(1) refer to "any damage suffered" in that sub-section: *Birse Construction Ltd* v. *Hastie Ltd*.

6.88   Any liability can be a liability in respect of the same damage: *Birse*

Construction Ltd v. *Hastie*. Judge Cyril Newman QC held that the 1978 Act provides a remedy regardless of the causes of action giving rise to liability, which allows apportionment between those found to be liable in respect of the same damage. The time at which the party from whom contribution is sought needs to be liable for the purposes of section 1(1) or 1(6) of the Act is the time at which contribution is being sought to be recovered and section 1(3) is not relevant to this test: *Co-operative Retail Services Ltd* v. *Taylor Young and Others*.

6.89 In *J. Sainsbury plc* v. *Broadway Malyan (a firm) and Ernest Green Partnership Ltd*, Sainsbury brought proceedings against its architect, Broadway Malyan, for breach of contract and negligence in design of a compartment wall with less than two hours fire resistance. A fire broke out and Sainsbury alleged that had the compartment wall been properly designed, the fire brigade could have prevented the fire spreading to the main sales area in the store. Sainsbury claimed damages for the costs of reinstatement of the whole building. Broadway settled its claim and then brought third party proceedings against Ernest Green Partnership, the engineers, claiming 50% of contribution under the 1978 Act. Ernest Green contested the reasonableness of the settlement on the grounds that Sainsbury's claim was for the loss of a chance to contain the fire and should be discounted by the percentage prospect of success of the chance.

6.90 In *J. Sainsbury plc*, the court held that the word "damage" in sub-sections (1) and (4) of section 1 of the Act was not only concerned with liability for loss and did not preclude a person from whom contribution was being sought of asserting that:

(1) the person claiming contribution paid too much, or
(2) that in assessing contribution the party liable to contribute should not be required to pay compensation for elements of the payment for which that person could never have been held liable had he been sued directly.

The court also stated the elements in a claim for contribution as:

(1) the facts relied upon against the party seeking contribution, would, if established, have rendered that party liable in law for some loss,
(2) the party has made a payment in respect of that loss,
(3) the party from whom contribution is sought, is liable in respect of the same loss.

6.91 The essential elements are *liability* and *common damage*. For example, A is the consulting engineer for a project and B the specialist services consultant. A enters into a collateral warranty with C but B does not. As a consequence of negligence on the part of both A and B in the design of the mechanical services, the heating installation is incapable of achieving the required outputs. The benefit of the collateral warranty has been assigned to D, the tenant in occupation. D brings proceedings against A under the collateral warranty but has no right to bring proceedings against B, either in contract or in tort as D's loss is purely economic. In these circumstances, whilst B is clearly blameworthy, A will have to pay the whole of the damages to D and will have no right of contribution from B. If however A and B's design faults had caused a heating boiler to explode destroying the building, whereby B had a tortious liability to D, then A would have rights of contribution against B.

6.92   In *Royal Brompton Hospital National Health Trust* v. *Hammond and Others (No 3)*, the Court of Appeal rejected a claim by architects for contribution against the main contractor on the grounds that the employer's claim against the architect was that the architect's behaviour had weakened or impaired the employer's prospect of success or increased the prospect of defeat in an arbitration brought by the contractor, whereas the employer's claim against the contractor was for damages for delayed completion. Accordingly the damage was not the "same damage" as required by the Act. The court held that it was necessary to distinguish between cases of negligent failure by an architect to condemn and require rectification of defective work where the damage might on both cases be regarded as defective building for which it could be said that the contractor and architect were liable in respect of the same damage. In the present case the damage caused by [the contractor's] breach of contract (if any) was the failure to provide the building on time. The damage caused by [the architect] was impairment of the ability to obtain financial recompense in full from [the contractor] for damage of a different kind, and contribution was therefore not available.

6.93   Section 1(3) of the Act provides that contribution may be recovered from someone who has 'ceased to be liable' thus preserving the rights of contribution where the party from whom contribution is sought could prior to the Act have pleaded a limitation defence or a settlement. The Act therefore reverses the decisions of *Wimpey & Co Ltd* v. *BOAC* and *Harper* v. *Gray and Walker*. (See also *Jameson*.)

6.94   Section 2(1) provides that the courts are to assess contribution on the basis of what is *just and equitable* having regard to the extent of the person's responsibility for the damage in question. Further, by section 2(3) the court must give effect to any limitation of liability clause contained in any relevant agreement.

6.95   In *Equitable Debenture Assets Corporation* v. *William Moss Group Ltd and Others*, the Official Referee gave separate apportionments of liability in respect of design and workmanship problems arising from defects in curtain walling. In respect of the design he apportioned 25% to the architect and 75% to the specialist sub-contractor. In respect of workmanship, he put 5% against the architect, 15% against the main contractor and 80% against the sub-contractor. Another example of an apportionment is the case of *Eames London Estates Ltd and Others* v. *North Herts District Council and Others*, where the apportionment in respect of defective foundations was $32\frac{1}{2}\%$ to the architect and $22\frac{1}{2}\%$ each to the local authority, the original developers and the specialist sub-contractor. See also *Saipem SpA & Conoco (UK) Ltd* v. *Dredging VO2BV and Geosite Surveys Ltd (the 'Volvox Holandia') (No 2)* for an illustration of an assessment of contribution and the available parties.

Limitation of action

Meaning

6.96   Limitation of action is a statutory remedy, which prevents a claimant from bringing proceedings after the expiration of specified time limits. The philosophy of the remedy is that defendants should not suffer the prejudice of stale proceedings and that the claimant should be encouraged to avoid delay. This philosophy is of particular relevance to the construction industry, which is well known for the longevity of its disputes. Limitation of action in respect of breaches of contract is governed by the Limitation Act 1980 which came into force on 1 May 1981. Limitation of action in respect of tortious liability is governed by the Limitation Act 1980 as amended by the Latent Damage Act 1986. Some commentators have considered that the Latent Damage Act also applies to breaches of contract where the contractual duty comprises skill and care, i.e. is analogous to the tort of negligence. It is the authors' view that this is not the effect of the Latent Damage Act by reason of the wording of sections 2 and 5 of that Act.

Limitation periods

6.97   Section 5 of the Latent Damage Act provides that an action founded on simple contract shall not be brought after the expiration of six years from the date on which the cause of action accrued. Section 8 provides that the limitation period in respect of a contract under seal shall be 12 years. Care needs to be taken therefore in executing collateral warranties for if the original contractual obligation is a simple contract and a party enters into a collateral warranty under seal, then the latter document will have extended the original obligations by a period of six years. Sections 14A and 14B of the Limitation Act (inserted by the Latent Damage Act) provide that the periods should be six years from when the cause of action accrued or (if this expires later) three years from when the claimant had knowledge of certain material facts. There is a long-stop of 15 years from the date of the negligent act or omission, after which date no action can be brought.

6.98   The Act does not interfere with the limitation period established by other statutes; for example, under the Civil Liability (Contribution) Act 1978 the period of limitation is two years from the date on which the right to contribution accrues, usually the date of quantification, and under the Defective Premises Act 1972 the period of limitation is six years after the completion of the building works or completion of rectification works, whichever is the later.

6.99   Section 13(1) of the Arbitration Act 1996 applies the Limitation Act to arbitral proceedings. Section 13(2) gives the court power to extend time by excluding from the calculation the period between commencement of the arbitration and the setting aside, in whole or in part, of any award. Section 13(3) provides that *Scott* v. *Avery* clauses (i.e. arbitration award as condition precedent to legal proceedings) shall be disregarded for the purposes of determining when a cause of action accrued for calculating the relevant limitation period.

Calculation of the limitation periods

6.100   The starting point of the calculation is the date of accrual of the cause of action. In contract, the general rule is that the cause of action accrues at the date of the breach of contract (unlike tort when the cause of action accrues from the date that the innocent party suffers damage). Usually there will be little difficulty in identifying the date of the breach of contract. However, it is important to note that in a construction contract, if it is an entire contract, the date of accrual of the cause of action in respect of defective building works is not the date when those works were carried out by the contractor but the date of practical or substantial completion and possibly the expiration of the snagging period. Further, designers and design and build contractors may have a continuing contractual duty to check their design and correct errors during the period of construction: *Brickfield Properties* v. *Newton* (see also *Chelmsford District Council* v. *T.J. Evers and Others*) where the writ was issued against a designer more than six years after the breach of contract but less than six years after the date of practical completion. The court refused to strike out the proceedings.

6.101   A cause of action cannot arise until there is a party who can sue and a party who can be sued: *Reeves* v. *Butcher*. It follows that even though a collateral warranty does not seek to create any greater liability than the original contractual arrangement, the execution of the warranty may operate to extend the limitation period. For example, A an architect designs a structure in June 1982 pursuant to a contract with his employer B which was entered into in 1981. As the development approaches substantial completion in December 1984, B sells the development to a tenant C and in January 1985 to facilitate this sale A enters into a collateral warranty in favour of C. A's design is defective. Under A's original contract with B which is a simple contract, the limitation period will have expired by June 1988. Under the collateral warranty with C, which is again a simple contract, the limitation period will not have expired until January 1991. If the collateral warranty were under seal, the limitation period would not expire until January 1997. Those giving collateral warranties therefore should be careful to abridge the limitation period (see paragraph 6.111).

6.102   Time stops running for the purposes of limitation upon the issue of a writ of summons or the service of a notice to concur in the appointment of an arbitrator: section 34 of the Act.

6.103   In *Lowsley* v. *Forbes*, the House of Lords held that the word 'action' in section 24(1) of the Act, meant a fresh action and did not include proceedings by way of execution on an existing judgment. However, section 24(e) barred the recovery of more than six years' arrears of interest.

6.104   Section 35(2)(b) of the Act (which operating together with section 35(3) appeared to preclude the substitution or addition of a new party to existing proceedings after the expiry of the limitation period) does not apply to claims which involve the addition or satisfaction of a new party but which do not involve a new cause of action, i.e. the Act did not interfere with the law relating to substitution of parties where the party being substituted was succeeding to a claim or liability already represented in the action: *Yorkshire Regional Health Authority* v. *Fairclough Building Ltd and The Percy Thomas Partnership*.

Deliberate concealment

6.105   Section 32 of the Act provides for a postponement of the commencement of the running of the limitation period if there has been fraud or deliberate concealment by the wrongdoer, or mistake. Both fraud and mistake are construed strictly, and accordingly are of limited relevance. In contrast, deliberate concealment is concerned with the situation where any fact relevant to the plaintiff's rights of action has been deliberately concealed from him by the party committing the wrongful act. In *Johnson* v. *Chief Constable of Surrey* the court held that the words 'any fact relevant to the plaintiff's right of action' in section 32(1)(b) of the Act must be construed narrowly and any new fact must be relevant to the claimant's 'right of action', to be contrasted with a fact that is relevant to the claimant's 'case' or 'right to damages'.

6.106   Section 32(2) provides that a deliberate commission of a breach of duty in circumstances in which it is unlikely to be discovered for some time amounts to deliberate concealment of the facts involved in that breach of duty; for example, mortar bridges permitting the transmission of damp from the outer leaf to the inner leaf of a brick cavity wall. Indeed, the authors were instructed in a case that involved the presence of cement bags in the cavity of the brickwork! It would appear, however, that the breach must be deliberate and not merely negligent: *William Hill Organisation Limited* v. *Bernard Sunley & Sons Limited* (a case concerned with the statutory predecessor of deliberate concealment, namely fraudulent concealment under the Limitation Act 1939).

6.107   The fact that the employer has engaged a clerk of works for supervising the construction works will not necessarily prevent the employer from relying on section 32: *London Borough of Lewisham* v. *Leslie & Co Limited* (another case dealing with fraudulent concealment).

6.108   Each case must be decided on its own particular facts. In *Gray and Others (the Special Trustees of the London Hospital)* v. *T.B. Bennett & Son, Oscar Faber and Others and McLaughlin & Harvey Limited*, a hospital development had been completed in 1963. In 1979, there was evidence of a bulge in a panel of brickwork as a consequence of which structural investigations were commissioned. These investigations revealed setting out errors in the concrete panels, resulting in an unsatisfactory fit of the brick cladding and wholesale mutilation of the concrete nibs in order to fit the brickwork. The employer brought proceedings against the contractors some 25 years after the construction of the development. The court distinguished the situation in *Gray* from that in *William Hill Organisation* and found that the breaches of contract in relation to the concrete panels and nibs had been deliberately concealed from the employer's supervisors, and therefore time for the purposes of limitation did not begin to run until the employer had discovered or could with reasonable diligence have discovered (section 32(1) of the Act) the concealment of the defective work; the earliest date for discovery was November 1979, when the employer noticed the bulge in the brickwork.

6.109   In *Sheldon and Others* v. *RHM Outhwaite (Underwriting Agent) Ltd and Others*, the House of Lords had to decide a point of fundamental importance in relation to postponement of the running of the limitation period under section 32(1)(H) of the Act; did statutory postponement operate, if, after accrual of the cause of action, the defendants had deliberately concealed from the plaintiff facts relevant to their right of action? The House of Lords held, reversing the Court of Appeal, that where the plaintiff's cause of action was deliberately concealed from the plaintiff by the defendant, within the meaning of section 32(1)(b) of the Act, the effect was that the limitation period did not begin until the plaintiff discovered the concealment or could with reasonable diligence have done so. For this purpose it made no difference whether the concealment was contemporaneous with the commencement of the cause of action or was subsequent to it.

6.110   In *Brocklesby* v. *Armitage and Guest*, the Court of Appeal held that it was not necessary for the purpose of entering limitation under section 32 of the Act, to demonstrate that the fact relevant to the claimant's right of action had been deliberately concealed in any sense greater than the commission of the act was deliberate, in the sense of being intentional, and that the act or omission did involve a breach of duty, whether or not the actor appreciated the legal consequence.

Abridgement of limitation periods

6.111   For reasons set out above, it is important for a party before entering into a collateral warranty to consider whether or not the document will extend the limitation period beyond the period created by the principal contract. If there is such an extension, then this can be provided for in the collateral warranty by an express condition abridging the new limitation period to correspond with the original period. Such conditions are valid: *Atlantic Shipping and Trading Company* v. *Louis Dreyfus & Co.* This case concerned a charter party which provided for the reference of all disputes under the contract to arbitration and also had a clause which stated, 'Any claim must be made in writing and claimant's arbitrator appointed within three months of final discharge and where this provision is not complied with, the claim shall be deemed to be waived and absolutely barred'. The court held that this clause was not open to objection on the ground that it ousted the jurisdiction of the court.

Acknowledgment and part payment

6.112 Section 29(5) of the Act provides that where there is a claim for a debt or other liquidated pecuniary demand and the person liable acknowledges the claim or makes any payment in respect of it, the claim is to be treated as having accrued on or before the date of the acknowledgment or payment. There can be successive acknowledgments or part payments, each one of which will give rise to a fresh calculation of the limitation period. However, once the limitation period has expired, an acknowledgment or part payment will not revive the claim.

6.113 An acknowledgment has to be in writing and signed by the person making it, and has to be an admission of liability in respect of a debt or other liquidated amount or of a sum that is capable of being ascertained from extrinsic evidence. If these requirements are satisfied, there will be an acknowledgment even though the debtor in the same document makes a statement that he will never pay the debt: *Good* v. *Parry*. On the other hand, a statement that monies had been paid on account and that money might be due was considered not to be an acknowledgment for the purposes of the Limitation Act: *Kamouth* v. *Associated Electrical Industries Limited.*

6.114 It is not considered that the act of entering into a collateral warranty will constitute an acknowledgment. However, it could if the appropriate words evincing an admission of liability were contained within the body of the document or in the recitals.

Chapter 7
Developers, Tenants, Purchasers and Funds

7.1 A funding institution, that is to say an organisation putting up money for a development, will have a dominant motive in mind when looking at collateral warranties: that is to obtain further protection in relation to the money that they are lending over and above the protection that they are already likely to have through a legal charge on the property and the other normal methods for securing money that has been lent. In a similar way, the purchaser of a freehold development will be seeking through a collateral warranty to give himself as much protection as possible in relation to the investment he has made: he will wish to have the right to make claims against those responsible for the design and construction of the development should something that they have done, or something that they have not done but should have done, cause the purchaser to expend money.

7.2 A tenant of leasehold premises in a similar way is looking to protect himself from a liability to repair the building, which he will normally have undertaken by entering into a lease. In respect of those issues, therefore, tenants, purchasers and funds have an interest in common: seeking to protect their own interests and pass as much risk as possible to the parties who have created the development: the architect, the engineer, the quantity surveyor, the contractor and, sometimes, the major (and on occasions, all) sub-contractors.

7.3 In developments, it is very common for the employer under the building contract not to be the owner of the land on which the building works are to take place. This arises for a variety of reasons, including creating an effective tax structure for a group of companies, joint ventures and project specific companies. However, where the employer is not the owner of the land, very difficult considerations of law arise (see *Alfred McAlpine Construction Limited* v. *Panatown Limited* and the discussion in paragraph 6.25).

7.4 In order to understand the pressures felt by developers, tenants, purchasers and funds when looking at the issues of collateral warranties, it is necessary to consider their positions separately in a little detail.

The position of a developer

The developer's problems

7.5 Where the employer under the building contract is not the same legal entity as the owner of the land, careful consideration has to be given to the best structure to protect the owner of the land. This problem arises out of the decided cases and, in particular, the House of Lords decision in *Alfred McAlpine Construction Limited* v. *Panatown Limited*.

7.6   In *Panatown*, Panatown was the employer under the building contract with McAlpine but the land was owned by UIPL, a sister company of Panatown. When things went wrong with the project, including allegations of defective work by McAlpine, an arbitration was started by Panatown against McAlpine. In that arbitration, McAlpine took the position that because Panatown had no proprietary interest in the land (and thus the building), Panatown were not entitled to recover damages from McAlpine. This is the classic 'no loss' argument. That contention was, in due course, accepted by a majority decision in the House of Lords. The decision of their Lordships appears to turn entirely on the proposition that UIPL had the benefit of a collateral warranty from McAlpine and, given that direct remedy in the hands of UIPL, there was no remedy available to Panatown against McAlpine. This also means that Panatown and UIPL would have been in a better position if there had been no warranty in place at all because Panatown would then have been able to pursue recovery of substantial damages from McAlpine.

7.7   However, both the 'broader ground' and the 'narrow ground' were given new life by *Panatown*. The broader ground is that the measure of damages must reflect the employer's performance interest (*St Martins Property Corporation* v. *Sir Robert McAlpine Limited*). The narrow ground is that the measure of damages is the loss suffered by the third party landowner, recoverable by the employer who must then account to the third party for the damages received (*Panatown, Dunlop* v. *Lambert* and *The Albazero*). Notwithstanding that 'new life', the considerations in these cases are extremely complex and not every case will fall within either the narrow or the broader ground. There is still a possibility of a legal black hole opening up to prevent recovery of damages.

7.8   Where there is an employer who is not the owner of the proprietary interest in the land on which the project is to be constructed, how is their position to be protected? Are those parties best served by not entering into collateral warranties at all?

7.9   The downside of the owner of the land not having collateral warranties is that any future litigation about such a project will still be subject to the uncertainties that exist in this area of the law. There is something to be said for the certainty of a contractual relationship created by a warranty as opposed to relying on the ever changing and developing common law. For that reason, collateral warranties will remain the norm on projects. However, the question inevitably arises: how are those warranties to be best drafted to seek to ensure that they provide to the owner of the land at least the same rights as those given to the employer under the building contract?

7.10   There are some pointers to the drafting of such warranties that come out of the *Panatown* decision, as follows.

7.11   The first is that the collateral warranty must create rights on the part of the landowner to enforce, for his own benefit, the construction contract itself (or at least specified provisions in that building contract). The reason is that in *Panatown*, the collateral warranty was set up as a reasonable skill and care obligation. Generally, that will not provide an adequate remedy to a building owner vis-à-vis the contractor.

7.12   The second is to ensure that, if it proves necessary, all of the employer, the landowner and the contractor can have their claims dealt with in the same proceedings. In practice, many building contracts have an arbitration agreement and most collateral warranties have no dispute resolution machinery. The result is that the building contract dispute will be heard in arbitration and the collateral warranty dispute in the courts. One obvious way to deal with this is to delete arbitration provisions from construction contracts, so that the court has jurisdiction to deal with disputes under that contract and relevant collateral warranties. Clearly, if the subject matter of the claims under the construction contract and the warranties are the same or substantially the same, then there are rules in the court for joinder of those actions into one set of proceedings. That will be more convenient and will avoid duplication of cost and the risk of inconsistent findings between the arbitrator and a court, there being considerable difficulties as to joinder in arbitration.

7.13   It is in this area where the Contracts (Rights of Third Parties) Act 1999 can assist, but so far there are no decided cases in this field and that is probably holding back the wide use of the Act in construction contract drafting. However, the building contract, using the Act, can contain an agreement that the landowner, who is not the employer and not a party to the building contract, has the right to enforce certain terms of the building contract. Such a provision could also, if the parties so wanted, exclude all other third party rights. Indeed such a provision could extend such rights to all companies within a certain group of companies, whether existing or not at the date of the building contract. That will provide the preservation of the right to enforce where there are intercompany transfers of the land from time to time after the building contract has been entered into. This subject is discussed further in Chapter 3 and in particular in paragraph 3.59 where a draft clause is suggested for this purpose.

7.14   There are, as yet, no cases explaining how an express exclusion of the operation of the Contracts (Rights of Third Parties) Act 1999 might affect the common law position discussed in *Panatown*. It is entirely conceivable that where there is no collateral warranty, such exclusion, which is presently commonplace, might prevent the landowner putting forward claims on both the narrow and broad grounds referred to in paragraph 7.7 above. In short, excluding the operation of the Act may have unintended consequences in relation to the rights of the building owner who is not also the employer under the building contract.

The position of a tenant

The tenant's problems

7.15   The law relating to landlord and tenant is a very complex area. It is, perhaps, therefore not surprising that there has been a great deal of litigation in relation to the meaning and effect of leases. Fortunately, in considering the impact of leases on collateral warranties, it is necessary to look at a relatively small aspect of landlord and tenant law: the obligation contained in a lease as to repair, known as the repairing covenant. It is essential in a lease to define whether the landlord or the tenant is to repair; in commercial leases, the obligation to repair is usually imposed on the tenant. However, whilst it is easy to say that the obligation to repair falls on a tenant, the nature of that obligation depends on the precise wording of the repairing covenant. Sometimes it is difficult to ascertain the precise meaning and effect of a repairing covenant in relation to particular circumstances that have arisen. However, some repairing covenants can require the rebuilding of the premises, for example:

'To repair and keep in repair the Premises and in addition when necessary to carry out all works of re-building, re-instatement and renewal of the Premises (whether in whole or in part) notwithstanding the cost of or reason for such works to the intent and effect that at all times during and at the end of the Term there shall be upon the land a high class building in good and substantial repair...'

7.16   At the other end of the scale, there can be provisos excluding from the tenant's repairing obligations, a liability to remedy latent defects. Such a provision is often sought by an ingoing tenant of a new development, but not often granted by the landlord (except perhaps when there is an over-supply of buildings and a lack of willing tenants or in buildings that are multi-tenanted where the repair costs are then recovered through the service charges). Such a clause might be in this form:

'Provided that nothing in this lease shall be construed as obliging the Tenant to remedy any Defect of whose existence the Tenant has within the first ... years of the Term notified the Landlord or any want of repair which is attributable to such Defect and which manifests itself within such period...'

7.17   Clearly such a clause would require a careful definition of 'Defect'. However, it is between these two extremes that most leases will fall and a typical simple form for a repairing covenant given by the tenant might be:

'To repair the Premises and keep them in repair.'

7.18   What is the tenant's position if he has a repairing covenant in his lease on a new building and he discovers that there are serious design and construction faults? Whilst the answer to this question will depend on the particular circumstances that have arisen and the precise form of the wording of the repairing covenant in the lease, the tenant is likely to be in difficulties in arguing that he does not have an obligation to put right those serious defects in design and construction at his own expense. It is to be doubted whether a covenant to repair would be sufficient to force a tenant to completely rebuild. The legal position in relation to a latent defect in a new building causing a state of disrepair for the purposes of a repairing covenant is, however, something that gives rise to complicated legal issues. What is tolerably clear is that if there is a lack of repair caused by a latent defect, then that lack of repair falls to be dealt with under the repairing covenant. The more difficult question is whether the latent defect itself falls to be rectified by the tenant. If the only realistic way of effecting the relevant repairs is also to rectify the latent defect, that is likely to fall within the repairing covenant of the tenant: *Quick* v. *Taff-Ely Borough Council.*

7.19   On the other hand, in *Ravenseft Properties Ltd* v. *Davstone (Holdings) Ltd*, a tenant was required to lay out substantial sums to remedy a latent defect. The court in that case, however, did not set out a principle that remedying latent defects would always fall within a tenant's repairing covenant, but Forbes J did make it clear that a latent defect was not necessarily outside the repairing covenant:

> 'The true test is, as the cases show, that it is always a question of degree whether that which the tenant is asked to do can properly be described as a repair, or whether on the contrary it would involve giving back to the landlord a wholly different thing from that which he demised.'

7.20   In another leading case, the tenant escaped liability for the cost of remedying a latent defect and the landlord had to bear the loss: *Brew Bros Ltd* v. *Snax (Ross) Limited.* In that case, the adjoining owner brought a claim in nuisance against the landlord of the premises: a wall was moving by reason of undermining of the foundations by a drain. The landlord joined the tenant into the proceedings claiming from him a full indemnity in respect of the adjoining owner's claim on the basis of the obligations arising under the repairing covenant in the lease. The lease was for 14 years and the defect arose just after the end of the first year of the lease. The Court of Appeal held that the repairing covenant included the drains and that both the landlord and the tenant were liable in nuisance to the adjoining owner. It was also held that the work required to make the premises safe was more than repair and was not therefore within the repairing covenant. As to what constituted 'repair', Sachs LJ said:

> 'It seems to me that the correct approach is to look at the particular building, look at the state which it is in *at the date of the lease*, look at the precise terms of the lease, and then come to a conclusion whether, on a

fair interpretation of those terms in relation to that state, the requisite work can fairly be termed repair. However large the covenant it must not be looked at *in vacuo*.'

7.21 It is also clear that there must be a state of disrepair before any question can arise as to whether it would be reasonable to remedy a design fault when doing the repair: *Quick* v. *Taff Ely Borough Council*. In *Post Office* v. *Aquarius Properties Ltd*, unusually, there was an inherent defect in the building that did not cause disrepair: defective retaining walls permitted flooding of a basement, ankle-deep. No damage had been caused to the building by the flooding: the building was in the same condition as when it was completed and let. As there was no damage, the Court of Appeal held that the tenants were under no obligation to the landlord to remedy the defect.

7.22 *Crédit Suisse* v. *Beegas Nominees Limited* is a case where the landlord was held liable in respect of defects in a building by reason of the particular wording of the covenant that they had given. Crédit Suisse was the tenant of newly constructed commercial premises. The outside walls were clad in aluminium and glass. Water leaked into the building through the cladding. The covenant given by the landlord was 'To maintain, repair, amend, renew ... and otherwise keep in good and tenantable condition' the structure of the building including its walls. It was held that those words were wide enough to require the landlord to put the building in good and tenantable condition even if the building had never been in such condition. In this case, Lindsay J said:

> '... whilst I accept the inevitability of the conclusion of the Court of Appeal in *Post Office* v. *Aquarius Properties Ltd* that one cannot have an existing obligation to repair unless and until there is a disrepair, that reasoning does not apply to a covenant to keep (and put) into good and tenantable condition. One cannot proceed from "no disrepair, ergo no need to repair" to "no disrepair, ergo no need to put or keep in the required condition". Leaving aside cases, such as this, where there is special provision for there to have been prior knowledge or notice to the covenantor, all that is needed, in general terms, to trigger a need for activity under an obligation to keep in (and put into) a given condition is that the subject matter is out of that condition.'

7.23 *Holding & Management* v. *Property Holding & Investment Trust* gives further authority for the proposition that if damage has already occurred, work designed not merely to remedy the damage which has occurred, but which also seeks to prevent a recurrence of the damage, appears to fall within the scope of a repairing covenant. It also seems that a repairing covenant may also cover the carrying out of preventative repairs if damage is imminent.

7.24 The application of the present law since *D. & F. Estates* is exemplified by the case of *Ernst & Whinney* v. *Willard Engineering (Dagenham) Ltd and*

Others. E & W took an assignment of a lease, containing a full repairing covenant, for a commercial office building. Before they moved into the building, they discovered that the air conditioning ductwork was defective (it leaked and there were restrictions impairing performance). The total cost of the remedial works was put at approximately £1.5m. E & W had sought to argue that the ventilation system was defective and that the defects caused physical damage to the building, which in turn had been caused by the failure of the defendants to carry out the work with reasonable skill and care. The judge took the view that he could not accept that argument because the building that E & W had leased, as their predecessors had leased it before them, was the building as built containing the ventilation system as it was, whether or not it was defective or whether its defects were apparent or latent. He went on to say that whether or not the system was deficient in terms of workmanship and performance was a matter that could only be measured by reference to the installation contracts, to which E & W were not parties. E & W failed in their claim. This is the kind of case to which the government had regard when parliament brought into the law the Contracts (Rights of Third Parties) Act 1999. If that Act had been in force at the time of this case, the result would have been different unless the effect of the statute had been excluded in Willard's contract (see Chapter 3).

7.25   It is, therefore, easy to see that a tenant on a repairing covenant under a lease is taking on a risk, the full extent of which may, of itself, be the subject of some uncertainty. It cannot be a surprise, therefore, that tenants look for every means of reallocating that risk to someone else in so far as they are able to do so. Given the way the law of tort has developed, they cannot seek redress against any member of the team who designed and constructed the development, and with whom they have no contractual relationship: that contractual relationship is usually created by collateral warranties. The Contracts (Rights of Third Parties) Act 1999 may also provide a remedy unless its application has been excluded (see generally, Chapter 3).

7.26   A tenant will be concerned also to try to secure his position through the collateral warranty in relation to rent of alternative accommodation if he has to vacate the premises while the defects are repaired.

Parties giving collateral warranties to tenants

7.27   The tenant will be looking for warranties that will provide a remedy for him in the event of defects in workmanship, materials and design. It follows that different considerations arise depending on the method of procurement of the project. If the building contract is a JCT 98 (not With Contractor's Design), under which the contractor will normally have no obligation as to design (unless the option for Performance Specified Work is incorporated), then the tenant will look for collateral warranties from the contractor (in relation to the construction quality risks), the architect and the engineers: structural, mechanical and electrical, and environmental (thereby covering the design element). However, increasing use is being made of clause 42.1 of that contract under which arrangements are set out for 'Performance Specified Work'. Under these arrangements, certain requirements as to performance of parts of the work can be specified and the contractor has an obligation to meet those requirements. For example, the functional requirements of air-conditioning might be so specified, leading the contractor inevitably into a need to carry out design. It follows that even on JCT 98, there can still be a design obligation on the contractor. Quantity surveyors are sometimes but not always asked to give collateral warranties to tenants and, in any event, they are not usually involved directly in either design or construction, at least in the context of a collateral warranty. However, the position may be different if a quantity surveyor is involved in fitting out works for the tenant, particularly if those works are being executed under the main contract, as opposed to a separate contract with the tenants.

7.28 Where the project is carried out under a design and build contract, and the contractor has responsibility, therefore, for every aspect of the design and construction of the project, the position is different. Clearly the tenant will look for a collateral warranty from the contractor. The contractor himself will have employed an architect and the various engineers to carry out the design for him. Should the tenant take collateral warranties from the architect and engineers in these circumstances? If the tenant's collateral warranty with the contractor has been carefully drafted, then the tenant, in the event of a breach of the warranty, can bring proceedings against the contractor and the contractor can, if he so wishes, join the architect and/or the engineer into those proceedings. In this way a chain of contracts and liability is set up.

7.29 On the other hand, it is not unknown for contractors to cease trading through insolvency; it is this fear that can cause tenants to seek collateral warranties from the architect and the engineers, as well as the contractor on design and build projects. In circumstances where the contractor is insolvent, the tenant who has a collateral warranty with the architect and the engineer will still be in a position to bring proceedings against them at least in relation to matters which fall within their duties – this is likely to be only design because there are few contractors who will appoint architects and engineers on design and build projects to carry out any supervisory duties in relation to the quality of the construction work itself.

7.30 Irrespective of the form of procurement of the project, it is becoming common for the major sub-contractors (for example, in respect of the piling/foundations, the structure, the waterproof envelope and the mechanical and electrical services) to be asked to give warranties to tenants. There are several good reasons for adopting this approach. Firstly, main contractors are not liable to third parties for the torts of their sub-contractors, who are in law, independent contractors – indeed this was one of the difficulties faced by the plaintiff in *D. & F. Estates*. On a project procured on a basis such as JCT 98, if the tenant has a collateral warranty with the main contractor, and a defect arises which is a breach of contract by the sub-contractor (which will also be likely to be a breach of the main contract), then in theory there is a chain of contracts setting up liability and the tenant should be able to recover from the main contractor, who in turn will recover from the sub-contractor. However, that chain of contracts could be broken by, for example, an insolvency of the main contractor, and prudence suggests that a tenant is well advised to take warranties from the major sub-contractors. If those sub-contractors have entered into employer/sub-contractor warranty agreements then a contract is created under which the sub-contractor accepts responsibility for reasonable skill and care in his design, selection of goods and materials, and in complying with a performance specification – in other words, the employer has a remedy against the sub-contractor under the warranty in respect of design and workmanship problems, but a tenant does not unless third party rights exist by reason of the Contracts (Rights of Third Parties) Act 1999 (see Chapter 3).

7.31   The most difficult type of project from a tenant's point of view in relation to collateral warranties is a construction management project. By 'construction management' is meant a project where the main contractor is only managing the project and does not enter into any of the trade contracts himself – those trade contracts are between the trade contractor and the developer; the effect of such an approach on collateral warranties for tenants is fairly horrendous. Unless the developer will give it (and he probably will not), the tenant does not have one person, such as a main contractor, to look to for his warranty; he therefore faces the prospect of obtaining warranties from the vast majority of the trade contractors. Unless the construction management project has been set up with this point in mind (and most are these days) the tenant may face an impossible task. This is all the more so where the building is multi-tenanted and every tenant is looking for a warranty from every trade contractor for the important elements of the construction of the building.

7.32   On management construction projects (where the management contractor enters into the trade contracts himself), similar issues to those set out above arise in relation to a project on, for example, JCT 98. However, each and every management contract and construction management contract needs to be considered carefully by tenants in order to ascertain precisely what it is that they need, or alternatively, where they are offered warranties from some parties, whether or not what they are being offered does in reality meet the risk in relation to the repairing covenant that they are taking on under their lease.

What tenants look for in collateral warranties

7.33   The tenant is looking for the designers to undertake an obligation to him in respect of design; that certain materials have not been specified for use in the building; that the designers have professional indemnity insurance and will maintain it; and that they can assign the benefit of the collateral warranty to third parties when they come to dispose of the lease and that those third parties can also assign the benefit of the collateral warranty when they dispose of their interest. From the contractor, the tenant will be looking for a warranty that the contractor has carried out the construction of the work in accordance with his contract with the developer; that he has not used in the construction of the building certain specified materials; and that the benefit of the collateral warranty can be assigned to third parties.

7.34   From a design and build contractor, the tenant will want in addition a warranty in relation to design and that the contractor has and will maintain professional indemnity insurance.

7.35   From sub-contractors and trade contractors, the tenant will be looking for similar warranties and the tenant should be particularly careful to be certain to obtain a design warranty from those sub-contractors and trade contractors who are designing. For example, whilst the engineer may design structural steelwork, it is the practice within the construction industry for the steelwork sub-contractor to design the joints in the steelwork.

The position of a purchaser

The purchaser's problems

7.36   The fundamental problem faced by a purchaser of property in the United Kingdom is the principle of *caveat emptor* (buyer beware). Put more simply, on a sale of land and premises, it is up to the purchaser to find out whether or not the building has been built properly and designed adequately and whether or not it contains any latent or patent defects. It is not possible to imply a term into the contract for sale of property to the effect that the property is free from defects. This fundamental principle is, therefore, a serious issue faced by every purchaser of real property. It is for this reason that purchasers invariably appoint surveyors/engineers/ architects to carry out a survey and report on the state of the building prior to the purchaser becoming legally bound to purchase the building. However competently those surveys and reports are carried out, they cannot, inevitably, uncover extensive latent defects, although they can, and do, regularly uncover patent defects. Indeed, carrying out a survey at a level required to uncover some latent defects would be likely to be very expensive indeed: for example, inspecting brickwork cavities with fibre optics to check for bridging of the cavity and, more importantly, the presence or absence of brick ties. It is to be doubted whether most vendors of commercial property would permit such surveys to be carried out in any event. Clearly, therefore, the purchase of property carries risk for the purchaser.

7.37   Interestingly, a verbal collateral contract was established between a vendor and a purchaser as long ago as 1901: *De Lassalle* v. *Guildford*, see paragraph 1.4.

7.38   However, the practise of conveyancing has moved on since 1901 and oral representations such as arose in the *De Lassalle* case are usually now expressly excluded by a contract term. Sales of property, both domestic and commercial, are now usually governed by The Law Society's General Conditions of Sale. The 1984 revision of those conditions provided that the purchaser acknowledged in making the contract that he had not relied on any statement made to him save one made or confirmed in writing. However, The Standard Conditions of Sale 1995, Third Edition are rather different in relation to such matters compared with the 1984 and 1990 Editions. They provide that if a statement in the contract or in the negotiations leading to it, is or was misleading or inaccurate due to an error or omission, a remedy is provided (clause 7); so that for example if there is a material difference between the description or value of the property as represented and as it is, the purchaser will be entitled to damages. It follows that in order to obtain compensation under such a provision, it will be necessary to prove that a statement was made (whether oral or in writing) and that it was misleading or inaccurate due to an error or omission. The 1995 Standard Conditions of Sale do not make any attempt to negative reliance by a purchaser on oral statements not confirmed in writing. The balance in the 1995 Standard Conditions is more in favour of a purchaser in this respect than were the 1984 Conditions.

7.39   Many cases have arisen on the answers given to preliminary enquiries in conveyancing transactions; difficult issues have arisen on the pre-1990 Conditions of Sale as to whether a claim could be brought under the Misrepresentation Act 1967 in respect of the loss suffered by the purchaser in reliance on the representation contained in the answer to the preliminary enquiries. That remains, at least in theory, a substantial weapon in the hands of the purchaser if he has asked appropriate questions and received informative replies. However, when the property market is booming, the answer to particularly searching preliminary enquiries is more likely to be 'the purchaser must make his own enquiries'. The purchaser then has to decide, on a commercial basis having made such enquiries as he can, whether or not he wishes to proceed with the purchase. Inevitably, in this respect, the vendor of property is in a better position than the purchaser.

Parties giving collateral warranties to purchasers

7.40   The same parties are likely to be involved in giving collateral warranties to purchasers as were discussed under paragraphs 7.27 to 7.32 above in relation to tenants. However, where the development is sold during construction, or, well after completion, the alternatives of novation or assignment of benefits may be more appropriate than warranties (see 11.13 to 11.16).

What purchasers are looking for in collateral warranties

7.41   Different considerations will arise depending on whether the purchaser is completing his purchase before or after completion of the building project. It is not at all uncommon for a freehold commercial property development to be sold during construction by one party to another (for example, between pension funds). Depending on the particular circumstances, a project in progress is probably best dealt with by way of novation agreements with all the members of the professional team and the contractor whereby the purchaser stands in the shoes of the vendor, the project otherwise continuing without interruption. Each project of this kind needs careful consideration on its own particular facts and with careful consideration of the types of contracts being used and the method of procurement of the project. In these circumstances, it may be unnecessary for there to be additional collateral warranties created for the simple reason that the purchaser will take over all the benefits of all the contracts of the professional team and the contractor, all of which should contain an enabling clause requiring the giving of warranties. However, where the original employer has obtained collateral warranties before the sale (for example from sub-contractors), then the benefit of those warranties will need to be assigned to the purchaser and the drafting of collateral warranties should make express provision for that possibility.

7.42   Purchasers after completion of the building project will be looking for all the same things as discussed in paragraphs 7.33 to 7.35 above in relation to tenants. In addition, they are likely to be looking for a right to obtain copies of plans, drawings, specifications, calculations, electronic information and similar documentation in relation to the design and construction of the development. This inevitably involves considering issues of copyright, which will have to be dealt with in the collateral warranty at the same time as obligations on the relevant parties to provide copies of the relevant documents. It is clearly important to a purchaser who may wish to carry out alterations to the development that he has a facility to obtain necessary information in relation to the construction of the development. If the purchaser is intending to let the building, he will be seeking to have the ability to provide to tenants, collateral warranties from the designers and contractors.

The position of the funding institution

The funding institution's problems

7.43   The funding institutions behind property developments come in many forms, including banks, merchant banks, consortia of banks, insurance companies and pension funds (including those of large corporations). Before entering into a funding arrangement, funding institutions will naturally satisfy themselves that their involvement will be secure from a financial point of view; this will be likely to involve requirements from them as to terms in the construction contract that create as near to price certainty as can be achieved, including restrictive reasons for extension of time for completion and variation in the work. That is particularly so on projects under the government's Private Finance Initiative. They will also require arrangements that deal with a situation where, if things do not go according to plan, they are as well safeguarded as can be achieved. There will always be a lengthy funding agreement between the funding institution and the developer, allocating the risk between those two parties and setting out each party's rights, duties and obligations, and often restricting the right of the developer to vary the building works without prior consent of the funder. If the architect, the engineers or the contractor get into difficulties, then vis-à-vis the fund that will usually be a problem for the developer to resolve and take the financial risk, not the fund. On the other hand, what is the funding institution's position if the developer gets into serious difficulties, for example a serious breach of the finance agreement with the funding institution involving perhaps non-payment of interest or the insolvency of the developer?

7.44 It is specifically in relation to the whole area of developer default that potential problems arise for a funding institution in funding a development. In order to safeguard the investment they have made up to that point, the fund will wish to have arrangements in place from the outset that enable the fund, in the event of serious default by the developer, to take over the whole project themselves, or possibly through a third party appointed by the fund, so that they can secure completion of the project with a minimum of extra expense, disruption and delay, thereby safeguarding their investment. These are usually called 'step-in rights'. They go well beyond creating rights in respect of quality in the design and construction; they are creating an express contractual right for the benefit of the funding institution in relation to its financial position as to security for the loan to the developer.

Parties giving collateral warranties to a funding institution

7.45 The parties giving collateral warranties to a funding institution will be substantially the same as those referred to in paragraphs 7.27 to 7.32 in relation to tenants. However, the fund will almost certainly wish to have a warranty from the quantity surveyor because they will wish to rely on the quantity surveyor in relation to his financial duties.

What funding institutions are looking for

7.46 Funding institutions are looking for at least the same matters as are dealt with in paragraphs 7.33 to 7.35 in relation to tenants. Funds may well seek wider duties than just design workmanship and materials. Sometimes they seek to have the same duties owed to them as are owed under the principal contract.

7.47   In addition to those, funding institutions will be looking for step-in rights that enable them to take over and complete the project, but only if they wish to do so, in the event of serious default by the developer. This will usually take the form in law of a novation agreement (see paragraphs 9.73 to 9.79). The effect will be, for example, that the client in the architect's appointment will no longer be the developer but will be the funding institution (or a third party appointed by the funding institution), but that will be the only change in the architect's appointment. In other words, the client will have changed, but none of the rights, obligations or duties between the client and architect will have changed. One client is simply substituted for another client.

7.48   The funding institution will also wish to have a right to use drawings, specifications, calculations, electronic information and the like produced by the design team so that the fund is free to make use of those documents for the purposes of the development. Again, this will involve dealing with the question of copyright in the warranty.

Obligations to enter into collateral warranties

7.49   In an ideal world, every contract between every party on a development project and agreements for lease/sale agreements would be entered into on the same day. This is, of course, unrealistic. However, the fact that it cannot be done does raise some potential difficulties in relation to collateral warranties. The architect is often the first person appointed by a developer when he has a project in mind. At that stage, although it is possible, it is unlikely that the developer has in mind a particular tenant or purchaser. The same situation will probably apply at the time of the appointment of the engineers and the contractor. The funding institution will usually come into the picture some time after the architect and usually before the contractor. Unless there is a pre-let of the property, it is likely that the tenant will be the last party in time to become involved with the project. It is often the tenant or the purchaser that has the most concern about the need for, and the contents of, collateral warranties.

7.50   Situations often arise where tenants will not agree to enter into leases unless and until they have collateral warranties in the form put forward by them (or at least agreed by them). By that stage, there will be no legal sanction on the architect, engineer and contractor to give collateral warranties in any form unless such an obligation, by way of an enabling clause, has been incorporated into their terms of engagement and the building contract.

7.51   Where no legal obligation to enter into collateral warranties has been imposed in the terms of engagement and in the building contract, the fact that there is no legal obligation on the architect, engineer and contractor to give collateral warranties, may well lead to an outright refusal by those parties to enter into warranties. The only pressures that may arise are the usual commercial pressures – there will be no legal obligation. If the architect, engineers and contractor do agree to discuss warranties, then there will inevitably be a long drawn out and expensive discussion as to the wording and extent of the warranties. At worst, this can result in a tenant withdrawing from the proposed transaction.

7.52   With all this in mind, provisions are sometimes seen in terms of engagement of architects and engineers and building contracts along the following lines:

> 'The Architect shall within 14 days of written notice from the Developer enter into a collateral warranty with any party proposing to enter into an agreement for lease of the Project in such form as that party shall reasonably require.'

7.53   Such provisions are wholly ineffective in law. For an obligation of that type to be effective, the terms of the collateral warranty have to be certain. There is nothing certain where no terms are set out. Such a provision purporting to impose an obligation to enter into a collateral warranty where the terms are not certain will not give rise to a right on the part of the developer to apply to the court, for example, for an order of specific performance requiring the architect to enter into a collateral warranty. These clauses are akin to an agreement to agree which is of itself unenforceable. The better approach is to have a clause something like the following:

> 'The Architect shall within 14 days of a written notice or written notices from the Developer enter into a collateral warranty and/or collateral warranties in the form of the draft collateral warranty annexed hereto as Appendix A with any party proposing to enter into an agreement for lease of the Project and/or any part of the Project.'

7.54   Appendix A then contains the full wording of the proposed collateral warranty for that particular architect or engineer or contractor. Such an obligation would be enforceable by the courts. The inevitable difficulty that arises is that the collateral warranty has to be drafted and in place prior to the appointment of the architect, engineer or contractor as the case may be. At that time, the tenant/purchaser is unlikely to be known. Each and every tenant and purchaser has their own views about collateral warranties. It may be that the one that has been drafted, and forms Appendix A in the example above, is acceptable, but equally it may not be acceptable to the tenant. However, at least in these circumstances the tenant can be offered, for certain, a warranty in the form of Appendix A; the fact that the architect, engineer and contractor know that they have to give a warranty in that form often enables amendments needed by a particular tenant to be more readily agreed than if there was no draft warranty agreed from the outset.

JCT enabling clauses

7.55   Many of the standard forms of contract now contain express enabling
clauses in relation to the requirement for the provision of collateral
warranties. For example, the Joint Contracts Tribunal for the Standard
Form of Building Contract have published enabling provisions with their
Standard Form of Agreement for Collateral Warranties, MCWa/F and
MCWa/P&T (2001 Editions), which are respectively where a warranty is
to be given by a contractor to a fund and a purchaser/tenant. These
precedents are drafted for use with various members of the JCT family of
building contracts (JCT Standard Form of Building Contract 1998 Edition,
JCT Intermediate Form of Building Contract 1998 Edition, and JCT
Standard Form of Building Contract with Contractor's Design 1998
Edition). The enabling provision for MCWa/F (2001 Edition) is as follows
(showing the appropriate clause numbering for JCT 1998 Edition):

> **'19B   Warranty by Contractor to a person ("Funder") providing
> finance for the Works**
>
> 19B.1.1   By a notice in writing by actual delivery or by registered post
> or recorded delivery to the Contractor at the address stated in
> the Articles of Agreement the Employer may require the
> Contractor to enter into a JCT Warranty Agreement ("War-
> ranty MCWa/F") with a Funder identified in the notice. The
> Warranty must be entered into within... days (not to exceed
> 14) from receipt of the Employer's notice.
>
> 19B.1.2   A notice under clause 19B.1.1 is only valid if the entries in
> clauses 1(a), 6(3), 6(4), 13 (and also in clause 9 where that
> clause is not to be deleted) of the Warranty to be entered into
> have been notified in writing by the Employer to the Con-
> tractor prior to the date of this Contract.
>
> 19B.2   Clause 19B.1 shall not apply where the Contractor, on or prior
> to the date of this Contract, has given a Warranty on the JCT
> Warranty Agreement (MCWa/F) to the Funder.'

7.56   This enabling clause contains some potential traps for the unwary
developer and will, if used at all, be amended by developers and funds to
remove those traps. The first trap is that the notice to the contractor
requiring a warranty to be entered into with a fund is only valid if certain
information as to the completion of various blank spaces in the warranty
itself has been given to the contractor prior to the date of the building
contract (clause 19B.1.2 of the building contract). This includes the persons who are to be deemed to have given similar warranties (for the net
contribution provision in clause 1(a) of the warranty), time periods for
notices (clauses 6(3) and 6(4) of the warranty), details of professional
indemnity insurance as to the limit of indemnity (generally and any
limitation in respect of claims arising out of pollution and/or
contamination) and the time period for which it will be maintained

(clause 9 of the warranty) and the limitation period after which claims cannot be made under the warranty (clause 13 of the warranty). In short, if that information has not been given by the employer to the contractor pre-contract, the notice to enter into collateral warranties will be ineffective. Secondly, the notice, to be effective, must be in writing and sent by registered post or recorded delivery or actually delivered (clause 19B.1.1 of the building contract). Thirdly, the same written notice, to be effective, must name the funder to whom the warranty is required to be given. These requirements may well have the effect of making the machinery more certain in the legal sense but they also create a potentiality for difficulty in relation to the obtaining of warranties by the notice method.

7.57   The JCT enabling clause for CoWa/P&T (2001 Edition) is as follows (using the clause numbering for JCT 98):

'19A   Warranty by Contractor – purchaser or tenant

19A.1   Clause 19A shall only apply where, before the Contract has been entered into, the Employer has in a statement in writing to the Contractor (receipt of which the Contractor hereby acknowledges) set out the maximum number of Warranties that he may require pursuant to clause 19A.

19A.2   On or after the date of Practical Completion of the Works (or of practical completion of a Section of the Works where the Contract is modified for completion by phased sections) the Employer, by a notice in writing by actual delivery or by special delivery or recorded delivery to the Contractor at the address stated in the Agreement, may require as follows: the Contractor shall enter into a Warranty Agreement on the terms of the JCT Warranty Agreement MCWa/P&T (a copy of which, with clause 1(a) and 1(a)(i) and clause 8 completed and also clause 5 where that clause is not deleted, has been given to the Contractor before the Contract has been entered into, receipt of which the Contractor hereby acknowledges) with the Purchaser/Tenant referred to in clause 19A.4.

[*Note: If, after the Building Contract has been entered into, the Employer requires any amendment to the terms of the Warranty Agreement MCWa/P&T given to the Contractor pursuant to this clause, this can only be done by a further agreement with the Contractor.*]

19A.3   The Warranty shall be entered into within _____ days (not to exceed 14) from receipt of the Employer's notice referred to in clause 19A.2.

19A.4   A notice under clause 19A.2 shall identify the Purchaser/Tenant with whom the Employer has entered into an agreement to purchase/an agreement to lease/a lease and where relevant shall sufficiently identify that part of the building(s) comprising the Works which has been purchased/let.

19A.5   The Employer may require the Contractor to enter into a

Warranty Agreement as referred to in clause 19A.2 with a person who has entered into an agreement to purchase/an agreement to lease/a lease as referred to in clause 19A.4 before the date of Practical Completion of the Works (or of practical completion of a Section of the Works where the Contract is modified for completion by phased sections); but on condition that the Warranty Agreement shall not have effect until the date of Practical Completion of the Works (or, as relevant, the date of practical completion of a Section).'

7.58   The enabling clause for the MCWa/P&T Warranty, as with the enabling clause for MCWa/F, requires that the notice to the contractor to enter into warranties is only valid if certain information as to the completion of various blank spaces in the warranty itself has been given to the contractor prior to the date of the building contract (clause 19A.2 of the building contract). This includes either the limit on recoverable damages in respect of each breach or the maximum liability in total for all the breaches, in addition to the liability for costs of repair, renewal and reinstatement, which is unrestricted (clause 1(a)(i) of the warranty), details of professional indemnity insurance as to the limit of indemnity (generally and any limitation in respect of claims arising out of pollution and/or contamination – clause 5 of the warranty) and the time period for which it will be maintained (clause 5 of the warranty) and the limitation period after which claims cannot be made under the warranty (clause 8 of the warranty). In short, if that information has not been given by the employer to the contractor pre-contract, the notice to enter into collateral warranties will be ineffective. Secondly, the notice, to be effective, must be in writing and sent by registered post or recorded delivery or actually delivered (clause 19A.2 of the building contract). Thirdly, the same written notice, to be effective, must name the purchaser/tenant to whom the warranty is required to be given. These requirements may well have the effect of making the machinery more certain in the legal sense but they also create a potentiality for difficulty in relation to the obtaining of warranties by the notice method.

7.59   Curiously, unlike the enabling clause for CoWa/F, this enabling clause does not require the naming prior to the date of the building contract of the parties who are to be deemed to have given similar warranties (for the net contribution provision in clause 1(c) of the warranty). These are the consultants and the sub-contractors, blank spaces for which are left in the warranty itself. This makes for flexibility but at the expense of creating arguments as to whether the warranty is in a form that is legally certain for the purposes of the effectiveness in law of the enabling clause.

7.60   The enabling clause also contains restrictions as to when the employer can require warranties from the contractor. Clause 19A.2 states the notice requiring the entering into of warranties by the contractor can only be given 'on or after the date of Practical Completion (or of practical completion of a Section of the Works where the Contract is modified for completion by phased sections)'. So what is to happen if the employer needs warranties prior to practical completion? There is no mechanism, for example, for the employer to obtain warranties in the event that practical completion is never reached, say, for example, on a determination of the contractor's employment prior to practical completion, or, a repudiatory breach of contract by the contractor accepted by the employer at a time prior to practical completion.

7.61   Clause 19A.5 deals with this by giving the employer power to call for warranties prior to practical completion. However, that is restricted by a condition: those warranties will not be effective until the date of practical completion of the works or the date of practical completion of a section (if relevant). Such a provision will not be likely to be acceptable to most developers because it restricts the freedom which they want in order to deal with their property transactions to best effect for them. For the same reason, funds, in looking at their security, are unlikely to find acceptable such a provision in a building contract.

Chapter 8
Insurance Implications

8.1   Very few professional practices of designers have sufficient resources within their own organisations to meet anything other than minor claims brought against them in respect of professional negligence. It is for this reason that professional indemnity insurance is so important to the construction professions and to contractors working on design and build projects. Contractors, architects, engineers and quantity surveyors should simply not be asked to give collateral warranties which may result in professional indemnity insurance cover not being available to meet a claim brought under the provisions of that warranty: for such a situation to arise cannot be in the interests of the construction industry professions, contractors or the tenant, purchaser or fund to whom the warranty has been given. It follows that the dimension of professional indemnity insurance should never be forgotten in the negotiation of a proposed collateral warranty. For a discussion of latent defects insurance, see Chapter 11.

Principles of professional indemnity insurance

Proposal form, disclosure and risk

8.2   Insurance policies are nothing more nor less than contracts; the usual rules as to the formation of contracts apply to contracts of insurance. In practice, the usual procedure is for the person seeking insurance to complete a proposal form – each company has their own standard form. The person seeking insurance will sign the proposal form and by so signing he is offering to accept insurance on the basis of the proposal form and the insurers' standard conditions (to which reference will be made in the proposal form, although the terms will not usually be set out). Those terms will contain a provision to the effect that the answers given on the proposal form are incorporated into and are the basis of the insurance. As soon as the insurers have accepted the proposal, they will then issue a policy.

8.3   Contracts of insurance are, however, different to other contracts in one important respect: they are based on the principle of the utmost good faith of the parties (known as the principle of *uberrimae fidei*). One of the aspects of this principle is that the purpose of a proposal form is to enable the insurer to assess the risk so that he can decide whether or not to accept the proposal and, if so, on what terms both as to the conditions and the premium. Clearly, insurers cannot make a proper evaluation of those matters unless full disclosure is made by the person seeking insurance of every matter that is relevant to the risk. When making a proposal to an insurer, it is necessary to disclose all facts that are material and not to make a statement that amounts to a misrepresentation of a material fact. Such non-disclosure or misrepresentation entitles the insurer to avoid the policy.

Non-disclosure

8.4   Disclosure means disclosing all facts that are material. Material facts are matters that would have affected the mind of a prudent insurer in deciding whether to take the risk and, if so, on what conditions and at what premium. This involves disclosing all material facts that are actually within the knowledge of the person seeking the insurance and this duty is not limited by what the person applying for the insurance thinks is relevant. It is clear that facts which show that a risk is not an ordinary risk, but a greater risk than the ordinary, are material for this purpose.

The cover

8.5   The intention of a professional indemnity policy is to provide an indemnity in respect of the designer's legal liability for damages in respect of claims brought against the designer for breach of professional duty. The policy is therefore written on the basis that it covers claims made during the period of insurance. The period of insurance of professional indemnity policies is usually 12 months. It follows that a professional indemnity policy will cover *claims made against the insured during that 12-month period.* Indeed difficulties can arise where the full extent of a claim is not known during one period of insurance but becomes known during a subsequent period of insurance at a time when a different insurer is on risk: see for example *Thorman and Others* v. *New Hampshire Insurance Co (UK) Ltd and the Home Insurance Company.*

Renewal

8.6   Professional indemnity policies run for a period of 12 months and can only be renewed with the same insurer if the insurer consents. At renewal, professional indemnity insurers may require the completion of a fresh proposal form and the same duty of the utmost good faith arises on the completion of a proposal form for renewal for the simple reason that it is in reality a proposal for a new insurance policy. If there is no new pro-posal form at renewal, then the insured has a duty to notify any material matters to the insurer before renewal. In any event, a new insurer may be appointed on renewal.

Generally

8.7   The matters set out above are very much a thumbnail sketch of some principles of professional indemnity insurance. They are intended merely as an introduction to the subject for the purposes of dealing with the particular points that follow in relation to collateral warranties and professional indemnity insurance.

Disclosure of collateral warranties

8.8   It is clear from the present position in the law of tort (Chapters 1 and 2) that in the absence of a collateral warranty, the tenant of a building put up by a developer would be unlikely to succeed in a claim in negligence brought against the architect or engineer or contractor of the developer in respect of negligent design. A collateral warranty entered into between the tenant and the architect, on the other hand, would enable that tenant to pursue his claim. It must follow that the existence of collateral warranties is a material fact that must be disclosed to insurers prior to professional indemnity insurance being taken out *and* at renewal. After all, against the background of the present law of tort, the existence of collateral warranties must be a matter which would affect the mind of a prudent insurer when considering whether to take the risk and, if so, on what conditions and at what premium (see paragraphs 8.2 to 8.4).

8.9   Indeed, the fact that there are so many different forms of collateral warranty that have been signed, and are being signed, suggests that each and every collateral warranty actually entered into should be produced to insurers at renewal – at the very least a schedule of collateral warranties should be appended to the proposal form with a statement that they will be produced to insurers if insurers wish to see them. This is patently a burdensome obligation both on the insured and on the insurer, but given the law in relation to material disclosure in professional indemnity policies, any other approach by the insured is clearly dangerous for the simple reason that it may be that the policy could be avoided by the insurer if disclosure is not made of these material circumstances. The simple fact that many brokers (but not all), most insurers (but not all) and most underwriters are not set up to deal with a volume of work of this sort on renewal is irrelevant to the principle that full disclosure must be given of material circumstances. In practice, some insurers are closely involved in the approval of warranties (but only from an insurance point of view) prior to their finalisation. This may assist in meeting the insured's duty to give disclosure of warranties to the insurer.

8.10   Another day-to-day problem is this: should the insured, when asked to sign a new collateral warranty, obtain his insurer's agreement before he agrees to execute a warranty? The strict answer to that question is that another warranty is, vis-à-vis the insurance policy, another material circumstance that will have to be disclosed on renewal; if on renewal disclosure is made and the insurer refuses cover at that stage, the insured runs the risk of being uninsured in respect of any claims arising under or out of that particular warranty. It is for this reason that a cautious view should be taken and all proposed collateral warranties should be shown to insurers, and their agreement to cover the risk obtained, before the warranty is executed. There are two methods of avoiding this burden.

8.11   The first is where particular standard forms of warranty have been approved by insurers for general use by people insured with them, and it will not then be necessary for the insurer's permission to be obtained on each and every occasion. The RIBA, RICSIS, ACE and RIASIS insurance schemes approved, at the time of its issue, the use of the standard form warranty to a company providing finance in the form CoWa/F (1992) and CoWa P&T (1993) (see General Advice given with both forms of warranty at paragraph 8). However, no consultant should assume that this is the case or that, even if it is, it will be the case forever. Clearly, that statement does not apply to insurers other than those stated in the General Advice. Indeed, it is the case that those warranties were last published in 1992 and 1993 respectively, so now, that statement should be checked with insurers before relying on it. Consultants should check on a routine basis with their own insurer what the position is at any particular time.

8.12   Some insurers, such as The Wren Insurance Association Limited, which is an architects' mutual insurer, has its own suggested form of collateral warranty. In practice, insurers' own standard forms are rarely acceptable to commercial developers in their unamended form. Whilst the Wren does have its own suggested form, it will look at any warranty submitted to it to ensure that there are no implications from a cover perspective. Furthermore, no-one should assume that because some insurers approve particular standard forms, other insurers will automatically approve the same warranties – it is not the case for the simple reason that each and every insurance policy is a separate contract between the particular insured and the particular insurer.

8.13   Finally, the mere fact that the Form of Agreement for Collateral Warranty (CoWa/F) to be given to a company providing finance (but not CoWa/ P&T) is said to have been agreed 'after discussion with the Association of British Insurers' does not mean that every insurer who is a member of that Association will give cover in respect of a liability arising under a warranty in that standard form; whether or not cover is given is a matter between the particular insured and the particular insurer by way of agreement or by way of endorsement on the policy. Furthermore, such approval as is given to particular standard forms by particular insurers is usually limited to cover where the warranty is entered into in its *unamended* form; for example, the approval of CoWa/F by the RIBA, RICSIS, ACE and RIASIS insurance schemes is clearly to that warranty, in its *unamended* form only. These standard form warranties are usually amended and that approval will clearly not apply to *amended* warranties.

8.14   The second way in which the difficulties of constantly referring warranties to insurers for approval can be overcome to some extent, is by agreeing with the insurer a suitable policy amendment by way of an endorsement. This is dealt with in more detail later in paragraphs 8.35 to 8.41.

Particular insurance problems

Operative clause

8.15   The purpose of professional indemnity insurance is to give an indemnity to the insured in respect of his loss – no more and no less than his loss: *Castellain* v. *Preston*. The policy of insurance achieves this effect by what has become known as the 'operative clause'. This is the clause of the policy that sets out the nature of the risk that the insurers are taking and in respect of which they agree to give the insured an indemnity in respect of any sum which the insured has to pay as a result of the loss occurring. The operative clause is therefore of the utmost importance in deciding whether or not a particular loss is covered by a policy. Having looked at the operative clause to see if cover is provided, it is necessary to look at the other terms of the policy to see whether there are, for example, any exclusions that would mean the insurer was not liable on the indemnity otherwise provided in the operative clause.

8.16   In looking at collateral warranties and insurance, therefore, it is essential to look first at the operative clause of the insurance policy. Unfortunately, there are innumerable different policy wordings produced by different insurance companies, mutuals and underwriters and, further, they vary from profession to profession and in relation to contractor's design and build cover. It is therefore impractical in a book of this sort to cover all the possible policy wordings, although some points are made below in relation to a small number of them. The over-riding principle must be for the insured to check that the operative clause of the policy does provide cover in respect of liability that may arise under collateral warranties.

8.17   Fairly typical policy wording for an operative clause is:

> 'The Insurer agrees to indemnify the Insured up to the limit specified in the Schedule in respect of any sum or sums which the Insured may become legally liable to pay as damages for breach of professional duty as a result of any claim or claims made upon the insured during the period of insurance arising out of the conduct of the practice described in the Schedule as a direct result of any negligent act, error or omission committed by the Insured in the said practice or business.'

8.18   The most important words in that clause in the context of collateral warranties are '. . . as a direct result of any negligent act, error or omission'. These words have given rise to considerable litigation in relation to particular circumstances that have arisen. Clearly, it is important in the context of collateral warranties to be reasonably confident that those words are apt to provide cover in respect of a claim arising under a contract as well as a claim for negligence as a tort. In practice, insurers usually accept that a breach of a duty under a contract (whether express or implied) to exercise reasonable skill and care will fall within the matters giving rise to indemnity under the operative clause above. Some support for the view that the word 'negligent' does not also condition 'error or omission' was given by Webster J in *Wimpey Construction UK Limited* v. *D. V. Poole*:

> 'A professional indemnity policy does not necessarily cover only negligence. In my view I must give effect to the literal meaning of the primary insuring words and construe them as to include any omission or error without negligence, but not every loss caused by an omission or error is recoverable under the policy. In the first place, which is common ground, it must not be a deliberate error or omission.'

8.19   Another type of clause found in professional indemnity policies, but not often in the construction professions, is:

> 'The Company will indemnify the Insured in respect of claims made against the Insured and notified to the Company during any period of Insurance against civil liability incurred in connection with the conduct of the Business carried on by or on behalf of the Insured.'

8.20   Such clauses as that above which give the indemnity in respect of 'civil liability' are least likely to give rise to difficulties in relation to collateral warranties. Civil liability is to be construed in contra-distinction from criminal liability and will be likely therefore to include claims in the tort of negligence and claims under a contract as well as other forms of action, for example breach of statutory duty.

8.21   There are further policies around where the operative clause gives an indemnity against 'liability at law for damages' in respect of claims arising out of the conduct of the business. On similar words, arguments have been advanced in cases that those words did not provide cover for claims arising from breach of an obligation to exercise reasonable skill and care under a contract. In other words, if that argument were right, there would be no cover for collateral warranties. The English case in which this point was argued is *M/S Aswan Engineering Establishment Co* v. *Iron Trades Mutual Insurance Co Ltd* where the operative words giving rise to the indemnity under the policy were:

> '. . . against all sums which the Insured shall become liable at law to pay as damages and such sums for which liability in tort or under statute

shall attach to some party or parties other than the Insured but for which liability is assumed by the Insured under indemnity clauses incorporated in contracts and/or agreements...'

8.22   Hobhouse J said of the submission that this clause limited claims under the policy to claims made against the insured in tort:

> '... If the words used have an ordinary and natural meaning that is reasonably clear that is the meaning which should be adopted and the court should not entertain an obscure or contrived argument to give these words some different meaning. This principle is reinforced where it is the insurance company that is seeking to reject the ordinary meaning and where the document is, as here, a standard form document produced by the insurance company itself. "Liable at law" on its ordinary meaning simply means legal liability.'

8.23   It should be noted, however, that a different conclusion has been reached on this point in two Canadian cases, albeit they are not binding on the English courts: *Canadian Indemnity Co* v. *Andrews & George Co*; *Dominion Bridge Company Limited* v. *Toronto General Insurance Company*.

8.24   There are policies in existence, particularly for design and build contractors, where the operative clause appears to limit cover expressly to 'negligence', for example:

> 'We ... agree to indemnify the Assured for any sum or sums which the Assured may become legally liable to pay ... as a direct result of negligence on the part of the Assured in the conduct and execution of the professional activities and duties as herein defined.'

8.25   The word 'negligence' is capable of having a particular meaning in law and that meaning is limited to claims in tort. If the word 'negligence' were to have that meaning in this operative clause, then the consequences for the insured would be very serious indeed for he would have no cover in respect of claims arising out of a breach of duty under a contract – either a collateral warranty or his terms of appointment unless there was also negligence as a tort. In the case of collateral warranties, it is unlikely that there could be a claim in tort, otherwise there would be no need for the warranty in the first place. Although it is conceivable that a judge construing this type of policy wording would interpret it liberally to include a breach of a duty under a contract to take reasonable skill and care, it is a risk that the insured might be well advised to avoid. If possible, therefore, it may be best to avoid such policy wording or, if it cannot be avoided, then to have the word 'negligence' defined for the purposes of the policy so as to include breach of any common law duty to take reasonable care or exercise reasonable skill, and/or breach of any obligation whether arising from express or implied terms of a contract or a statute or otherwise to take reasonable care or exercise reasonable skill.

Exclusions

8.26   Every professional indemnity policy contains a list of matters in respect of
        which the insurer will not be liable under the policy; these commonly
        include the excess, a claim brought about by dishonesty, fraud or criminal
        act, a claim brought outside a specified geographical area, libel and
        slander (unless there is a policy extension to cover these matters), and
        personal injuries caused to a third party unless they arise out of breach of
        professional duty. In the policies of particular professions, exclusions will
        be found which are peculiar to that profession; an example is a restriction
        as to cover for surveys and inspection unless the survey or inspection has
        been carried out by someone who has one of the specified qualifications.
        However, for the purposes of collateral warranties, there is one exclusion
        clause that can give rise to particular and serious problems. Typical
        clauses of this type are:

> 'The giving by the Assured of any express warranty or guarantee which
> increases the Assured's liability but this exclusion shall not apply to
> liability which would have attached to the Assured in the absence of
> such express warranty or guarantee.'

or

> 'Any claim arising out of a specific liability assumed under a contract
> which increases the Insured's standard of care or measure of liability
> above that normally assumed under the Insured's usual contractual or
> implied conditions of engagement or service.'

8.27   It is clear that most collateral warranties will be caught by such an
        exclusion clause in the policy. It follows that the policy would not give an
        indemnity in respect of claims made under collateral warranties in those
        circumstances. These exclusion clauses have been common form clauses
        in professional indemnity policies for a number of years and they came
        about because insurers became concerned that pressures were being
        placed on professional people to enter into contracts which provided for a
        higher level of duty than that to be expected of a reasonably competent
        person of that qualification holding themselves out as having those par-
        ticular skills. The basis of professional indemnity insurance is that the
        standard of care to be expected from the professional man is of the type
        considered in *Bolam* v. *Friern Hospital Management Committee* (approved in
        Whitehouse v. *Jordan*), subject to the qualifications put on the test in other
        cases (see paragraph 5.4). Clearly if a higher standard than that is to be
        expected of a professional man, by a contractual arrangement, then it is a
        matter that insurers are entitled to have disclosed to them as a material
        fact. The types of exclusion clauses set out above are, then, important to
        insurers but they have the unfortunate side-effect of excluding from
        indemnity under the policy claims under collateral warranties.

8.28   It used to be argued (prior to the changes in the law of tort) that all a simple collateral warranty did was to put in writing the duties that were in any event owed in tort – on that basis, these exclusion clauses were not so important. Now that the position has changed in tort, these exclusion clauses are of the utmost importance to people who want to rely on professional indemnity insurance either directly or indirectly: architects, engineers, quantity surveyors, contractors, funding institutions, tenants and purchasers. The only way to deal with these matters, where there is such an exclusion clause, is to agree an appropriate amendment to the exclusion clause with insurers or, alternatively, to have a policy endorsement so that collateral warranties are not caught by the exclusion, subject to the wording of the endorsement. This aspect is considered further in paragraphs 8.35 to 8.41.

Other matters of concern to insurers

Economic loss/consequential loss

8.29   When there are defects in the construction or design of a building, the losses suffered by the occupier or owner can extend well beyond the direct costs of repair. There could be loss of production in a factory by the production having to be either stopped or disrupted while remedial works are carried out; there could be the costs of removal to other premises while remedial works are carried out and the costs of renting those other premises; there could be loss of profit; there could be a claim for the cost of management time spent by the occupier/owner in dealing with all the consequences of the defective design or construction. All of these matters might loosely be called economic or consequential loss. They can, and often do, exceed the direct cost of remedial works.

8.30   In so far as the insurance policy is concerned, cover is an indemnity given in respect of 'damages' and that word in the context of an insurance policy means what the insured is legally liable to pay to the third party. It is for this reason that insurers are concerned as to the wording of collateral warranties in relation to consequential and economic loss. Insurers would prefer to see economic and consequential loss expressly excluded in collateral warranties but this is of course an approach that does not appeal to purchasers, tenants and funding institutions. Further consideration is given as to what damages can be recovered under collateral warranties in Chapter 6; there is a further discussion of the possibilities of limiting liability in collateral warranties by their wording, in paragraphs 9.14 to 9.21 and 9.92 to 9.99.

Assignment

8.31   The whole question of assignments of collateral warranties has been a matter of great concern to insurers. They felt that unlimited assignments, say from outgoing tenants to incoming tenants every time the tenancy changes, will inevitably lead to a greater risk of claims. That concern does not seem to have materialised over the last ten years or so. On the other hand, it must be the case that assignments increase the possibility of claims compared with a situation where assignments are not permitted or are restricted. It is because of that anticipation by insurers of that increased likelihood of claims that insurers will often insist on a restriction in a collateral warranty on rights of assignment. However, it is the case that insurers generally during the latter part of the 1990s gradually began to accept that they could be more generous in the number of assignments that they would allow. Typical requirements of insurers are considered in paragraph 8.35 below.

Fitness for purpose and other express guarantees

8.32   It is the case that the operative clauses of many insurance policies will only provide cover for 'negligent act error or omission'; this is based on the standard to be expected of a reasonably competent professional man carrying on that particular profession. In order to establish a liability on that sort of basis, it is necessary to look at what an ordinary competent person exercising that particular skill would do and to compare that with the actions of the person against whom the negligence has been alleged. In practice this is done by seeking the views of other persons exercising that particular skill and asking them whether the action that was taken by the person whose actions have been questioned are above or below the standard to be expected of an ordinary competent person carrying on that type of work. That is the function of expert witnesses and without such expert evidence, claims against professional people could not proceed at all: *Warboys* v. *Acme Investments*.

8.33   However, where there is an express or implied warranty that a design or piece of construction will be fit for a particular purpose, then the test in law is very different to that for professional negligence. Then, the questions that have to be asked are what was the purpose; was the design or construction fit for that purpose; if it was not, then there is liability even where there has been no negligence. Similar arguments arise in relation to guarantees of performance whether in relation to design or construction.

8.34   Insurance will not extend cover for such fitness for purpose obligations or guarantees, unless there is also negligence. Given that position, a person who enters into such obligations is likely to find themselves without the benefit of their insurance cover, unless there is also negligence in relation to the same facts.

Policy endorsements for collateral warranties

8.35   Professional indemnity insurers who understand the difficulties for designers and contractors created by the commercial need to give collateral warranties, have been prepared to agree policy wording that is of great assistance. The aim of such policy wording is to set out in the policy, or by way of an endorsement, the circumstances in which insurers will extend cover for collateral warranties; on that basis, the designer or contractor can then decide, when faced with a proposed collateral warranty, whether or not that particular warranty falls within the cover provided by the insurer or whether the warranty will have to be put to the insurer to see whether cover can be given by the insurer. This has several advantages to the designer/contractor and the insurer: many collateral warranties can be signed without the need to bother brokers/insurers with the wording; in those cases, the designer/contractor can have some confidence that his policy will cover those particular warranties; if proposed warranties are put forward which do not fall within the insurer's permitted restrictions, then the designer/contractor is in a good negotiating position with the third party because he can use the lack of insurance cover if he were to sign the proposed warranty as an entirely proper reason for seeking amendments to it; finally, the insurer will not be constantly bombarded with requests for approval of non-standard warranties, and the number that have to be put to him for approval will be considerably reduced.

8.36   There are typical formats for the kind of wording that insurers may be prepared to agree so as to overcome the difficulties created by the sorts of exclusion provisions referred to in paragraphs 8.26 to 8.28. The two types in fairly common use are set out here.

8.37   The first type of collateral warranty extension is known as EO17(A1):

'It is hereby noted and agreed that notwithstanding Exclusion "X", Underwriters hereon will extend cover to include liability assumed by the Assured under any Collateral Warranties, Duty of Care Agreements or similar Agreements given by the Assured in the conduct and execution of the Professional Activities and Duties as defined in the Policy

PROVIDED ALWAYS THAT this extension shall be subject to the following exclusions:

(i)   liability arising from the provision by the Assured of an express term guaranteeing or warranting the fitness for purpose or similar for the works.

(ii)   liability under any Warranty or Agreement if and to the extent that the period of such liability exceeds the period of the Assured's liability under the Contract to which the Warranty or Agreement is supplemental.

(iii)   liability arising under any express guarantee that the works will satisfy any particular performance specification or any express guarantee relating to the period of the project.

(iv)   any express contractual penalty of a financial nature or liquidated damages.

(v)   any liability to an assignee of the Warranty or Agreement if:

(a)   the Warranty or Agreement was originally given to funders, financiers or bankers and the benefit of the Warranty or Agreement has been assigned more than twice since the Warranty or Agreement was entered into by the Assured

(b)   the Warranty or Agreement was given to anyone other than funders, financiers or bankers and the benefit of the Warranty or Agreement has been assigned more than once since the Warranty or Agreement was entered into by the Assured.

However these Exclusions shall not apply to liability which would have attached to the Assured in the absence of such Warranties or Agreements.

In the event that the Assured gives a Warranty or Agreement assuming liabilities beyond the scope of cover provided by virtue of this extension then cover hereunder shall be limited to the extent of this extension.

Furthermore Underwriters may extend this Policy to include increased liability assumed by the Assured under any Warranty or Agreement which goes beyond the above exclusions subject to any additional information as may be required and at terms and conditions to be agreed by Underwriters.

Subject otherwise to the terms conditions limitations and exclusions of the Policy.'

8.38   This collateral warranty extension is intended to limit the liability of the insurer to indemnify the insured professional to such liability as would have arisen under the contract to which it is collateral. This relates to limitation periods, excluding fitness for purpose and guarantees, increased liability under the collateral warranty and exclusion of cover in respect of more than two assignments for funders and one in respect of others, such as tenants and purchasers. Such restrictions on assignments if reflected in the drafting of the collateral warranty are most unlikely to be acceptable to a commercial developer, particularly in relation to purchasers and tenants. Professional firms with such collateral warranty extensions need to be sure that they do not enter into collateral warranties with more assignments than they have cover for under their insurance policy. Clearly, the right course for such a firm is to approach insurers on a case-by-case basis to obtain agreement to more assignments not invalidating the cover. It is no doubt because of the reality of commercial developers not looking kindly on restrictions on assignment that another approach is available in the insurance world, which is the second type of collateral warranty extension to policies. Certainly insurers are more relaxed about assignment of collateral warranties than they were when the first edition of this book was published in 1990.

8.39   The second type of collateral warranty extension is known as EO17(B1):

'It is hereby noted and agreed that notwithstanding Exclusion "X", Underwriters hereon will extend cover to include liability assumed by the Assured under any Collateral Warranties, Duty of Care Agreements or similar Agreements given by the Assured in the conduct and execution of the Professional Activities and Duties as defined in the Policy

PROVIDED ALWAYS THAT this extension shall be subject to the following exclusions:

(i)   liability arising from the provision by the Assured of an express term guaranteeing or warranting the fitness for purpose or similar for the works.

(ii)   liability under any Warranty or Agreement if and to the extent that the period of such liability exceeds the period of the Assured's liability under the Contract to which the Warranty or Agreement is supplemental.

(iii) liability arising under any express guarantee that the works will satisfy any particular performance specification or any express guarantee relating to the period of the project.

(iv) any express contractual penalty of a financial nature or liquidated damages.

However these Exclusions shall not apply to liability which would have attached to the Assured in the absence of such Warranties or Agreements.

In the event that the Assured gives a Warranty or Agreement assuming liabilities beyond the scope of cover provided by virtue of this extension then cover hereunder shall be limited to the extent of this extension.

For the avoidance of doubt Underwriters note that the Collateral Warranties, Duty of Care Agreements or similar Agreements may be freely assignable but Underwriters liability will be limited to the period of liability that would have attached to the Assured under the Contract to which the Warranty or Agreement is supplemental.

Furthermore Underwriters may extend this Policy to include increased liability assumed by the Assured under any Warranty or Agreement which goes beyond the above exclusions subject to any additional information as may be required and at terms and conditions to be agreed by Underwriters.

Subject otherwise to the terms conditions limitations and exclusions of the Policy.'

8.40   The major difference between this second collateral warranty extension and the first one set out in paragraph 8.37 is in relation to assignment of warranties. In the first, there are restrictions; in the second, there are no restrictions on assignment, save that cover will not extend to such assignments if the liability can be incurred at a time later than that under the contract to which the collateral warranty is supplemental.

8.41   The difficulties of providing effectively in collateral warranties for restriction (as opposed to prohibition) on assignment are considered in paragraph 9.84.

Problems on changing insurers

8.42   All of the matters raised in this chapter need to be carefully considered on renewal of professional indemnity insurance with the same insurers but they need to be considered all the more carefully on changing insurers. Alternatively, the broker may be the same and the insurance scheme may be the same but the insurance company with whom the scheme is placed or, indeed, the underwriters participating in the cover may be different. In such changing circumstances, the agreement of an insurer in one year of insurance to the wording of particular collateral warranties will not bind any insurer at the time when a claim is made in a different policy year, perhaps many years after the collateral warranty was entered into, unless full disclosure has been given to insurers, or the wording of the warranty falls within the express terms of the professional indemnity policy as discussed in paragraphs 8.36 to 8.40 above. Brokers will often say that changing insurers can be dangerous in a professional indemnity context; in the age of collateral warranties, it is potentially even more dangerous.

Chapter 9
Typical Terms

9.1   Although there are a few standard forms of collateral warranty (see Chapter 10), which are not widely accepted in their unamended form, there are types of terms that often arise in collateral warranties. Those types of terms are considered here following a discussion of the general considerations that should apply to the drafting of collateral warranties.

General considerations

The principal contract

9.2   A draftsman should always bear in mind that the intention of a collateral warranty is to create a contractual relationship that is collateral to the obligations created by the principal contract. That principal contract can take a great many different forms: Standard Form of Agreement for the Appointment of an Architect, one of the several ACE Forms, the RICS Conditions of Engagement, the Main Construction Contract and all the various forms of sub and trade contracts. The guiding principle of the draftsman should be that the collateral warranty should not properly seek to impose any greater or more extensive obligations than those which are created under the principal contract; one exception to this principle is where a sub-contract between a contractor and a sub-contractor contains no obligations as to design but in circumstances where the sub-contractor is in fact designing. Clearly that design obligation should be the subject of a clause in a collateral warranty so as to create a contractual cause of action in favour of the beneficiary in the event of default in the design obligation.

9.3   The draftsman should be particularly careful to provide words and obligations in the collateral warranty that are entirely consistent with the words and obligations contained in the principal contract. Equally, he should avoid the temptation to put additional and onerous obligations in a collateral warranty that do not appear in the principal contract. To do so is likely to lead to extensive argument about the precise wording of the warranty with each party who is asked to sign the warranty and, in extreme cases, producing a situation on the project where the commercial reality of the building process is in danger of being lost. This can sometimes happen where the warranties that are to be obtained are decided by the funding agreement between the developer and the fund prior to the involvement of the professional and construction teams for the project. The developer needs to have his finance in place, for without the finance there will be no project, but it is a mistake in those discussions between the developer and the fund not to have regard to the commercial realities of the construction market place; the expectations of some funding agreements in relation to the collateral warranties that are to be obtained by the developer from the professional team, the main contractor and sub-contractors sometimes bear no relation to what is in fact achievable subsequently with those other parties.

Purpose

9.4   The draftsman should never lose sight of the purpose of the collateral warranty that he is drafting: simply put, it is to create a contractual relationship in circumstances where there are no relevant duties in tort owed by one party to the other. It follows that the language of tort is inappropriate for use in a collateral warranty. Some draftsmen used to seek to create a tortious type of obligation by means of a contract term:

> 'The Architect hereby agrees that it owes to the Tenant a duty of care in tort.'

9.5   In circumstances where the House of Lords have decided that no such duty is owed in tort, except perhaps for pure economic loss but not physical damage, it is very difficult to see how the courts could be persuaded to accept that such a clause has the effect of creating tortious duties. In relation to design, for example, the usual need will be to create a contractual obligation on the designer to use reasonable skill and care; the use of language appropriate to tort is wholly inappropriate in the setting up of contractual relationships. In any event, in the example given above, it is very difficult to attribute any meaning at all to a confirmation that a duty of care in tort is owed by one party to another, when, absent the clause, there is no such relevant duty.

Contract or tort

9.6   The issue has often arisen as to whether or not there can be, as between two parties, liability in both tort and contract in relation to the same set of facts. A further issue that arises is whether, if there is a contract, which may contain clauses excluding liability or limiting the consequences of liability, there can also be a duty in tort which is not subject to the same restrictions that are imposed in the contract in relation to liability under the contract. There was a trend in the 1980s to move towards looking only to the contract. For example, in *Tai Hing Cotton Mill Ltd* v. *Liu Chong Hing Bank Ltd*, in the House of Lords, Lord Scarman said:

'Their Lordships do not believe that there is anything to the advantage of the law's development in searching for a liability in tort where the parties are in a contractual relationship. This is particularly so in a commercial relationship... their Lordships believe it to be correct in principle and necessary for the avoidance of confusion in the law to adhere to the contractual analysis.'

9.7   Although this point was not directly before the House of Lords in *Murphy* v. *Brentwood District Council*, their Lordships seem to have had in mind that there could be liability in both tort and contract. In *Greater Nottingham Co-operative Society Ltd* v. *Cementation Piling and Foundations Ltd*, the Court of Appeal faced a contention by the employer that the existence of a collateral warranty in the old Grey Form (as with the 1963 Edition of the JCT Form) between them and the nominated sub-contractors gave rise to an argument that there was close proximity for the purposes of seeking to establish a duty of care in tort owed by the sub-contractor to the employer; the sub-contractor, on the other hand, argued that the existence of the collateral contract restricted liability in tort. In this case, it had been conceded by the employer that the collateral contract did not have provisions that were appropriate for the particular circumstances that had arisen and they therefore had pursued their claim in tort. The Court of Appeal adopted the analysis of Lord Scarman in *Tai Hing* set out above, and Purchas LJ said:

'... in considering whether there should be a concurrent but more extensive liability in tort as between the two parties arising out of the execution of the contract, it is relevant to bear in mind:
(a) the parties had an actual opportunity to define their relationship by means of contract and took it; and
(b) that the general contractual structure as between the Society, the main contractor and Cementation as well as the professional advisers produced a channel of claim which was open to the Society.'

9.8   The court went on to find that the economic loss suffered by the employer in this case was not recoverable in tort; one of the reasons for that finding was the fact that there was a collateral contract.

9.9   However, the trend of the 1980s has not been followed in the 1990s. Indeed, the late Judge Newey QC took a less restrictive view in *Hiron* v. *Legal & General Assurance, Pynford South Limited and Others*. He said:

'... while it may be that the courts are moving towards a restrictive view of duties in tort where there is a contract between the parties, it seems clear that at present the mere existence of a contract does not preclude liability in tort... I think that the correct approach is, where there was a contract between plaintiffs and defendants, it should be treated as a very important consideration in deciding whether there was sufficient proximity between them and whether it is to be just and reasonable that the defendant should owe a duty in tort.'

9.10   A similar approach was taken by Judge James Fox-Andrews QC in *Wessex Regional Health Authority* v. *HLM Design Limited*, where he said:

> 'I am satisfied that where there is a contractual relationship between a person and someone professing to special skills for which professional qualifications are necessary and the contract relates to the exercise of those skills and the case falls within the principles of *Hedley Byrne*, as explained in *Caparo* and *Murphy* there may be a concurrent duty to take reasonable care to prevent or avoid economic loss so long as it is fair and reasonable.'

9.11   In both *Conway* v. *Crowe, Kelsey Partner and Another* and *Lancashire and Cheshire Associations of Baptist Churches* v. *Howard and Seddon Partnership*, it was said that it was neither unfair nor unreasonable for there to be a cause of action still extant in tort at a time when the defence of limitation was available to the defendant in respect of the same claim put as a claim under a contract.

9.12   The House of Lords considered the issue of concurrent liability in *Henderson* v. *Merrett Syndicates Limited*. This was the well-known action where Lloyd's Names brought proceedings to try to recover losses they had suffered. There was a mixture of contractual and tortious claims. Lord Goff, having considered the issue of concurrent liability against its historical development, said:

> 'Attempts have been made to explain how doctors and dentists may be concurrently liable in tort whilst other professional men may not be so liable, on the basis that the former cause physical damage whereas the latter cause pure economic loss... But this explanation is not acceptable, if only because some professional men, such as architects, may also be responsible for physical damage. As a matter of principle, it is difficult to see why concurrent remedies in tort and contract, if available against the medical profession, should not also be available against members of other professions, whatever form the relevant damage may take.'

9.13   It follows from these cases that in drafting collateral warranties, consideration needs to be given to the following points:

- The collateral warranty must deal with all the points that the parties feel are important; any omission in the collateral warranty may be subjected to the kind of argument and analysis set out in the cases above as to whether that omission can be made good by a tortious duty to take reasonable care.
- The existence of a collateral warranty will influence how the court approaches the question as to whether or not there is a concurrent duty in tort and the nature and extent of the duty.

- A collateral warranty that contains a provision that expressly preserves rights in tort, if any, may have the effect of assisting the court when looking at whether or not there are co-extensive duties in tort.
- In like manner, a tortious duty could be expressly excluded by a provision in the collateral warranty. This is virtually unheard of in commercial collateral warranties.

Exclusion clauses and limitation of liability clauses

9.14   The Unfair Contract Terms Act 1977 renders totally ineffective a term that seeks to exclude or restrict liability for death or personal injury resulting from negligence (section 2(1)). 'Negligence' has a wide meaning for the purposes of the Act (section 1(1)) and includes:

- any obligation, arising from the express or implied terms of the contract, to take reasonable care or exercise reasonable skill in the performance of the contract;
- any common law duty to take reasonable care or exercise reasonable skill (but not any stricter duty).

9.15   It follows from these provisions in the Act that it is not possible in a collateral warranty to seek to exclude or restrict liability for death or personal injury resulting from, for example, failure by a designer to exercise reasonable skill and care in his design.

9.16   However, the provisions of the Act that deal with seeking to exclude or restrict liability for breach of contract (otherwise than for death or personal injury) only arise in circumstances where one of the parties deals as a consumer or on the other's written standard terms of business (section 3(1)). It is unlikely that one party will be dealing as a consumer in the context of collateral warranties on commercial buildings.

9.17   Difficult questions can sometimes arise as to whether or not a particular document is, for the purpose of the Act, 'the other's written standard terms of business'. Firstly, it might be argued that collateral warranties are not, in any event, written standard *terms of business* for they are ancillary or collateral to those terms which are contained in the principal contract. Secondly, many collateral warranties are the product of negotiation between the parties and such warranties will not have their exclusion clauses looked at under the provisions of the Act. Thirdly, what is the position where there is in a standard form of warranty, agreed by various trade and/or professional organisations, a clause, for example, putting a limit on the amount of money payable as damages for breach or otherwise restricting the scope of recoverable damages? There is some support in the case of *Walker* v. *Boyle* for the view that some standard forms of contract are 'industry' forms derived from negotiation by representative bodies of both parties and that, accordingly, such contracts would not be the other's standard terms of trading for the purposes of the Act.

9.18   If the draftsman is preparing a clause seeking to exclude or restrict liability for loss or damage, other than death or personal injury, by a contract term, and it falls to be considered under the Act, then, for that clause to be effective in law, it has to satisfy the requirement of reasonableness in the Act (section 2(2)). The Act states that test as 'the term shall have been a fair and reasonable one to be included having regard to the circumstances which were, or ought reasonably to have been, known to or in the contemplation of the parties when the contract was made' (section 11(1)).

9.19   Where consideration is given at the drafting stage as to whether or not a restriction of liability to a specified sum of money is reasonable under the provisions of the Act, some help can be derived from section 11(4) of the Act which provides:

> 'Where by reference to a contract term or notice a person seeks to restrict liability to a specified sum of money and the question arises whether the term or notice satisfies the requirement of reasonableness regard shall be had in particular to:
> (a)   the resources which he could expect to be available to him for the purposes of meeting the liability should it arise and
> (b)   how far it was open to him to cover himself by insurance.'

9.20   However, the Unfair Contract Terms Act may be inapplicable to non-standard negotiated collateral warranties. In these circumstances, the giver of the warranty may well wish to put a global financial cap on his liability and that will be the subject of negotiation. If the Unfair Contract Terms Act applies (or may arguably apply), then the considerations in paragraph 9.19 above will be important in the context of looking at a global cap. The high level of potential damages flowing from liability on PFI (Private Finance Initiative) projects will often trigger a discussion about capping of liability on such projects.

9.21   In like manner, the giver of warranties will try to limit liability to the cost of making good defects in design and workmanship whilst at the same time seeking to exclude all other liability. These kinds of matters are often the subject of intense negotiation on large projects.

Contra proferentem

9.22   It goes without saying that in drafting it is important to avoid ambiguities both within clauses and between clauses; the *contra proferentem* rule, which is discussed at 1.69, is a powerful tool for resolving ambiguities.

Limitation of action

9.23   Consideration should always be given by the draftsman of a warranty to the position in relation to limitation. The law in relation to limitation of action is considered in paragraphs 6.96 to 6.114.

Housing Grants, Construction and Regeneration Act 1996

9.24   The question arises whether collateral warranties are 'construction contracts' within the meaning of the Housing Grants, Construction and Regeneration Act 1996. This point has not arisen for decision in the courts. If they are 'construction contracts', then the Act brings into the contract certain terms, among which are a statutory right to adjudication (section 108), provisions as to payments (sections 109 and 110), withholding payments (section 111) and suspension of performance (section 112). This Act applies to England, Wales and Scotland.

9.25   This is not the place for a fully detailed discussion of the law relating to 'construction contracts' or the plethora of court decisions that have been the result of this Act. Suffice it to say that the most common route in the Act, but not the only one, to define a 'construction contract' is 'the carrying out of construction operations' (section 104(1)(a)). 'Construction operations' are defined in section 105 of the Act as including, among other things, 'construction, alteration, repair, maintenance, extension, demolition or dismantling of buildings, or structures forming, or to form, part of the land (whether permanent or not)' (section 105(1)(a)). There are further extensive definitions of 'construction operations' in section 105(1) and exceptions in section 105(2). For more detail, see the discussion in paragraphs 1.72 to 1.76.

9.26   What appears to be clear is that a typical collateral warranty to a tenant or a purchaser that warrants compliance with a building contract by a contractor cannot be 'construction operations' under the Act.

9.27   However, a typical collateral warranty to a fund, which will be likely to contain 'step-in rights' by which the fund is entitled to stand in the shoes of the employer, may be in a different category. Until those step-in rights are exercised, the collateral warranty does not appear to be 'construction operations'; it has the same character as a tenant or purchaser warranty. However, once those step-in rights are exercised, then arguably the warranty becomes a contract for the carrying out of 'construction operations' and, therefore, prima facie, subject to the provisions of the Act for a statutory right to adjudication, provisions as to payments, withholding payments and suspension of performance, in the same way as the principal building contract to which the warranty relates.

9.28   In the event that such a collateral warranty, which is a construction contract, does not comply with the provisions of the Act, then the provisions of a statutory scheme are incorporated into the warranty as implied terms (sections 108 to 111 and 113). This raises the possibility of there being a non-compliant warranty and a compliant building contract (such as JCT) with the consequence being whether or not the provisions of the scheme apply. The alternative approach is that in such circumstances, on a proper legal analysis, the provisions of the warranty merely make it clear that the construction operations are governed by the building contract and that the warranty is therefore part of a construction contract that is compliant with the Act.

Typical terms

9.29   One of the difficulties that all the parties to the construction process face is that there are no universally accepted standard forms of collateral warranty. Many of the problems in the agreement of warranties arise because there are a very large number of bespoke forms of warranty in the market place – many funds, developers, retail stores and many tenants have their own standard forms, or at least requirements, which reflect their own experience, fears, knowledge and lack of knowledge. It is therefore difficult and probably unproductive to look in detail at the precise wording of many of the clauses that appear in practice. In any event, the wording varies enormously from collateral warranty to collateral warranty. It is however possible to put into categories the types of term that are commonly found. A commentary on the BPF Form of Agreement for Collateral Warranty for use where a warranty is to be given to a company providing finance for a proposed development, is at 10.72 to 10.124 and a commentary on MCWa/F is at 10.125 to 10.146.

Design

9.30   An approach sometimes adopted is that a designer gives a warranty that he has and will perform his design agreement with his client in all respects in accordance with that design agreement. Although this has the benefit of being truly collateral to the principal contract, it may have the effect that the designer owes all the duties under his design agreement to the third party in addition to the client. This could result, for example, in a design and build contractor owing all the duties that he owes to the employer under the contract to a tenant including, by way of further example, an obligation as to completing on or before the completion date and an obligation to pay liquidated and ascertained damages. The effect, therefore, of such a wholesale incorporation of the principal contract into a warranty may lead to wholly unintended results. However, funds will want more than just a design warranty.

9.31   The better way of proceeding in relation to design is to repeat in the collateral warranty the clause, or part of the clause, that appears in the principal contract in relation to design. For example, the architect's duty in relation to the standard of his design is to be found at Condition 2.1 of the Standard Form of Agreement for the Appointment of an Architect:

> 'The Architect shall in performing the Services and discharging all the obligations under this Part 2 of these Conditions exercise reasonable skill and care in conformity with the normal standards of the Architect's profession.'

9.32   In like manner, the ACE Conditions of Engagement, Agreement B(1) 1995, 2nd Edition 1998, which is the contract between the engineer and his client on a project where an architect has been appointed, provides:

'The Consulting Engineer shall exercise all reasonable skill, care and diligence in the discharge of the services agreed to be performed by him.'

9.33   The transposition of such obligations from the principal contract into the collateral warranty is not difficult and should provide a basis for drafting and agreement of this clause of the collateral warranty to be given by architects and engineers with the minimum of difficulty and discussion.

9.34   It is not at all uncommon for warranties to be put forward to consultants that seek to increase the allocation of risk to them. A typical example is:

'The Architect warrants that he has exercised and will continue to exercise in the performance of the Services all the reasonable skill and care to be expected of a competent and qualified architect experienced in the provision of services for works of a similar nature, value and complexity to the works to be carried out under the Building Contract.'

Under such a collateral warranty, the test for breach is subjected to statements as to the relevant standard against which the architect will be judged, namely, someone who is experienced in similar work, not merely an ordinarily competent member of a profession.

9.35   Sub-contractors usually fall into one of two categories in relation to a design warranty: this is dependent upon whether or not design is part of the obligations undertaken under the sub-contract.

9.36   If there is no design obligation in the sub-contract, then the sub-contractor should be required to give a collateral warranty to the employer/developer in any event and irrespective as to whether he is to be asked to give warranties to other parties. If the sub-contractor is a nominated sub-contractor under the JCT Standard Form of Building Contract, 1998 Edition, or a named sub-contractor under the JCT Intermediate Form of Building Contract, or a works contractor under the JCT Management Contract, there are standard forms of employer/sub-contractor agreements and these are to be found respectively as NSC/W, ESA/1 and Works Contract/3. In respect of the standard of design to be expected of the sub-contractor/trade contractor, all three agreements are in substantially the same form:

'The sub-contractor warrants that he has exercised, and will exercise, all reasonable skill and care in
(1)   the design of the Sub-Contract Works in so far as the Sub Contract Works have been or will be designed by the Sub contractor, and
(2)   the selection of materials and goods for the Sub-Contract Works in so far as such materials and goods have been or will be selected by the Sub-Contractor, and
(3)   the satisfaction of any performance specification or requirement in so far as such performance specification or requirement is included or referred to in the description of the Sub-Contract Works.'

9.37   Again, the incorporation of such obligations in warranties to parties other than the employer/developer should not be too difficult or contentious.

9.38   Where the sub-contractor is carrying out his design as part of his obligations under the sub-contract, the sub-contract should deal with the question of design. In the absence of an express provision dealing with design, the liability of a designing contractor will be likely to be a fitness for purpose obligation (see paragraph 5.17). The difficulties in relation to fitness for purpose in the context of design are discussed below in paragraph 9.42. However, the Standard Form of Sub-Contract for domestic sub-contractors for use with the JCT Standard Form 'With Contractor's Design' (1998), DOM/2, contains an express provision in relation to design:

> 'To the extent that the Sub-Contractor has designed the Sub-Contract Works (including any further design which the Sub-Contractor is to carry out as a result of a Variation required by the Contractor) the Sub-Contractor shall have in respect of any defect or insufficiency in such design the like liability to the Contractor, whether under statute or otherwise, as would an Architect or, as the case may be, other appropriate professional designer holding himself out as competent to take on work for such design who acting independently under a separate contract with the Contractor had supplied such design for or in connection with works to be carried out and completed by a building contractor not being the supplier of the design.'

9.39   The effect of such a provision (which is a mirror image of the provision in the JCT 'With Contractor's Design', 1998 Main Contract) is that the liability in respect of design is to be the same as that of an architect or other professional designer, namely, reasonable skill and care. The effect therefore of this clause is to prevent the implication of a term that would otherwise be implied to the effect that the design shall be fit for its purpose. Again, there is little difficulty in providing a similarly worded clause in a warranty to be given by a sub-contractor or main contractor to parties other than the original employer.

9.40   Difficulties are sometimes perceived where architects and engineers are engaged by contractors in relation to a design and build project of the contractor. Usually, the main contractor, as under the JCT 'With Contractor's Design', 1998, will have dealt with the question of design expressly in the contract between him and the employer/developer. In relation to design, therefore, the contractor is in a position, if he so agrees, to give warranties in relation to the design to the fund, a purchaser and tenants. The question therefore arises whether or not it is necessary for the architect and engineer of the contractor to give warranties also to the fund, purchaser and tenants. Such warranties will often be needed in relation to the fund and a purchaser (particularly a purchaser prior to completion of the building works) by reason of the desire of a fund and a purchaser to protect their position if, for example, the contractor becomes insolvent. A similar issue arises in relation to design. Usually there will be a chain of contracts, providing a route for a claim: for example, the tenant through the collateral warranty to the contractor and the contractor through his engagement of the architect to the architect. If the contractor becomes insolvent, then this chain of contracts will be broken and the tenant will have no right in tort against the architect. It is for this reason that warranties are usually sought from architects and engineers on design and build contracts.

9.41   A further important point that arises in relation to warranties from architects and engineers on design and build projects is the question of supervision or inspection of the contractor's work. It is in practice rare for architects and engineers to be employed by a contractor in that capacity on a design and build project. However, the Standard Form of Agreement for the Appointment of an Architect and the ACE Conditions of Engagement both provide for the architect and engineer respectively to carry out certain inspection functions on the site. If the architect/engineer used their standard terms of engagement for their appointments on design and build projects and subsequently collateral warranties are entered into with third parties by reference to those standard terms, a warranty as to inspection may be incorporated into the warranty, even though it was never intended by the contractor or the architect/engineer that inspection duties should be part of their function. It is therefore important for architects and engineers on design and build projects to be certain that they are not being fixed by the warranty with inspection duties to third parties in circumstances where they are not in fact carrying out such duties under their appointments.

Fitness for purpose

9.42   It is commonplace to see obligations as to fitness for purpose in respect of design as provisions in draft warranties. Such provisions as to fitness for purpose in design are not usually sensible provisions to be incorporated in collateral warranties to be given by professionals for the following reasons:

- Designers' professional indemnity insurance policies, including those of design and build contractors, are written on the basis of reasonable skill and care (see 8.17). Furthermore, professional indemnity policies usually exclude any liability assumed under a contract which increases the standard of care or measure of liability above that which normally applies under the usual conditions of engagement (see 9.32 to 9.34). It is to the benefit of every party to a collateral warranty that its provisions do not prevent the designer from having recourse to his professional indemnity insurers should a claim arise. To put it another way, there is no commercial benefit in drafting and securing harsh provisions if, when liability is established, there is no money available from insurance to meet the liability.

- The basis of the appointment under the principal contract of designers is that of reasonable skill and care – designers warrant that they will use reasonable skill and care, not that they guarantee to produce a particular result. There is a distinction to be drawn here between a term as to fitness for purpose that could be implied as a matter of fact to give effect to the actual intentions of the parties and a term to be implied in law (to give effect to the presumed intention of the parties). An express obligation as to reasonable skill and care does not exclude the former.

9.43   Sometimes there is an obligation found in collateral warranties to the effect that a contractor or sub-contractor warrants that materials will be fit for their purpose in so far as they have been or will be selected by the contractor/sub-contractor. Such a clause creates no particular difficulties where it is in relation to a contract for the supply of work and materials and the contract includes no design obligations: it is nothing more nor less than the position at common law (see for example, *Young & Marten* v. *McManus Childs*). Where the term is to be included in a contract where there is design in addition to work and materials, then consideration must be given as to whether or not such an obligation seeks to impose a fitness for purpose obligation in relation to design; it is a vexed question in construction law as to where design stops and construction takes over. Where there is any doubt, it is likely to be commercially sensible for all the parties to make such a fitness for purpose obligation in relation to materials subject to a duty to use reasonable skill and care in their selection.

Workmanship

9.44   Main contractors are often asked to warrant to third parties that they will carry out and complete the project in accordance with the building contract. The extent and scope of such warranties requires careful consideration. No contractor should be asked to give warranties to third parties that go beyond his obligations as to quality contained in the building contract. This warranty can be drafted by reference to the wording of the building contract in respect of quality. For example, a contractor giving a warranty to a tenant as to materials and workmanship in circumstances where the main contract is the standard form of building contract, JCT 98, might be asked to do so in the following terms (which are substantially based on clause 2.1 of JCT 98):

> 'The Contractor warrants to the tenant that he has and will use in the construction of the Works (as defined in the Building Contract) materials and workmanship of the quality and standards specified in and/ or required under the Building Contract.'

9.45   The use of such a warranty avoids the dangers of giving a general warranty to a tenant that the contractor will carry out and complete the works in accordance with the building contract; such an obligation may raise a great many questions as to which of the duties in the building contract the contractor is also intended to owe to the tenant.

Excluded materials

9.46   Every warranty contains a provision that certain materials will not be specified for use and/or used in the construction of the project. These used to be called 'deleterious materials' but it is the case that the usually excluded materials are not 'deleterious' as such; the use of the word 'deleterious' has all but disappeared from warranties. Some manufacturers of materials that are habitually included in such lists in warranties are very concerned about these lists causing problems both in the technical sense (i.e. is a useful product being banned for no good or sensible reason?) and in the commercial sense (do these lists damage the businesses of the manufacturers?). Indeed, Rockwool Limited, manufacturers of mineral wool, began a campaign in December 2001 to try to highlight the risks of specifying lists of excluded materials in construction documentation (circular letter from their solicitors, Morgan Cole, dated 3 December 2001). As well as highlighting the need to get that documentation drafted on a proper technical basis, Rockwool Limited have suggested that there could be legal proceedings brought by manufacturers in respect of restraint of trade where such documentation is drafted on an unsuitable basis leading to exclusion of their products from possible selection on projects. Draftsmen should, therefore, take great care in compiling and drafting these lists of excluded materials, including taking technical advice from their clients and their consultants where necessary.

9.47   It has to be very seriously doubted whether the use of such lists of excluded materials in warranties is a sensible and logical way to specify what materials are to be used in the works. The most usual and proper place to specify the quality of the project, and the quality of the materials, is in the specification within the primary contract itself, not in a contract collateral to that primary contract. If certain materials are not to be used on the project, then those materials can easily be expressly referred to in the specification as being materials that should not be used. If that were done, then there would be no need for an excluded materials provision in collateral warranties. However, it is still the accepted regime that such lists should appear in collateral warranties and the lack of logic in the usual approach is not an issue that very often gets raised in the negotiation of warranties.

9.48   There are other reasons for not having a list of particular excluded materials in a warranty. Firstly, such a list, being a list of materials that must not be used, can never, by definition be a complete list and, further, technical data changes over time.

9.49   Secondly, there is the danger that, on particular wordings, anything that is not specified as being excluded can be used.

9.50   Thirdly, and most importantly, there is no reason why the third party
        should not rely on the general and positive obligations created by war-
        ranties as to design and quality of materials and workmanship. Could it
        really be seriously suggested that an engineer would not be in breach of a
        duty to use reasonable skill and care in circumstances where he had
        specified the use of high alumina cement concrete for structural beams in
        the roof of a swimming pool exposed to high humidity? The same point
        can be made in relation to many of the most commonly stated excluded
        materials. There is little point in trying to have a definitive list of specific
        negatives when there is available an over-riding positive general duty to
        exercise reasonable skill and care. That is the best reason for not having a
        list of excluded materials.

9.51   However, notwithstanding what is said above, the presently per-
        ceived conventional wisdom (probably coupled with pressure from
        tenants and commercial conveyancing solicitors) is to have an exclu-
        ded materials provision. It generally remains the position that a war-
        ranty is usually required of the designer that he has not and will not
        specify certain materials. The basic list is that contained in, for exam-
        ple, the Form of Agreement for Collateral Warranty CoWa/F and
        CoWa/P; it is:

        • High alumina cement in structural elements.
        • Wood wool slabs in permanent formwork to concrete.
        • Calcium chloride in admixtures for use in reinforced concrete.
        • Asbestos products.
        • Naturally occurring aggregates for use in reinforced concrete which do
          not comply with BS 882: 1983 and/or naturally occurring aggregates for
          use in concrete which do not comply with BS 8110: 1985.

9.52   The list of excluded materials varies dramatically from project to project.
        Some of these lists are of an extreme length, and some of the items are so
        lacking in definition as to be unhelpful and pointless in a practical
        building sense, let alone from the point of view of a lawyer trying to
        construe their meaning. In any case, where the lawyer dealing with
        collateral warranties does not understand the technical aspects of these
        lists, he should seek advice from someone who does. There are real
        dangers in lawyers incorporating long lists of materials given to them by
        their clients without careful consideration of both the technical and legal
        implications. Some examples of the kinds of things that arise in warranties
        are given below, although this list is not exhaustive. In looking at this list,
        it is necessary to bear in mind the comments set out in paragraphs 9.46 to
        9.50 above:

        • 'Woodcrete and chipcrete'.
        • 'Mundic blocks' – It is probably the case that such blocks are limited to
          certain areas of the south west of England, but the so-called 'mundic
          reaction' between certain types of aggregates containing pyrites and

cement may be a much wider problem in concrete itself than in mundic blocks.

- 'High alkali cement not conforming with certain British Standards when used with aggregates containing reactive silica' (known in the building trade as 'concrete cancer').

- 'Urea formaldehyde' (a foam insulation product), or sometimes 'urea formaldehyde foam or other products which may release formaldehyde fumes'.

- 'Calcium chloride' – Such a chemical appearing on its own is bordering on the absurd; as a matter of strict chemical analysis, calcium chloride will almost certainly be present in a great many building products. What is probably intended, but is not effectively achieved, is to prohibit the use of calcium chloride as an additive in concrete for use in structural elements where steel is present. A formulation now in use is 'Chloride based accelerating admixtures (e.g. calcium chloride) used in concrete, mortar or grout', but even that wording does not mention steel.

- 'Materials which are composed of mineral fibres, either manmade or naturally occurring, which have a diameter of three microns or less and between 5 and 1000 microns in length and which contain any fibres not sealed or otherwise stabilised to ensure that fibre migration is prevented' – Such a definition is unhelpful in a legal document; it is presumably intended to deal with a description of fibres that can be dangerous to health, for example, by inhalation, leading to cancer or other illness but is that description in the drafting the right test for such things? However, it is rather more sensible and more certain in a legal document to specify precisely which materials are regarded as being excluded in this sense; if the definition is simply intended to catch particular products, then it will be better both from a legal and a practical point of view to spell out the precise excluded materials that are not to be used. Such descriptions by size of fibres in microns will not be readily understood by a majority of people in the building industry and professions. At the same time, such a generic description may well catch products that should not be caught by the words. Rockwool Limited take the view that such drafting is unnecessary, wrong and in restraint of trade in relation to their mineral wool products (see paragraph 9.46 above).

- 'White asbestos (chrysolite), brown asbestos (amosite, otherwise known as asbestiform cummingtonite-grunerite) or blue asbestos (crocidolite) or any asbestos or asbestos containing products as defined in the Asbestos Regulations 1969 and/or 1987 and/or any statutory modification or re-enactment thereof' – It is hard to understand how this adds very much to the obligation of the designer to use reasonable skill and care or the general duty to comply with the law under the Asbestos Regulations.

- 'Vermiculite unless it is established as fibre free' or 'Vermiculite products'.

- 'Brick slips where they do not comply with BS 5628, Part III and BS 3291, Part V' – this is a further example of a provision which is wholly unnecessary; any designer who provides for brick slips which do not comply with current British Standards will be in breach of his overriding duty to exercise reasonable skill and care.
- 'Lead' or 'Lead in drinking water supply pipes' – The former is inappropriate; there are very few buildings without lead flashings somewhere; in any event, the chemical element 'lead' will be present in many other building products. For example, does the draftsman really think that a building services engineer or a contractor will specify for use lead water supply pipes in the twenty-first century? Sometimes 'lead-based paint' is also excluded.
- 'Any materials containing lead which may be ingested, inhaled or absorbed, except copper alloy fittings containing lead permitted in drinking water pipe work by any relevant authorities' – This is a variation on the previous item. It suffers from the disadvantage that it is not clear precisely what kind of materials and fittings are intended to be banned
- 'Lightweight thermal-type or air entrained low density concrete blocks' – It may be that such materials are positively specified in the building contract itself.
- 'Calcium silicate bricks' – This exclusion is not necessary because it is well understood in construction where it is appropriate to use such bricks.
- 'Deeply recessed joints when used in conjunction with perforated bricks.'
- 'When manufactured from galvanised material, wall ties, cramps, straps, restraint and support angles and lintols in cavity walls.'
- 'Polyurethane foam or polyisocyanurate board' – These are sometimes banned, together with urea formaldehyde foam (see above). All these materials are quite often listed as excluded materials in warranties, but included on the same project expressly by the specification of the contract works; an example is composite metal cladding for roofs, walls and insulated doors, which will contain foam of one or another kind.
- 'Pitch polymer damp-proof courses in horizontal locations.'
- 'Materials referred to as being hazardous to health and safety in *Hazardous Building Materials: A Guide to the Selection of Alternatives* edited by S.R. Curwell and C.G. March in the edition current at the date of this warranty' – This is an attempt to incorporate wholesale into a warranty part of the contents of a book that runs to well over 100 pages. It is very much to be doubted whether the joint editors and the authors of *Hazardous Building Materials* intended their book to be used in this way. The book is a useful and helpful contribution to wider understanding of the hazards of certain building materials, the alternatives that are available, and the hazards, if any, of the possible alternatives. However, as a matter of law, it is almost impossible to give any useful legal construction to a clause in a contract that seeks to incorporate parts of a

book when the parts that are incorporated are not readily identifiable, there being no precisely identified list in the book of 'materials referred to as being hazardous to health'; indeed, in one part of the book, general guidance is given as to potential hazards of materials when in position, when disturbed and in the environment in waste disposal, each ranked separately as to none reasonably foreseeable, slight/not yet quantified by research, moderate, and unacceptable. Does the draft clause intend to treat as excluded all materials save those where no hazard is reasonably foreseeable when judged against the material in position or when it is disturbed, or when it is disposed of? In short, this clause is an attempt to do something useful, but in practice is lacking in legal definition and clarity.

9.53   In the JCT Collateral Warranties (2001 Edition), another, but very similar, formulation has been adopted that had also been seen prior to their publication in bespoke warranties. It is not to use materials '... other than in accordance with the guidelines contained in the edition of the publication *Good Practice in Selection of Construction Materials* (Ove Arup & Partners) current at the date of the Building Contract...' (clause 2 of MCWa/F, but MCWa/P&T, SCWa/f and SCWa/P&T are similar). The edition current at the date of writing this book is 1997, so it is already four years old. The purpose of the publication was to encourage good practice in the selection of materials, not to assist in defining what should not be used for the purposes of a collateral warranty. This draft, therefore, can be subjected to the same criticisms as above in relation to the use in warranties of the publication by S.R. Curwell and C.G. March.

9.54   The environmental factor is now appearing in collateral warranties, with certain types of material, some of which are often used in the construction industry, being excluded:

- 'Lindane, tributyltin oxide and pentachlorophenol' – These are materials in certain types of timber preservation, where it may be that (leaving aside the manufacturing and distribution risks) the greatest risk occurs during and shortly after application of the preservative to the timber. Another formulation is ''Wood preservatives which have not been used in compliance with all appropriate current British Standards, in particular BS 52628, BS EN 335-1, 335-2 and 335-3.'
- 'Any of the products containing cadmium that are referred to in the Environmental Protection (Controls on Injurious Substances) (No 2) Regulations 1993.'
- 'Materials which produce or are likely to produce CFCs, save in accordance with the standard set out in the Montreal Protocol 1987' or sometimes just 'chlorofluorocarbons'.

Sweep up provision

9.55   There is often a sweep up provision at the end of the excluded materials clause; again from a drafting point of view, it is almost certainly more satisfactory to rely on the positive duty to exercise reasonable skill and care (see paragraph 9.50) than it is to have a sweep up clause as a negative obligation tucked on to the end of the excluded materials provisions. However, a not untypical sweep up clause found in collateral warranties is:

> '... any other substance or method of use or incorporation which is or may reasonably be suspected to be unstable, inadequate, dangerous, combustible or otherwise unsuitable for building purposes or for the type of building or conditions which it was used or is the subject of statutory control or does not conform to British Standards.'

9.56   Such a clause is so widely drawn and inconsistent with the standard forms of building contract and sub-contract that it should not properly be incorporated into any warranty. In particular, such clauses are often in conflict with the provisions in contracts and sub-contracts that require contractors and sub-contractors to comply with instructions they are given under those contracts. Further, not every building material in use is the subject of a British Standard or Code of Practice. It does not follow that such materials are deleterious or should be excluded from use.

Inconsistency

9.57   Conflicts between excluded materials clauses in collateral warranties and the provisions in the contracts to which the warranties are collateral are a potentially serious source of problem. What is to happen if a material banned in the collateral warranty is instructed to be used by the architect in circumstances where the contractor has an obligation under the main contract to comply with architect's instructions? For this reason, the excluded materials should also appear in the specification for the project (in the main contract and the sub-contracts) and, for an abundance of caution, there should ideally be a provision that the architect is not permitted to instruct the use of those excluded materials.

9.58   Sometimes, there is an additional provision requiring the giver of the warranty to notify the employer if materials which were not deleterious at the time of incorporation have become generally known to be deleterious prior to practical completion. Such an apparently desirable clause can also lead to difficulties. If such a notice is given by reason of the collateral warranty, what is to happen to the project that is actually being built? Is work to stop? Who is to bear the cost of replacing the materials, and the costs of the inevitable delay? All these points need to be borne in mind when looking both at warranties and at the contracts to which they are collateral. They also provide an added reason for keeping the excluded materials clause in a warranty in short and precise form, avoiding a lack of particularity in the drafting and avoiding the incorporation of general obligations, which in turn may lead to difficulties in relation to the principal contract.

9.59 A final point is that whilst a designer, who is not also building, can warrant that he will not specify particular materials, he cannot properly warrant that he will see that the contractor does not incorporate such materials into the construction (even though the designer has not specified them).

9.60 On the other hand, of course, it may be possible for a design and build contractor to give such a warranty, but not where he does not have complete control over the specification. The contractor on the JCT 98 'With Contractor's Design' Form should watch this point in relation to the contents of the 'Employer's Requirements' and the right of the Employer under that form of contract to require a 'Change'.

Copyright

9.61 There will usually be provisions dealing with copyright in drawings, specifications, calculations and the like. Recently, it has become usual for these provisions to deal also with electronic information (such as CAD design and other stored electronic 'documents') in addition to drawings, specifications and calculations. Occasionally, a fund or tenant will seek to have the copyright himself. Such a provision is in direct conflict with the provisions of most of the standard form conditions of engagement. The simplest way to avoid this problem is for the copyright to remain with the designer but for a licence to be given in the warranty agreement in respect of the copyright but limited to the purposes of constructing the development and any alteration, rebuilding and/or extension to it; it would not be sensible for a designer to give a licence in his intellectual property in the absence of such a limitation. For example, the Standard Form of Agreement for Appointment of an Architect (SFA/99, updated April 2000) reserves ownership in the copyright to the Architect '...in the work produced by him in performing the Services and generally asserts the right to be identified as the author of the artistic work/work of architecture comprising the Project'; unless the wording of the warranty contains the appropriate limitation on the licence for use on the named project, the licensee may well be able to construct another building using the same design without payment of any further fee and/or licence fee.

9.62 The copyright/licence provision needs careful consideration in any event both as to its scope and its effect; this is another area of the drafting of collateral warranties where careful regard should be had to the copyright provisions of the contract to which the warranty is collateral. Sometimes, there is not only a provision in the warranty granting a licence in respect of copyright but also an indemnity provision indemnifying the fund/

tenant in respect of any proceedings, damages, costs and expenses which he may incur by reason of the giver of the warranty infringing or being held to have infringed any patent rights. The giver of such an indemnity must check that he is not giving an indemnity in respect of something over which he has no control. For example, a wide-ranging indemnity is given by the contractor to the employer under the JCT 98 'With Contractor's Design' contract at clause 9.1. The indemnity is given in respect of various matters including infringement of patent rights and infringement of copyrights. However, there is a savings provision in the same contract at clause 9.2 so that where the contractor has supplied and used patented articles or inventions and the like in compliance with the employer's instructions, royalties, damages and the like which the contractor has to pay to the owner of the patent are paid to the contractor by the employer. It is important for contractors to see that there is no undermining of a provision such as that in any collateral warranties that they give.

9.63   The Copyright, Designs and Patents Act 1988 creates moral rights in relation to copyright – such rights are to be distinguished from the copyright author's economic rights. One of the moral rights created by the Act is to give to the designer of the building a right to be identified as the designer 'by appropriate means visible to persons entering or approaching the building... and the identification must... be clear and reasonably prominent' (section 77(7)(b) of the Act). A person does not infringe that right unless the right has been asserted by the author of the design (section 78 of the Act). It is therefore the case that if the building owner/fund/purchaser/tenant do not wish the designer of the building to have the right to be identified by a prominent notice on the building, then they will need a provision by which the designer waives his moral right under section 77(7)(b) but designers would be well advised not to consent to any waiver of moral rights which goes beyond the right to be identified on the building.

9.64   Sometimes the granting of the licence is made conditional on the designer having been paid his fees in full; there may be further provisions that the licensee is entitled to call for the originals and/or copies of any and all of the relevant documents. The scope of such an obligation needs to be carefully considered by the designer, and designers will wish to have a provision that they are paid the cost of the provision of copy documents.

9.65   The construction professional should be alert to the need to have a provision in the collateral warranty enabling him to charge and be paid for searching for and copying documents in relation to the licence that he is giving.

Insurance

9.66   In warranties given by a party who is carrying out design, there is often a requirement as to professional indemnity insurance; these provisions vary enormously in their content and effect. A typical term might require a warranty that professional indemnity insurance is in force at the date of the warranty, that the premiums have and will be paid, and that the insurance will be maintained into the future, sometimes without limit in time and sometimes with a limit in years (usually six years for a contract under hand and twelve years for a deed).

9.67 These kinds of provisions give rise to very real practical and legal problems. The first is that which arises from the nature of the professional indemnity insurance market. That market changes from time to time both in the average levels of premium required and in terms of the capacity and, consequently, the amount of the insurance (the limit of indemnity). It should also be remembered that professional indemnity insurance is annually renewable and is made on the basis that it covers claims made during the period of insurance. Given the inevitable vagaries of the professional indemnity insurance market place and the annual basis of this type of insurance, it is very difficult for a designer to give warranties in relation to future insurance, even one or two years ahead, let alone six or twelve years ahead. It follows that in looking at these types of clauses, the designer should have in mind the difficulty of forecasting the professional indemnity market as to availability of insurance, the size of the market and its cost; what is the designer to do if, six years after entering into this kind of obligation, the cost of the insurance becomes prohibitive to him in the context of his business at that time?

9.68 Secondly, this type of provision assumes that the designer's firm or company will continue in business into the future in the same legal form; where it is a firm, the partners who entered into the warranty (and who are liable for breaches) may cease to be partners and/or may retire or die; where the designer is a company, it may cease to trade without going into liquidation or receivership, or it may go into liquidation or receivership. These points should be borne in mind by parties seeking from designers onerous insurance provisions in warranties.

9.69 Thirdly, where there is a prohibition, as there is in some policies, preventing disclosure of the existence of insurance to a third party, insurers' express permission should be obtained to entering into such a warranty. Depending on the terms of the insurance, the disclosure of its existence can have the effect of rendering the policy void. Further difficulties arise in relation to these provisions on renewal and on changing insurers (see paragraph 8.42).

9.70 Fourthly, what is to happen if the designer is in breach of his obligation to insure in the warranty? Very few clauses contain any provision to deal with that aspect. Clearly the designer would be in breach of contract under the warranty and that would give rise to a claim on the part of the other party for damages. Those damages are the sum of money that would put the receiver of the warranty in the position he would have been in but for the breach. That sum might be the amount of the premium for the professional indemnity insurance. It is, however, extremely unlikely that a third party would be able to take out professional indemnity insurance on behalf of the designer in any event, although it might be possible to obtain some latent defects insurance and it might be possible to argue that the costs of that insurance are the damages that are recoverable for breach of the warranty as to insurance. Short of that, it follows that the consequences of breach of a provision of this kind are not generally going to be helpful to the party in receipt of the benefit of the warranty as to insurance.

9.71   To try to deal with all these competing factors, an unobjectionable provision might be for the designer to warrant that there is professional indemnity insurance in existence at the date of the warranty and that the premium has been paid; the designer could further warrant that he will use his best endeavours to obtain professional indemnity insurance in succeeding years provided that such insurance is available in the market place at reasonable rates; these provisions could be extended by the designer undertaking to notify the other party or parties in the event that he cannot obtain insurance at reasonable rates so that a decision can be made as to how best to protect the positions of the parties. This kind of provision is to be found in the Forms of Agreement for Collateral Warranty published by the British Property Federation and the Joint Contracts Tribunal (CoWa/F, CoWa/P&T, MCWa/F, MCWa/P&T, SCWa/F and SCWa/P&T).

9.72   Sometimes warranties require the warrantor to confirm that the warranty has been shown to the warrantor's professional indemnity insurers who have confirmed in writing that the warrantor will be held covered under that warranty. It is highly unlikely that insurers would so confirm in that form. Those who require these kinds of provisions should bear in mind that professional indemnity insurance is an annual policy and covers claims made during the period of insurance. There must, therefore, be some doubt about the value of obtaining these kinds of confirmations when the warranties are likely to be in place for 12 years. However, the following formulation might possibly be acceptable to some insurers:

> 'This agreement has been shown to the Consultant's insurers and they have noted it, subject otherwise to the terms, conditions and exceptions of their current professional indemnity policy.'

Novation and step-in rights

9.73   Funds and purchasers will wish to try to secure their investment. In particular, they may wish to have the right to continue with the project if the employer/developer is unable to continue, perhaps through receivership or liquidation or a serious breach of the funding agreement or the agreement to purchase. In order to try to give effect to this concern, funds and purchasers often require a provision in the warranty that they have the right to take over the project. What this means in practice is that in each warranty with each member of the professional team and the main contractor, those parties will agree that, if requested by the fund or purchaser, they will enter into a direct agreement with the fund in respect of the completion of the design/construction for the project. In principle there is nothing objectionable to such an arrangement, but there are major points that need to be carefully considered in relation to each of the different types of provisions that appear in practice.

9.74 The first point is the approach to the drafting; on a proper view, this arrangement is a novation agreement whereby, for example, in the building contract, the contractor continues as before and the fund or purchaser as the case may be are substituted for the employer/developer on the basis that the fund/purchaser take on all the rights and obligations of the employer. In other words, there is a substitution of parties; this would mean, for example, that where the contractor had existing claims against the employer (say in respect of unpaid interim certificates), the fund/purchaser would be liable. It is not unusual for funds and purchasers to try to avoid a commitment as to past liabilities by providing that they shall only be liable on the agreement from the date of the novation. Clearly such a provision is not one that designers and contractors would wish to agree. This is, of course, a question of commercial balance but it does seem that if the fund/purchaser wish to have the right to take over the agreements, then they should also agree to be liable in respect of past matters.

9.75 What is to happen to the existing agreement between the designers and the developer, and, the contractor and the developer, when the fund or purchaser exercise their right to step into the shoes of the developer? The terms of the principal contract should be carefully considered in this context: for example, if the employer/developer falls out because of liquidation, there are provisions in the building contract to deal with that matter. The consequence of entering into a fresh agreement with a third party is that such a step is likely to be a repudiatory breach of the principal contract. There must, therefore, be a provision to the effect that the designer/contractor will not be in breach of his principal contract if the fund/purchaser exercises its right to step into the shoes of the developer/employer.

9.76 It follows from the need for a novation agreement and the need to avoid repudiatory breaches of the principal contract that these kind of arrangements need to be tripartite, that is to say, in the case of a contractor, the contractor, the developer/employer and the fund/purchaser should all be parties to the agreement. This facilitates the drafting of a novation provision whereby one party is substituted for another, and also enables a provision to be incorporated so that if the right to novation is exercised, the designer/employer can agree that that will not be a breach of the principal contract.

9.77 The most satisfactory way to resolve these problems is for there to be an obligation on the designer/contractor and on the employer/developer in a tripartite warranty agreement to enter into a novation agreement if the fund so requests, by which novation the fund will be substituted for the developer as if he had always been a party to the principal contract. In those circumstances, there will often be a release by the designer/contractor of the obligations of the developer (the fund having assumed all those obligations), but there will not usually be a release of the obligations owed by the developer to the fund. As a matter of style, the novation agreement can either be contained within the body of the warranty or by appending a draft novation agreement to the warranty and by a clause in the warranty requiring the other parties to enter into that agreement within a period of time after notice in writing from the fund/purchaser. Consideration should be given in drafting the funding agreement and in the contract for purchase to deal with these issues.

9.78   Problems are often caused by the draftsmen of these kind of provisions who seek to amend in the warranty the terms of engagement of architects and engineers and the building contract with the contractor; their intention is to provide that the architect, the engineer and the contractor will not exercise the rights under their principal contracts to terminate or determine their engagement without giving prior notice to the fund and/or developer. Sometimes a draftsman will seek to achieve this by simply deleting the termination and determination provisions in the principal contract. That is a very dangerous course of action, which could lead, for example, to a contractor having a purported obligation to continue with the project for the liquidator of a developer. The simplest way to deal with these problems is to provide that the architect, engineer and contractor give to the fund/purchaser a copy of any notice of termination/determination that they give to the developer/employer under or in respect of a breach of the principal contract. Convoluted provisions to deal with this issue are only likely to lead to lack of clarity and uncertainty.

9.79   It is increasingly the case in large commercial developments that there is more than one fund. Sometimes two or three funds are sharing the financing. They cannot sensibly all have rights to step in given in the same terms; for example, if there were a developer insolvency, there might then be arguments as to which fund had stepped in first. Usually, this can be resolved by having an agreement between the funds as to how this is to be managed, such as a priority between the funds being pre-agreed in the event that step in becomes possible.

9.80   Finally, novation will usually be a wholly inappropriate provision to be included in a tenant warranty.

Assignment

9.81   This subject is so complex that it merits virtually the whole of the contents of Chapter 4 and it is necessary to have in mind, when drafting, all the issues considered in that chapter. The following is a summary of issues, which draftsmen may find helpful.

No provision for assignment

9.82 If there is no provision for assignment, then the benefit of a warranty is freely assignable without the consent of the other party. Such an assignment can be legal or equitable (see paragraphs 4.5 to 4.21).

Prohibition on assignment

9.83 Where there is in a contract an express prohibition on assignment, then such prohibition is likely to be effective in law. It follows that if there were a purported assignment in such cases, it is likely that that assignment would be invalid (see paragraphs 4.38 to 4.49).

Restriction on assignment

9.84 Problems can arise where a contract provision seeks to impose a limit on the number of assignments that can be made; this can occur where the draftsman is attempting to make the provision on assignment in the warranty consistent with the provisions in the professional indemnity insurance policy as to the cover under that policy where there are assignments of collateral warranties; for example, the insurance policy wording might restrict cover to two assignments. If that provision is simply incorporated into the warranty, then every time there is an assignment the assignee steps into the shoes of the assignor and he has the right to make two assignments – in other words, the counting may never start. The simplest practical legal solution to this difficulty is to provide in the warranty that assignment is prohibited save where the express consent in writing of the original giver of the warranty has been obtained. The provision could continue to recite that the giver of the warranty shall not withhold his consent where the assignment is a first or second assignment. Such a provision puts the control in the hands of the original contracting party who needs to have that control – the insured in this example. The approach used in the JCT warranty, MCWa/F, is that there can be two assignments by way of a legal assignment, provided written notice is given to the contractor. The first is to 'P1' and the second from 'P1' to 'P2' but no further assignments are permitted.

9.85 Sometimes, a provision is inserted into a warranty that the agreement is personal to the parties. If a contract is a personal contract then the benefits of that contract cannot be assigned either by a legal assignment or by an equitable assignment. The issue here is whether a contract becomes a personal contract simply because it is agreed between the parties that it is. There has to be doubt as to whether such a provision in warranties is effective (see paragraph 4.36). It may be preferable, therefore, to prohibit assignment by an express term rather than seeking to do it by this indirect method.

9.86 Where there is a sub-lease, the first tenant remains liable on his covenants to the landlord notwithstanding the covenants of the sub-tenant. It may be the case that the original tenant has assigned the benefit of collateral warranties by absolute legal assignment to the sub-tenant. If the sub-tenant then defaults on his repairing obligations (say through bankruptcy), the original tenant will remain liable on his repairing covenant to the landlord but he will not have the benefit of the collateral warranties. It is for this reason that it is sometimes suggested by funds and developers that the givers of warranties should agree to enter into fresh warranties every time there is a new tenant of the same premises. The effect of that request is to ask the givers of the warranties to insure part of the commercial risk inherent in the operations of landlord and tenant. This new warranty approach is likely to be strongly resisted by all parties who are asked to give warranties; consideration might be given to a provision in the assignment requiring re-assignment of the benefit to the first tenant if certain circumstances were to arise. However, such a provision must be capable of being carried through in accordance with any restrictions on assignment that there may be in the warranty.

'Other parties' clause – net contribution provisions

9.87   The Civil Liability (Contribution) Act 1978 deals with contribution between people liable in respect of any damage whether tort, breach of contract or otherwise (section 6(1)). Under the provisions of this Act where two or more people are in breach of separate contracts with the same third person, producing the same damage, those two or more persons can claim contribution against each other. This is discussed in detail in paragraphs 6.85 to 6.95.

9.88   Following the restriction of available remedies for construction defects and design problems in the law of tort, contracts, in the form of collateral warranties, are now being used to fill the gap. There is serious concern in the construction professions and among their insurers, that if they give warranties and other professional people do not, they will be liable to the third party for the full amount of the loss and unable to claim contribution from those other parties who did not give collateral warranties – those other parties can never be liable in respect of the same damage, there being no relevant remedy available in tort. The givers of warranties have therefore sought methods to try to make sure that they are not liable for the full amount of the loss in circumstances where other parties ought properly to be liable at the same time. The creation of enforceable provisions to this effect is difficult.

9.89   The most effective provision is likely to be that a warranty does not come into force and effect at all, a condition precedent, unless certain named other parties have entered into collateral warranties with the same third party to the same effect. This is sometimes called the 'Three Musketeers' clause – all for one, one for all. Unsurprisingly, developers, funds, purchasers and tenants do not accept these kinds of provisions.

9.90   Another method is for the liability under the warranty to be limited to such losses as is just and equitable having regard to the extent of the warrantor's responsibility and on the further basis that certain named parties are deemed to have given a warranty in similar form. This is the approach in the BPF warranties (CoWa/F and CoWa/P&T). A similar approach is adopted in the 2001 Editions of the JCT Collateral Warranties (MCWa/F, MCWa/P&T, SCWa/F and SCWa/P&T). This is the so-called 'net-contribution' provision. It is an attempt through contract terms to achieve the position that would apply under the Civil Liability (Contribution) Act in relation to multi-party proceedings in the court.

9.91   It seems that it is very difficult to draft such a clause to ensure legal certainty because it inevitably involves consideration as to whether or not other parties would have been liable and, if they would have been so liable, what apportionment would have been made by a judge if they had all been before the court. These issues will have to be considered against the background that they involve the proper legal construction of a clause in a contract, and, a claim made *under* that clause, not as damages for *breach* of contract (where different considerations will apply). The effectiveness of such provisions has yet to be tested in the courts, although they are now commonly put forward, but not so commonly included in executed warranties because they are generally not acceptable to developers or funds. One of the difficulties is that an attempt to try to give a court power to make that apportionment, in circumstances where the other parties were not before the court, would present some difficult issues; it is conceivable that a court would have some difficulty in giving effect to such a provision. However, the approach of the courts to such matters may be dominated by the current purposive construction used by judges to ascertain the legal meaning and effect of contracts, and that such a construction will assist in establishing the effectiveness of such provisions. No doubt, such provisions will be tested in the courts in due course in an appropriate case.

Limiting of liability

9.92   The parties to a contract can, in English law, agree a term excluding liability or limiting the consequences of liability, subject only to some statutory regulation. The most important statutory regulation is the Unfair Contract Terms Act 1977. It is the case that the courts when construing exclusion clauses are more likely to find favour with a clause that purports to limit the consequences of liability, rather than purporting to exclude it altogether (see for example *George Mitchell (Chesterhall) Limited v. Finney Lock Seeds Ltd* and *Ailsa Craig Fishing Co Ltd v. Malvern Fishing Co Ltd and Another*). In drafting a clause seeking to impose a money limit on a claim for damages, section 11(4) of the Unfair Contract Terms Act is helpful (see paragraph 9.19). This provides that in looking at whether a particular term is reasonable under the Act, it is open to the court to have regard to two things. Firstly, the resources which that party could expect to be available to him for the purposes of meeting the liability should it arise; secondly, how far it was open to him to cover himself by insurance. This section gives some support to the view that it might be reasonable to limit liability to the limit of indemnity in the professional indemnity policy (provided, of course, that that limit of indemnity was a reasonable limit having regard to the type of work being undertaken).

9.93   The whole question of consequential loss is a matter that gives rise to concern among the givers of warranties. It is one thing, they say, to be liable for the direct costs of remedial works following defective design or workmanship, it is another to have to pay all the other economic consequences that may flow from such a breach of contract. Those other consequences can include loss of rent, the tenant moving out of the building while repairs are carried out, the costs of disruption/loss of profit in their business and the costs of returning after the remedial works. Sometimes, possibly often, those indirect costs can exceed the cost of the remedial works themselves.

9.94   In contract, the damages that can be recovered are those that would put the innocent party in the position he would have been in, in so far as money can do it, but for the breach of contract. The kinds of damages that can be recovered are governed by the question of remoteness and this is discussed in detail in paragraphs 6.12 to 6.20. In principle, the kind of indirect losses discussed in the previous paragraph may well be recoverable as damages for breach of a warranty obligation. Various attempts have been made, therefore, to try to restrict that potential liability. One method is to exclude economic and consequential loss. However, those words have led to difficulty of interpretation in the court. For example, in *British Sugar plc* v. *NEI Power Projects Limited and Another*, the court decided that the effect of the contract was a contractual provision:

> 'The Seller will be liable for any loss, damage, cost or expense incurred by the Purchaser arising from the supply by the Seller of any such faulty goods or materials or any goods or materials not being suitable for the purposes for which they are required *save that the Seller's liability for consequential loss is limited to the value of the contract.*'

9.95   The Court of Appeal unanimously decided in relation to that particular contract that the parties simply agreed to limit the defendants' liability for loss and damage not directly and naturally resulting from the defendants' breach of contract to an amount equal to the value of the contract.

9.96   Notwithstanding that decision, there is a tendency for the draftsmen of warranties to provide expressly for what is covered, rather than seeking to limit the scope of recoverable damages. For example, an architect might seek to limit his liability under the warranty to the direct cost of remedying defective work, all other costs, losses, damages and expenses being excluded. Developers, funds, purchasers and tenants are not keen on these limitation provisions.

9.97   Again, in looking at limitation of liability, the principal contract should not be ignored in drafting the warranty. For example, the Standard Form of Building Contract, JCT 'With Contractor's Design', 1998, contains a provision limiting the liability of the contractor in respect of design (except where the contract is for the provision of dwellings, in which case the Defective Premises Act 1972 prohibits the excluding of liability). Clause 2.5.3 of that contract limits the liability for 'loss of use, loss of profit or other consequential loss arising in respect of the liability of the Contractor' for an insufficiency in design to a sum which has to be stated in the appendix to the contract. In the event that such a limitation is agreed under the main contract, then that limitation should properly follow through into the collateral warranties as well.

9.98   It is common for professional indemnity policies that provide cover for collateral warranties by way of a policy endorsement to exclude cover where 'the period of such liability exceeds the period of the Assured's liability under the Contract to which the Warranty or Agreement is supplemental' (see paragraph 8.37). It may well be, therefore, that such a provision should be included by the insured in any warranty that he gives. Sometimes this problem is met by a clause that provides that the damages recoverable under the warranty shall not be any greater than if the receiver of the benefit of the warranty had been, jointly, a party to the principal contract. The wording needs to be drafted with great precision – use of the words 'greater liability' will usually not create sufficient clarity.

9.99   During the construction of a project, there can be agreements in relation to, for example, what remedial action should be taken in respect of defective work. Indeed, clause 8.4 of JCT 98 provides expressly for defective work to be the subject of remedial work and then left in place, rather than simply removal of the defective work, which used to be the only remedy available to the architect under earlier editions of the JCT contract. That example leads to consideration in warranties of a provision that any agreement under the principal contract will also bind the receiver of the benefit of the warranty. This problem can be met to a certain extent by making sure that the main operative clauses of the warranty are truly collateral, but every giver of warranties should consider whether a wider clause should be expressly incorporated to cater for this issue.

Limitation of action and indemnity provisions

9.100   The law in relation to limitation of action is considered in paragraphs 6.96 to 6.114. In so far as drafting is concerned the following points are important.

9.101   It is unlikely that simply entering into a collateral warranty constitutes an acknowledgement of liability for the purposes of limitation (see paragraphs 6.112 to 6.114); however, it is axiomatic that there cannot be a breach of a collateral warranty until a date after the date of the warranty. If the giver of the benefit of the warranty wishes to impose through the warranty some agreement as to limitation of action then it is open to him to do so for the reason that limitation periods can be altered by agreement. Sometimes the giver of the benefit of the warranty will seek to have a clause incorporated that says that he will have 'no longer lasting liability' than that arising under the principal contract. It is thought that those words may not have the desired effect for the simple reason that they may not be sufficiently clear in relation to limitation. A party desiring to effect such a limitation, therefore, should expressly set out in the clause the limitation period by reference to the principal contract; in other words, whilst the warranty may have been entered into at a later date, and no cause of action can arise before that date, the date by which the breach is to be taken for the purposes of the limitation period (6 or 12 years depending whether under hand or under seal) will be the same date that would arise on a breach of the principal contract.

9.102   The givers of indemnities under collateral warranties should be aware of the consequence that has on the limitation position. This is that the cause of action does not accrue on an indemnity until the liability of the person seeking to be indemnified has been established: *County & District Properties* v. *Jenner*; *Green & Silley Weir* v. *British Railways Board*. This principle of law can have the effect of substantially extending the period of time after which proceedings can be successfully brought.

Delay

9.103   On a building project, the building owner will usually be able to recover liquidated and ascertained damages from the contractor in the event that he is late in completion, subject to any provisions for, and the granting of, any extensions of time for completion. The whole issue of delay vis-à-vis the contractor is therefore dealt with in this way. On a design and build project, any delay in the design of the contractor (or the people to whom he sub-lets the design) will also be the subject of liquidated damages and extension of time machinery. However, where the main contractor is not designing, such as where JCT 98 (without contractor's design) is used, then it is the case that the architect, the engineer and any designing nominated sub-contractors can cause delay to the main contract. In each of those cases, the main contractor will usually be entitled to an extension of time in respect of the delay caused due to the issue to him of late information.

9.104   Although nomination is very rarely utilised nowadays, in these circumstances, the standard form of nominated sub-contractor/employer warranty, NSC/W, has certain provisions. The first is that the sub-contractor agrees to supply the architect with information in due time so that the architect will not be delayed in issuing that information to the main contractor such that the main contractor may have a valid claim to an extension of time for completion or a valid claim for direct loss and/or expense. The effect of such a warranty is that, whilst the employer will not be able to obtain liquidated damages from the main contractor by reason of the extension of time that has to be granted by the architect, the nominated sub-contractor renders himself liable to that claim from the employer, together with any claim for direct loss and/or expense that the employer has to pay to the main contractor as a result of the sub-contractor's late information.

9.105 Further, under NSC/W, the sub-contractor warrants that he will so perform his obligations under the sub-contract that the architect will not be under a duty to issue an instruction to the main contractor to determine the employment of the sub-contractor. Again, breach of this warranty to the employer may render the sub-contractor liable to the damages suffered by the employer under the main contract in circumstances where the sub-contract has been determined. Thirdly, the sub-contractor warrants under NSC/W that the contractor will not become entitled to an extension of time for completion of the main contract works by reason of delay on the part of the nominated sub-contractor in performing his work. Again, breach of this obligation will entitle the employer to recover main contract liquidated damages from the sub-contractor.

9.106 Consideration should be given as to whether clauses of this type should be incorporated in collateral warranties and/or in the principal contracts (for example, the architect and engineer).

Disputes resolution procedure

9.107 Most collateral warranties are silent on the procedure to be adopted in the event of a dispute. On the other hand, most of the contracts to which they are collateral contain arbitration agreements or agreements that the courts have jurisdiction.

9.108 In relation to a tenant's claim under a warranty at a time after completion, this will give rise to no difficulties and the dispute can simply be heard in the court; however, in relation to a dispute that arises under a warranty at a time prior to the final certificate under the building contract there are many potential difficulties. For example, procedural problems will be created where there is a dispute under the main contract between the contractor and the employer as to, say, an employer's set-off of liquidated damages in circumstances where the contractor believes he is entitled to an extension of time. At the same time, the funder exercises his right to require novation, believing the employer to be in material breach of the funding agreement. The building dispute will be the subject of an arbitration clause and the dispute about the right to novate will be likely to be subject to the jurisdiction of the courts; in both cases, the resolution of the dispute turns upon the same facts and the same evidence and there is a clear risk of inconsistent findings between the court and the arbitration. There are three ways to deal with this matter:

- The first is to make no provision for disputes resolution procedure in the collateral warranty and for all the parties to keep their fingers crossed.

- The second is to have an arbitration clause in the warranty together with multi-party machinery both in the warranty and in all the relevant principal contracts but limited to disputes on issues which are substantially the same as or connected with issues raised in the related dispute under the other contract. Such a multi-party arbitration clause should give the arbitrator power to make directions and all necessary awards in the same manner as if the procedure of the High Court as to joining one or more defendants or joining co-defendants or third parties was available to the parties and to him. Clearly, such multi-partite machinery will not usually be relevant in warranties to tenants but may well be relevant to warranties given to the fund and the purchaser so as to avoid the risk of multiplicity of proceedings.
- The third method is to have provisions in all the principal contracts and in the collateral warranties that English law applies and the English courts shall have jurisdiction. The court then has power to order joinder of parties to such separate disputes. This will not, of course, overcome the risk of multiplicity of proceedings where there are arbitration clauses in some of the principal contracts. In this connection, the position of a developer/landowner who is not the employer under the building contract is important: see paragraph 7.10.

9.109   Sometimes, developers and funds will seek to overcome the risk of multiplicity of proceedings by arranging for the arbitration clause in the building contract (and in the other principal contracts) to be deleted and, in their place, an English law and English courts clause is inserted. This used to be a potentially dangerous route. Under JCT 98, for example, an arbitrator has the power to open up, review and revise any certificate, opinion, decision, requirement or notice given during the course of the works. It used to be doubted that the court had such power, whether litigation is commenced where there is an arbitration clause in the contract or where the arbitration clause has been deleted (*Northern Regional Health Authority* v. *Derek Crouch Construction Co*). However, in *Beaufort Developments* v. *Gilbert Ash*, the House of Lords decided that *Crouch* had been wrongly decided and that the court does have the power to open up, review and revise any certificate, opinion, decision, requirement or notice. In any event now, the JCT forms provide a choice that can be made prior to executing the contract as between arbitration and the courts for dispute resolution.

9.110   Finally, it is inappropriate to seek, as some draftsmen do, to delete the arbitration machinery in the principal contract by a clause in the collateral warranty. The proper place for amendments to the principal contract is in the principal contract and not in the contract that is collateral to it.

Notice

9.111   It is useful to include provision as to how notices are to be given in the procedural sense: by post, recorded delivery, special delivery, by hand, by fax, by telex, at what address and when service is to be taken as having taken place.

9.112　Although strictly not a term of the contract, the basis on which the execution of the collateral warranty is to be carried out is important for two reasons. Firstly, where the warranty is to be executed under hand, the draftsman must provide for consideration (see paragraph 1.48). Secondly, warranties under hand are subject to a six-year period of limitation of action whereas contracts executed as deeds are subject to a twelve-year period of limitation of action (see paragraph 6.97). The receivers of the benefits of warranties will usually be looking for twelve years, and, therefore, documents to be executed under seal. As to the statutory provisions for execution, see paragraphs 1.53 to 1.55.

Contractors and sub-contractors

9.113　A question arises as to whether or not there should be a provision in sub-contracts in relation to collateral warranties entered into by the main contractor. There are two important points here. The first is whether or not the limitation periods are different under the sub-contract and the main contractor's collateral warranty; the danger point for the main contractor is expiry of the limitation period under the sub-contract at a time when the limitation period under the collateral warranty has not expired. Secondly, making it clear in the sub-contract that any damage that the main contractor suffers under a collateral warranty is damages in the legal sense that are recoverable under the sub-contract.

9.114　As to the limitation point, a useful device is to have an indemnity in the sub-contract. The cause of action does not accrue on an indemnity until the liability of the person seeking to be indemnified, the main contractor, has been established. The second point, foreseeability of damage, can be dealt with by a provision in the sub-contract. Putting these two points together, the type of provision that main contractors might wish to incorporate in their sub-contracts is:

> 'The sub-contractor hereby acknowledges that any breach by him of the sub-contract may result in the Contractor committing breaches of and becoming liable in damages under the Main Contract and other contracts made by him in connection with the Main Works (including but not so as to derogate from the generality of the foregoing collateral contracts with funding institutions, tenants and/or purchasers of the Main Works and/or any part thereof) and may occasion further loss or expense to the Contractor in connection with the Main Works and/or further or otherwise and all such damages, loss and expense are hereby agreed to be within the contemplation of the parties as being probable results of any such breach by the Sub-Contractor. The Sub-Contractor shall indemnify and save harmless the Contractor from and against all such damages, loss and expense as aforesaid.'

Guarantees of obligations under warranties

9.115   It is becoming increasingly common for the obligations entered into by warrantor companies in collateral warranties to be guaranteed by the parent company of the warrantor. This is usually carried through by having a three-party collateral warranty: the warrantor company, the beneficiary of the warranty and the parent company of the warrantor (the guarantor). The collateral warranty has, in addition to all the usual terms, the terms of a guarantee by which the guarantor guarantees the obligations of its subsidiary company. This is shown diagrammatically in Fig. 1.2 in Chapter 1 in relation to a funder.

Chapter 10
Practical Considerations

10.1   A great many practical issues are raised on a regular basis by those
         involved in collateral warranties. Does a warranty have to be given at all?
         What the tenant wants is unfair; who is to pay the legal costs of nego-
         tiation of non-standard warranties? Do warranties have to be signed by
         the parties? How should the negotiations be conducted and how should
         professional indemnity insurers be involved? Should different terms be in
         the warranties depending on whether they are being given to funds,
         purchasers or tenants? The aim of this chapter is to look at those matters
         and then to provide a list of the standard forms that are available, with a
         commentary on the Form of Agreement for Collateral Warranty, known
         as CoWa/F, for use where a warranty is to be given to a company pro-
         viding finance for a proposed development, prepared and approved for
         use by the British Property Federation, the Association of Consulting
         Engineers, the Royal Incorporation of Architects in Scotland, the Royal
         Institute of British Architects and the Royal Institution of Chartered
         Surveyors. There is also a commentary on the 2001 version of MCWa/F:
         The JCT Form of Agreement for Collateral Warranty for use where a
         warranty is to be given to a company providing finance (funder) for the
         proposed building works by a contractor.

Does a warranty have to be given?

10.2   A warranty does not have to be given by anyone unless they are under a
         binding and certain contractual obligation to do so, subject, of course, to
         the commercial pressure that can be brought to bear. It follows that in the
         absence of such a binding and certain contractual obligation, such as that
         in paragraph 7.53, no warranty whatsoever has to be given. This is par-
         ticularly so where after practical completion of a development, proposed
         tenants seek warranties from the architect, the engineer and the con-
         tractor. This gives rise to points that are the opposite sides of the same
         coin. Firstly, building owners would be well advised to deal with the
         whole question of collateral warranties at the outset of the project and see
         that they have bound all the principal parties to the construction process,
         through their contracts, to give certain collateral warranties when
         required by the building owner; this should be the case even though the
         names of the fund/purchaser or tenants are unknown at that stage.
         Secondly, it may be unwise for architects, engineers and contractors to

raise the question of warranties, prior to entering into their principal contracts, with a view to avoiding any binding obligation to enter into warranties.

Contractual obligation to give warranties

10.3   The type of clause that could be incorporated into the principal contract to create this obligation is discussed in paragraphs 7.49 to 7.61. Where such clauses are incorporated by the building owner into tender documents or proposed contract documents, then the tenderer would be well advised to check the provision particularly carefully, including the warranty itself. In contractors' organisations, the estimating department may not be best placed to pick up the risks of collateral warranties and/or their wording and the dangerous consequences that can flow from badly drafted warranties or warranties that have been well drafted but with onerous conditions. Contractors' estimators would be well advised to have such provisions carefully vetted by in-house or external lawyers.

10.4   Where a binding contractual obligation has been entered into to give warranties, then it is likely that on the failure of that party to give the warranty, the other party to the principal contract could apply to the High Court for an order of specific performance requiring the warranty to be entered into. For such an application to be successful, the wording of the contractual obligation will have to be clear and unambiguous, and the terms of the warranty will have to be certain in law. A provision that simply purports to require a warranty to be given, but on terms that are not described, will be insufficient to enable an application for specific performance to be successfully made to the court.

Commercial pressure

10.5   In most cases, it is the commercial pressure to sign warranties that is one of the dominant factors in the decision as to whether or not they should be given. In other words, if warranties are not agreed to be given by a party, then that party may not obtain the project; at a much lower level, there may be such a good working relationship between the parties, with ongoing work, that the pressure to give warranties is entirely reasonable; however, in the absence of a binding obligation, the party who is asked to sign warranties is in a much better negotiating position on the terms of the warranty and, perhaps, the extent of the parties to whom warranties are to be given. There is little point in the party giving a warranty permitting commercial pressure to force him into a position where he enters into very onerous obligations, which are not collateral to the principal contract, or obligations that put at risk his professional indemnity insurance cover or the future financial viability of his business.

Commercial balance

10.6   The reality is that the purpose of a contract is to allocate risk between the parties; in theory, that allocation of risk is a matter of negotiation between the parties. Usually, in collateral warranties, there is not a balancing of risks – most of the risks are undertaken by the person giving the warranty and the negotiation is about the nature and extent of the risks that that person is prepared to undertake.

Legal costs and consideration

10.7   On lengthy non-standard collateral warranties, the legal costs involved in the original drafting and subsequent vetting, negotiation, re-drafting of and producing the final version can be heavy. Indeed, on smaller projects, or sub-contracts on larger projects, the legal costs can be out of all proportion to the value, in terms of profit, of the principal contract. There is no reason in law why the collateral warranty obligations should be given by a professional person or a contractor free of actual consideration in money. Although it is unusual to agree such a 'signing fee', the reality is that the developer, fund and tenant are gaining real benefits through a warranty. A building will be more readily marketable with warranties. The commercial reality suggests a fairly substantial fee might be appropriate. Regard might be had to the financial benefit to the other party; what the premium might be on ten years non-cancellable building insurance (which might not be needed if there are collateral warranties); what the market will bear; the future liability risks and any increased professional indemnity insurance premiums, now and in the future by reason of collateral warranties.

10.8   If there is a contractual obligation to enter into a warranty, and there has been no prior agreement as to legal costs or some provisions set out in the obligation as to legal cost, then each party will have to bear their own legal costs in seeking advice. However, if there is no legal obligation to enter into a warranty, the person asked to give the warranty is in a rather better position. If the commercial situation enables him to do so, he can say to the other party, 'I am not obliged to give you a warranty but if you wish me to consider giving one, then I must ask you to pay my reasonable legal costs that I will incur in dealing with it and a signing fee of £. . .'. The other party, particularly where it is a developer with prospective tenants who will take a lease if they are given warranties, may be susceptible to that kind of suggestion. Such an agreement as to legal cost can speed the process to agreement of the warranty dramatically; it is simply the case that when one party is picking up the legal costs of two parties, he is likely to be keen to see resolution with the minimum of time and therefore expenditure of costs.

Negotiating and insurance

10.9   Where the party asked to give a warranty in relation to design has professional indemnity insurance, then particular regard should be had to the matters set out in Chapter 8. Some further points are worth making in relation to negotiating warranties where there is a professional indemnity insurance aspect.

10.10   Where there is an endorsement to the professional indemnity policy in the form of, or similar form to, that set out at in paragraph 8.37 or 8.39, the person giving the warranty must keep the scope of the warranty within the parameters set by the endorsement to his policy. Indeed, the fact that those parameters are so set by the policy is a very useful tool in negotiations by that party with the person seeking the warranty. He should argue that there is little point in him entering into a collateral warranty that would take him outside the cover of his professional indemnity insurance policy – it is in the interests of all parties to warranties that insurance cover is available in the event of a claim. It is no answer for a developer, fund, purchaser or tenant to say that insurance is a matter wholly for the person giving the warranty; whilst that is true as a matter of law, it flies in the face of the commercial reality in the event of a claim.

10.11   Where the insurance policy has no endorsement in relation to collateral warranties, or where a warranty cannot be agreed within the terms of the endorsement, then it is absolutely essential to refer all such warranties to insurers for approval prior to entering into them. In order to minimise the workload on insurers, it is sensible to get a draft into its final form before putting it to insurers, rather than constantly referring all the interim drafts; insurers, when faced with such non-standard warranties, will either refuse cover, agree to cover without any extra premium, or agree to cover with extra premium payment. The insured should clarify with the insurer whether that extra premium payment is a one-off in that year of insurance or whether the additional premium is likely to reflect in premiums for subsequent years of insurance; in any event, it is necessary to disclose warranties that fall outside any agreed endorsement to the policy to insurers at the time that insurance is being taken out and at any renewal whether with the same insurer or a different insurer.

Warranties must be executed

10.12   Unilateral promises, save where given under seal, are not binding in English law, although they can be in Scotland. In order, therefore, for there to be a binding collateral warranty, it must be signed and/or sealed by all the parties to it. Where there is a novation agreement, with three parties, then all three parties must sign and/or seal the warranty.

10.13   As to the statutory provisions in relation to the execution of documents, see paragraphs 1.53 to 1.55.

The givers, receivers and contents of warranties

10.14   In order to ascertain what warranties are required from whom and what terms they should contain, it is necessary to review the intention of the developer and the method of procurement of the project.

The intentions of the developer

10.15   In setting up the collateral warranty arrangement for a project, it is essential to look first at what the developer's intentions are. Is there to be a fund providing finance for the project and what are their requirements as to collateral warranties, if any? Is there to be a provision for the possibility that the development might be sold during construction or on completion? If it is to be sold during construction, then consideration should be given to inserting an obligation in all the principal contracts (e.g. architect, engineer, quantity surveyor and contractor) requiring that party to enter into a novation agreement, in the form of a draft that should be attached, in the event that the building owner so requires; this is in addition to considering warranties. Is the building to be occupied by the building owner or let by him on full repairing leases to tenants or sold to a purchaser on completion for his occupation or for letting to tenants of the purchaser? There is little point considering who should be required to give warranties, and on what terms, to whom without looking first at the way the project is to be dealt with.

10.16   Where the developer landowner is not also to be the employer under the building contract, then regard should be had to the points raised in paragraphs 7.5 to 7.14 above.

10.17   Once the developer's intentions are established, it is a relatively simple task to prepare a list of the principal contract and collateral warranty arrangements that will be required. Sometimes this can be best explained by preparing a chart of the contractual arrangements in diagrammatic form (such as that in Fig 1.2 in Chapter 1). Before this can be done, however, it is necessary to look at the method of procurement of the project.

Method of procurement

10.18   The method of procurement will affect what warranties are required from whom and the contents of those warranties.

JCT 98 (no design by the contractor) or the Intermediate Form of Contract or similar

10.19   On these types of projects, the building owner will engage the architect and the engineer to provide the design. A main contractor will be engaged by the building owner to carry out the construction of the design provided by the architect and the engineer. There may be, but rarely these days, nominated sub-contractors who will carry out design for the building owner (usually on the terms of the standard form NSC/W) and who will carry out the construction of their work under a sub-contract with the main contractor.

10.20   For the purposes of collateral warranties, similar issues arise in relation to management contracts.

JCT 'With Contractor's Design' 1998

10.21   Under this form of contract, the building owner sets out his requirements in a document which, in the contract, is called 'the Employer's Requirements'; the design and build contractor puts forward his proposals in a document called 'the Contractor's Proposals'. Both these documents taken together form the description of the work to be done for the purposes of the contract. It follows that on the face of it, the contractor is undertaking all the design and all the construction work and only he should be required to give warranties. In practice, the contractor will usually sub-let the design to architects and engineers and in some cases (such as mechanical and electrical services) to sub-contractors. In theory there will be a chain of contracts for liability through collateral warranties given by the contractor so that, for example, the tenant can sue the contractor who in turn can join in the architect or engineer or sub-contractor responsible. However, if the contractor goes into liquidation, there will be a break in the chain of contracts and the tenant will be unable to proceed against the architect, the engineer or the sub-contractors in tort. It is for this reason that warranties are often required on design and build projects from the architect and engineer of the contractor, as well as some of the sub-contractors. Clearly a decision needs to be made as to which of the sub-contractors are to be required to give warranties – all of them would be inappropriate.

Construction management

10.22   Usually, the building owner will engage the architect and the engineer and the construction manager; each trade contractor will then enter into a contract with the building owner but on the basis that the construction manager has the right and obligation to supervise and direct the trade contractors. These projects are a potential breeding ground for collateral warranties. Clearly the architect and the engineer can be asked to give warranties in respect of design and other matters without too much difficulty. As to the trade contractors, it may be inappropriate to ask for warranties from all of them but clearly the major trade contractors will be asked to provide warranties. As to whether or not a warranty should be taken from the construction manager, this will depend on the duties that he has under his construction management contract. If those duties include, for example, inspecting the work of trade contractors to ascertain that they are carrying out work in accordance with the trade contract then collateral warranties may well be appropriate. Each construction management contract should be looked at individually to see whether the duties undertaken are duties that are appropriate to the giving of collateral warranties.

Standard forms of collateral warranty

10.23   The most widely used standard forms of collateral warranty available in England and Wales and Scotland are shown in Appendices 4 to 9 of this book, with the exception of the JCT Subcontractor Warranties (SCWa/F and SCWa/P&T) and the Scottish Building Contract Committee Subcontractor Warranties (SCWa/F/Scot, SCWa/P&T/Scot and Employer/Sub-Contractor Warranty Agreement). These forms are now listed below, together with short comments, followed by a commentary on two of these forms.

CoWa/F: The BPF, ACE, RIAS, RIBA, RICS, Form of Agreement for Collateral Warranty for use where a warranty is to be given to a company providing finance for a proposed development (Third Edition, 1992)

10.24   This form is the subject of a detailed commentary starting at paragraph 10.72.

10.25   It is used by professional firms giving a warranty to a fund. It is shown in Appendix 4 of this book.

10.26   It can be used in England and Wales and Scotland.

CoWa/P&T: The BPF, ACE, RIAS, RIBA, RICS, Form of Agreement for Collateral Warranty for use where a warranty is to be given to a purchaser or tenant of premises in a commercial and/or industrial development (Second Edition, 1993)

10.27   It is used by professional firms giving a warranty to purchasers and tenants. It is shown in Appendix 5 of this book.

10.28   It can be used in England and Wales and Scotland.

MCWa/F: The JCT Form of Agreement for Collateral Warranty for use where a warranty is to be given to a company providing finance (funder) for the proposed building works by a contractor (2001 Edition)

10.29   It is used for main contractors giving a warranty to funds. It is shown in Appendix 6 of this book.

10.30   With the pad of forms, there is an enabling clause for use with JCT contracts. There is a commentary on that enabling clause starting at paragraph 7.55.

10.31   It is designed for use only in England and Wales.

10.32   There is a commentary on this form starting at paragraph 10.125.

MCWa/P&T: The JCT Form of Agreement for Collateral Warranty for use where a warranty is to be given to a purchaser or tenant of building works or part thereof by a main contractor (2001 Edition)

10.33   It is used for main contractors giving a warranty to purchasers and
tenants. It is shown in Appendix 7 of this book.

10.34   With the pad of forms, there is an enabling clause for use with JCT contracts. This is commented on starting at paragraph 7.57.

10.35   It is designed for use only in England and Wales.

SCWa/F: The JCT Form of Agreement for Collateral Warranty for use where a warranty is to be given to a funder providing finance for the works by a sub-contractor (2001 Edition)

10.36   It is used for sub-contractors giving a warranty to funds.

10.37   There is an enabling clause provided with the standard form. That
enabling clause is to be inserted in the main building contract. It provides
for the main contractor to give the notice requiring the collateral warranty
to be entered into with the fund. This is a route that can lead to difficulties
because the employer/developer has no control over the obtaining of
those warranties whilst he has a vital interest in obtaining them. Indeed,
the JCT suite of contracts and enabling clauses does not provide for the
employer to have a right to call for the main contractor to obtain such
warranties from subcontractors; there is a legal 'black-hole' as it were.
Employers and funds, which are otherwise content with the JCT suite of
contracts and warranties, will need to consider whether these arrangements for the obtaining of warranties are sufficiently secure for them.

10.38   A method to resolve this mechanistic problem is to require, in the main
contract, that the contractor obtains a sub-contractor/employer warranty
as a condition of sub-letting. That sub-contractor/employer warranty
should contain an obligation on the sub-contractor, at the request of the
employer, to enter into collateral warranties (in defined forms, appended
to the sub-contractor/employer warranty) with funds, tenants and
purchasers. This gives control to the employer.

10.39   It is designed for use only in England and Wales.

SCWa/P&T: The JCT Form of Agreement for Collateral Warranty for use where a warranty is to be given to a purchaser or tenant of building works or part thereof by a sub-contractor (2001 Edition)

10.40   It is used for sub-contractors giving a warranty to purchasers and tenants.

10.41   The same comments apply as are made in paragraphs 10.37 and 10.38 above.

10.42   It is designed for use only in England and Wales.

MCWa/F/Scot (Funder): The Scottish Building Contract Committee Form of Agreement for Collateral Warranty for use where a warranty is to be given by the main contractor to a company providing finance (funder) for proposed building works let or to be let under certain SBCC main contract forms (November 2001)

10.43   It is used for main contractors giving a warranty to funds. It is shown in Appendix 8 of this book.

10.44   With the pad of forms, there is an enabling clause for use with certain SBCC contracts.

10.45   It is designed for use only in Scotland.

MCWa/P&T/Scot (Purchaser and Tenant): The Scottish Building Contract Committee Form of Agreement for Collateral Warranty for use where a warranty is to be given by the main contractor to a purchaser or tenant of the whole or part of the building(s) comprising the works which have been practically completed under certain SBCC main contract forms (November 2001)

10.46   It is used for main contractors giving a warranty to purchasers and tenants. It is shown in Appendix 9 of this book.

10.47   There is an enabling clause for use with certain SBCC contracts.

10.48   It is designed for use only in Scotland.

SCWa/F/Scot: Form of Agreement for Collateral Warranty for funding institutions by a sub-contractor (November 2001)

10.49   It is used for sub-contractors giving a warranty to funding institutions.

10.50   There is an enabling clause.

10.51   It is designed for use only in Scotland.

SCWa/P&T/Scot: Form of Agreement for Collateral Warranty for purchasers and tenants by a sub-contractor (November 2001)

10.52   It is used for sub-contractors giving a warranty to purchasers and tenants.

10.53   There is an enabling clause.

10.54   It is designed for use only in Scotland.

SBCC Standard Form of Employer/Sub-Contractor Warranty Agreement (November 2001)

10.55   This is a warranty to be given by a sub-contractor to an employer. It provides for warranties as to design, selection of materials and complying with specifications in much the same way as the JCT Employer/Sub-contractor Warranties (see paragraph 9.36).

10.56   It provides some time-keeping warranties to the employer in relation to the provision by the subcontractor of design and the like so as not to cause delay to the main contract. If the sub-contractor is in breach of those warranties, then the employer may be able to recover directly from the sub-contractor monies that he has paid to the main contractor in consequence of that breach by the sub-contractor leading to claims under the main contract.

10.57   It permits assignment by the employer within three years of practical completion where the employer parts with his interest by sale, lease, assignation of his lease or by sub-letting or otherwise disposing of his interest in the works.

10.58   It is designed for use only in Scotland.

The Wren Insurance Association Limited: Collateral Warranty to a Purchaser/Tenant/Financier of the Development

10.59   This is a suggested form produced by The Wren Insurance Association Limited, an architects' mutual insurer. It is a warranty for a professional firm to a purchaser, tenant and financier. It is not in widespread use but the following comments are made on parts of the warranty.

10.60   The form creates a reasonable skill and care duty to the beneficiary in respect of the architect's duties under the principle appointment. That can only lead to a liability that is limited to a net contribution. By that is meant such sum as is just and equitable for the architect to pay having regard to his responsibilities and on the basis that all the '...other consultants, advisers, contractors and sub-contractors...' are deemed to have provided warranties to the beneficiary. The effectiveness of such provisions is discussed at paragraph 9.87. The wide descriptions of those from whom warranties are deemed to have been taken (including 'advisers' and 'sub-contractors') may lead to a risk of the provision being uncertain and, therefore, unenforceable.

10.61   There is a caveat on liability that the architect '...owes no greater duties or obligations in time or in nature and shall have no greater liabilities...' under the warranty than he has under the principle appointment.

10.62   The licence in respect of copyright, and the obligation to produce copies of documents, are both subject to payment of fees due to the architect.

10.63   Assignment is restricted to an absolute legal assignment, subject to the architect's express consent, such consent not to be unreasonably withheld. This would not be acceptable on most commercial developments.

10.64   There is a limitation of the time within which claims must be brought under the warranty. The draft provides for a choice of 6, 10 or 12 years from practical completion under the building contract. Even though that is so provided, the attestation provision in the draft is under hand. Most beneficiaries will want a warranty under seal, so the form would have to be appropriately amended to achieve execution as a deed.

10.65   Third party rights under the Contracts (Rights of Third Parties) Act 1999 are expressly excluded.

10.66   It is designed for use only in England and Wales.

Model Form of Employer/Specialist Contractor Collateral Warranty Agreement published by the Confederation of Construction Specialists by CCS Services Limited (1991)

10.67   This warranty is designed to create obligations from a sub-contractor to an employer where the sub-contractor is assisting in the design of the subcontract works.

10.68   There is a reasonable skill and care obligation in relation to design, including the selection of materials and goods and the satisfaction of a performance specification.

10.69   There are 'reasonable steps' type obligations in relation to materials being of good quality and not to use certain excluded materials. The warranty has provisions as to copyright and direct payment by the employer of the sub-contractor in default of the main contractor's payment to the sub-contractor.

10.70   The agreement is subject to the 'law of England'.

10.71   This warranty is not thought to be in regular use.

Commentaries

Commentary on CoWa/F: BPF, ACE, RIAS, RIBA, RICS, Form of Agreement for Collateral Warranty to a company providing finance (Third Edition 1993)

10.72   CoWa/F is for use where a warranty is to be given to a company providing finance for a proposed development. This form is designed for use by architects, engineers and quantity surveyors who are asked to give warranties to a fund. It has been prepared and approved for use by the British Property Federation, the Association of Consulting Engineers, the Royal Incorporation of Architects in Scotland, the Royal Institute of British Architects and the Royal Institution of Chartered Surveyors. This warranty (Third Edition, 1992) is published by the British Property Federation Limited and the copyright is owned jointly by the BPF, the ACE, the RIAS, the RIBA and the RICS with whose kind permission the warranty together with the 'General advice' and their 'Commentary on clauses' is reproduced in Appendix 4 to this book.

10.73   This warranty is not, of course, appropriate for tenants or purchasers but its sister form, CoWa/P&T, is suitable for that purpose. These two forms have similar provisions, save that CoWa/F gives a right to the fund, in certain circumstances, to step in, that is to step into the shoes of the 'client'.

10.74   At the launch of its forerunner, the First Edition, in May 1990, Mr Michael Mallinson, the President of the BPF, said, 'I can see consultants insisting that they will not sign any other form. Since the form has been agreed by the Association of British Insurers and follows discussion with the Committee of London and Scottish Banks, I hope that financial institutions will accept it and not insist on their own forms. I hope that any other forms of collateral warranty proposed by funding institutions can now be buried – perhaps cremated would be a better word...'. That wish has not been fulfilled in practice. The form, even if used, is routinely amended.

10.75   This form is only appropriate where a warranty is to be given to a company providing finance for a proposed development.

10.76   The form is only appropriate for members of the professional team – the architect, the engineer and the quantity surveyor – called the firm in this form.

10.77   The form is drafted only for use where the architect/engineer/quantity surveyor has been appointed by the developer (called the client in this form) in circumstances where the client is receiving funding from a funding institution (called the company in this form). It follows that this form is not appropriate for design and build projects where the contractor has appointed the professional team.

10.78   The form is a tripartite agreement to which each of the firm, the client and the company (the financier) are parties. The client is a party to the warranty for the purposes of the variation arrangements (which arise for example where the finance agreement between the company and the client has been terminated) and so that the client can undertake that warranty agreements in the same form will be obtained from other parties (see the commentary later in this chapter on clauses 5, 6 and 12).

Recitals

10.79   There are three recitals setting out that the company has entered into a finance agreement with the client; that the client has appointed the firm under a contract; and that the client has entered, or may enter, into a building contract.

10.80   There is no particular problem with these recitals save that the form of the building contract that is proposed to be entered into is important because it triggers the definition of 'practical completion' in clause 9. Some building contracts do not use the words 'practical completion' which are very much a creature of the JCT form. An example from another contract is the British Property Federation Edition of the ACA Form of Building Agreement, where the words used are 'taking-over'. Parties proposing to use this form of warranty should therefore be aware of the need to know of the form of the proposed building contract, if it has not been entered into at the time the warranty is given, and to make an amendment, if necessary, to the words 'practical completion' in clause 9.

Consideration

10.81　The warranty sensibly provides (immediately above clause 1) for consideration by means of the usual formula, namely, in consideration of the payment of £1 by the company to the firm, receipt of which the firm acknowledges. Although such a provision is probably not necessary where the warranty is to be executed as a deed, it is most certainly necessary where it is executed under hand. It follows that having this provision in the standard form will avoid the risk of the consideration point being missed where the warranty is to be executed under hand.

10.82　However, the express words contain no provision in respect of consideration vis-à-vis the client. Consideration is discussed in paragraph 1.48 but in essence the usual analysis involves either detriment to the promisee or benefit to the promisor. Indeed, there has to be consideration for each promise not for the contract as a whole. In relation to the client, therefore, it is necessary, in the absence of an express statement as to consideration, to search the warranty to try to find the appropriate detriment and/or benefit. In so doing, the propositions as to consideration where there are three parties to an agreement set out in paragraph 1.50 need to be borne in mind. It is unfortunate that it should be necessary to carry out this analytical exercise, which in the event produces an uncertain result: it is hard to find the appropriate detriment and/or benefit for the clients' promises vis-à-vis the firm. There is therefore doubt whether in that respect there is consideration in this warranty. This issue could be put beyond doubt by amending the consideration provision to say that each party pays every other party £1, receipt of which is acknowledged, and in consideration of the mutual undertakings set out in the warranty.

Clause 1

10.83　The purpose of this clause is to fix the firm with a duty in contract to the company; there are alternative provisions in this clause appearing in square brackets. This is to enable the warranty to be amended to correspond with the contract to which this warranty is collateral. For example, where the architect is engaged on the Standard Form of Appointment of an Architect, the words '[care and diligence]' should be deleted in the warranty. Where the engineer is appointed on the basis of the ACE Conditions, the words '[and care]' should be deleted. The RICS Standard Form of Agreement for the Appointment of a Quantity Surveyor does not contain a provision as to the standard of the duty to be exercised. It will be that of reasonable skill and care by implication of a term to that effect.

Therefore, where this form is used for a quantity surveyor, clause 1 of the warranty should be amended by deleting the words '[care and diligence]'. If some other conditions of engagement are used, then clause 1 should provide the same words by way of duty as appear in those conditions. In any event, those completing this warranty should check that the words setting out the duty in the appointment are reflected here in the warranty. For example, some appointments say that the skill and care is that to be expected of a professional firm experienced in the size and quality of the work to be undertaken. It would be useful to reflect that wording in the warranty.

10.84   This clause provides a very wide ranging duty to the company: reasonable skill and care in the performance of *all* the firm's duties under their Conditions of Engagement. Taking the Standard Form of Appointment of an Architect (updated April 2000) as an example, those duties include each of the services to be provided (which could include appraisal, strategic brief, outline proposals, detailed proposals, final proposals, production information, tender documentation, tender action, mobilisation, operations on site up to and after practical completion, being Work Stages A to L), not making any material alteration to the approved design without consent, informing the client that the expenditure or the building contract period are likely to be materially varied, visiting the site and the exercise of suspension and determination rights (see clauses 2.7, 2.8 and 8.1 to 8.8). It should be noted that the architect's duties include administering the building contract under Work Stages H to L – that of itself includes various duties such as certification of sums due to be paid to the contractor, the certification of loss and expense (even though the ascertainment may have been delegated to the quantity surveyor) granting extensions of time, certifying practical completion and giving non-completion certificates. In respect of all these functions, therefore, an architect will, on this form of warranty, have entered into a contractual obligation with the fund to perform all those duties with reasonable skill and care. This warranty is not limited to the exercise of reasonable skill and care in the design.

10.85   There are two provisos to the firm's liability. The first is a net contribution clause (clause 1(b)).

10.86   This clause provides that warranties no less onerous than this warranty are deemed to have been entered into between the company and other parties, which are to be named or described in the blank space in clause 1(a). Firms will wish to have the contractor as one of the other parties. In view of the potential for reducing the amount of outlay in the event of a claim, there is some guidance on this clause in the 'Commentary on Clauses'. However, that commentary does not give much detail as to which other firms and companies should be listed.

10.87   The question that arises is what is the effect of this clause in relation to the firm's liability, and if it is liable, the recoverable damages? It is clear that the clause does not create a condition precedent to the firm's liability that those other warranty agreements have been entered into, but it does purport to limit the recoverable damages to that 'proportion of the Company's losses which it would be just and equitable to require the firm to pay having regard to the extent of the firm's responsibility' for the losses. For a general discussion on such provisions, see paragraph 9.87.

10.88   If several firms are liable in respect of the same damage, then the firm would be entitled to bring contribution proceedings under the Civil Liability (Contribution) Act 1978 against other Firms (section 1(1) of the Act). Here, we have an attempt by contractual provision to carry out a similar exercise to that which a court would do under that Act when all the relevant parties are before the court, in order to try to ascertain the limit of liability sought to be created by the wording of the warranty. That would be no easy task and a court might have some difficulty in accepting that it should perform such an exercise, given that the other parties are not before the court. There appear, as yet, to be no decided cases on this approach.

10.89   The second sub-clause is 1(b) which provides that the firm is entitled, in proceedings brought under the warranty, to rely on limitation and defences that it has available under the principle appointment. In short, the company is not in a better position under the warranty than the client is under the appointment.

Clause 2

10.90   This is the provision that sets out which materials are not to be used as a warranty from the firm to the company. It is intended to be deleted only where the warranty is given by a quantity surveyor, on the basis that he does not usually have any duties in relation to specification.

10.91   The obligation is not to specify certain stated materials (at (a) to (e) in the clause) but it is subject to a reasonable skill and care test. The warranty does not purport to require the firm to see that such materials are not incorporated into the works. There is a savings provision so that this duty will not apply where the client has authorised the use of those materials (either in writing or, if oral, confirmed in writing). Clause 2 begins with the words 'without prejudice to the generality of Clause 1' and the intention of these words is to provide that the detailed materials provisions in clause 2 do not cut down the over-riding general duty, set out in clause 1, to exercise reasonable skill and care in the performance of the duties. It follows that an express approval by the client to the use of any of the specified materials does not bind the company if the company can establish a breach of clause 1 by the firm.

10.92   In the form at clause 2(f), a space is left for completion by the parties, if they wish, and the marginal note reads 'further specific materials may be added by agreement'. It is likely that this space will be used, not only for further specific materials, but also for a sweep up clause of the type discussed in paragraphs 9.55 and 9.56; all the issues discussed there should be considered if a sweep up clause is inserted in this form of warranty.

Clause 3

10.93   This provides that the company cannot give any instructions to the firm except where the company is stepping into the shoes of the client. As to that exception, see the commentary on clauses 5 and 6 below.

Clause 4

10.94   Here, the firm acknowledges that it has been paid all the fees and expenses due and owing to it by the client at the date of the warranty agreement. That part of this clause probably has two effects: firstly, it gives a firm that has not been paid what it is owed at the time it is asked to sign the warranty agreement the opportunity to refuse to sign unless and until it has been paid. Secondly, if the firm were to sign in circumstances where they had not been paid up to date, then the firm cannot recover their outstanding fees from the Company under the provisions of clause 7 following the company taking over from the client under either of clauses 5 or 6. It follows that every firm should be particularly careful to check that they have been paid up to date prior to signing this form of warranty. No provision is made for what is to happen if there is a *dispute* about fees at the time the firm is asked to sign the warranty.

10.95   This clause further provides that the company has no liability to the firm in respect of fees and expenses unless and until they have taken over from the client under clauses 5 or 6.

Clauses 5, 6 and 7

10.96   These provisions set out the circumstances in which the company is entitled to take the place of the client under the contract between the firm and the client and what the consequences of such action are to be. It should be noted that the company is given these rights either for itself or its 'appointee'. The effect of this provision is intended to be that the company can require the firm to accept a third party of the company's choice. In practice, where the company is a bank or pension fund it is highly unlikely they would wish to be involved in the detailed work that is necessary to be the client for the purposes of conditions of engagement. In such circumstances, a bank/pension fund are likely to wish to appoint a project manager or similar to perform these functions. The firm is given no right of reasonable objection to the appointee.

10.97   There are two circumstances in which the company can 'require... [the Firm to]... accept the instructions of the Company or its appointee':

(1)   where the finance agreement (between the company and the client) has been terminated (clause 5). The written notice of the company is, for the purposes of this clause, to be conclusive evidence that the finance agreement has been so terminated.

(2)   where the firm has a right to terminate or treat as having been repudiated the contract with the client or to discontinue the services to be provided by the firm (clause 6). The procedure is that where such a right arises on the part of the firm, they agree in the warranty that they will not exercise that right unless they have first given at least 21 days notice in writing to the company. The company may during that 21 days give notice that either they or their appointee are going to give instructions in the future. If the company does give that notice, then the firm's rights to terminate the contract with the client or treat it as having been repudiated or to discontinue the services to be provided by the firm cease.

10.98   Whether action is taken by the company under clause 5 or clause 6 is entirely at their discretion; it follows that the firm cannot force the company to take any action under either of these clauses and that clauses 5 and 6 are for the benefit for the company.

10.99   However, if the company does take action under clauses 5 or 6, then clause 7 applies. Clause 7 provides for several matters:

(1)   The company or its appointee becomes liable for payment of the firm's fees and that includes any fees outstanding at the date of the notice, but see the commentary on clause 4 at 10.94 above.

(2)   The company or its appointee take on the performance of the client's obligations under the contract between the client and the firm (but on the words there has to be doubt as to whether the company takes over the *existing* obligations of the client, other than fees, or whether the effect of this provision is simply to deal with future obligations).

(3)   The appointment continues in full force and effect.

(4)   The firm becomes liable to the company or its appointee under the contract between the client and the firm in place of the firm's liability to the client.

(5)   Where the company appoints an appointee in the notice, then notwithstanding the fact that the appointee is primarily liable for the payment of the firm's fees, the company is liable to the firm 'as guarantor'.

10.100   The provisions in clauses 5, 6 and 7 are a variation of the contract, not a novation. Variations (as novations) require consideration, unless, perhaps, the document is executed under seal. This variation needs consideration where the warranty is not executed as a deed and there is doubt as to whether there is consideration vis-à-vis the client.

10.101   All these provisions in clause 7 are preceded by the words 'It shall be a condition of any notice given by the Company under Clauses 5 or 6…'. These words are capable of two interpretations given the proposition that the precise requirements as to the contents of a notice are not set out. The first is that if a notice is given, then the matters set out in clause 7 apply; the second is that a notice is not a valid notice under the warranty unless it recites on its face all the matters set out in clause 7. The latter is probably the better view. If that is right, it follows that a company giving a notice under clause 5 or 6 should recite the provisions of clause 7 in the notice in order to put this point beyond doubt. The procedure for giving notices is in clause 12.

10.102   However, other difficulties arise on the wording of the provisions of clauses 5, 6 and 7 in relation to the company appointing an 'appointee'. It is not possible for a party to be a party to a contract in the absence of agreement between the parties. The appointee is not a party to the warranty at the outset, unlike the company. The appointee is not the agent in law of the company; the language of clause 7 is inconsistent with a view that the appointee takes on the obligations of a principal, namely the obligations of the client to the firm and the obligation to pay the firm (*Beswick* v. *Beswick*); indeed the company is said to be a guarantor of payment of the fees to the firm by the appointee. Against this background, it is submitted that the appointee provisions of this warranty agreement do not work in law to create the intended rights and duties between the appointee and the firm and they appear to be unaffected by the Contracts (Rights of Third Parties) Act 1999.

10.103   An incidental difficulty arises where the warranty and/or the appointment of a firm have been executed under seal by the original parties. Such a deed cannot be varied unless the variation itself is contained in a further deed. On any view, the appointment of an appointee is a variation, and, the notice under clauses 5 and 6 is not a deed. A further difficulty is that, in any event on the provisions of clauses 5, 6 and 7, without more, the appointee cannot be bound under seal. One way to avoid all these potential difficulties in relation to an appointee is to amend the warranty so that where the company wishes to appoint an appointee, the firm has an obligation on receipt of notice from the company, to enter into a variation agreement in the form of a draft attached to the warranty agreement – the form of that draft agreement could be easily derived from clauses 5, 6 and 7 and the attestation clause in the draft could make it clear whether the variation agreement was to be executed under hand or under seal. The fact that the identity of the appointee is not known at the date of the collateral warranty agreement would not prevent this procedure from being effective.

10.104   It follows that firms in receipt of notices under the provisions of clauses 5, 6 and 7 of this warranty should give them very careful consideration on all the points mentioned above and in particular, where there is an appointee, whether or not that appointee is bound in law to the obligations that are set out in clause 7 and whether or not the firm will be in breach of its contract with the client if it acts on the notice.

10.105   The language of these clauses is that of 'obligations', 'instructions' and 'liability'. In novation agreements, the usual language seeks to deal with the benefits and burdens of each of the three parties. Looking at the provisions of these clauses from the standpoint of benefits and burdens, and the express words in clause 7, it is the case that the client has no discharge in respect of the burdens he has undertaken to the firm. This might be of assistance to a firm seeking to recover fees unpaid at the date of the notice in circumstances where they cannot recover them from the company by reason of the provisions of clause 4. A further potential problem arises in relation to the provision in clause 7 that the firm and the client remain bound by their contract; in clauses 5 and 6 provision is made for the firm 'to accept the instructions of the Company ... to the exclusion of the Client...'. That deals with what the position of the firm is to be, but it does not provide (and neither does clause 7) that the client shall issue no further instructions to the firm. This could lead to difficult issues where the company is disputing the operation of clauses 5 and 6 for bona fide reasons. The point could be put beyond doubt by simple amendments.

10.106 There is no provision for the client to be served with a copy of any notice given under clauses 5 or 6. Given the major changes in the responsibilities of the parties that are triggered by that notice, this is a curious omission.

Clause 8

10.107 This clause provides for the company to have a licence to copy and use certain documents, the copyright in which remains with the firm. The licence granted is limited to 'any purpose related to the Premises'. 'Premises' is a new word – in the first edition of this warranty it was 'development'. The word 'premises' is not defined and may lead to some difficulty in interpreting the proper legal meaning of this clause. It may be that this is a drafting error: 'Premises' is defined in CoWa/P&T but not CoWa/F. On the wording of the clause the licence also extends to the use of the documents for the construction, completion, maintenance, letting, promotion, advertisement, reinstatement, refurbishment and repair of the development. The clause provides that the firm will not be liable to the company or its appointee for any use to which the documents are put where that use is for a purpose other than that for which the documents were prepared by the firm; this is clearly an important provision from the firm's point of view. The licence extends to a right on the part of the company to copy the drawings and documents; there is, therefore, no right on the part of the company to require the firm to produce copies at the firm's expense.

10.108 One potentially serious consequence for the firm of the licence contained in this collateral warranty is that if the client becomes insolvent at a time when they have completed their design, but not been paid, the company can use that design to complete the building without paying the firm's fees because the company has no liability to the firm in respect of fees unless they trigger the provisions of clauses 5 or 6 (which they may be disinclined to do when they will obtain little benefit, having the design already, and may incur substantial liability to the firm for fees).

10.109 This clause also contains reference to an 'appointee'. Indeed, the licence given by this clause is given to the company *and* its appointee. There is an issue as to whether or not the appointee in clause 8 is the same as the appointee in clauses 5, 6 and 7. As a matter of legal construction it does not follow that it is the same appointee.

Clause 9

10.110   For a commentary on this professional indemnity insurance provision, see paragraphs 9.66 to 9.72. The discussion referred to in this clause 9 is nothing more nor less than an agreement to agree and is unenforceable in law. The same points in relation to the reference to 'appointee' in this clause arises as in paragraph 10.109. The 'appointee' in this clause is the appointee under clauses 5 or 6.

10.111   There is a need to complete the period of years that the insurance must be in place and the amount of the limit of indemnity in the blank spaces provided in this clause.

Clause 10

10.112   This clause sets out why the client has agreed to be a party to the collateral warranty – for the purposes of clauses 5 and 6 (see 10.97 above). The wording of the clause is more akin to a recital than a condition and a question may arise as to whether the clause does as a matter of law give an acknowledgement that the firm will not be in breach of its contract with the client if it acts for the company under clauses 5 and 6, or whether the clause simply states that that is one of the purposes for which the client has agreed to be a party – the purpose not in reality being fulfilled. On the other hand it may be that a court would be disinclined to adopt such a strict legal construction. The point could be put beyond doubt by inserting after the word 'acknowledging', the words 'which the Client hereby does so acknowledge'.

Clause 11

10.113   This clause deals with assignment. It is only to be used when English law applies.

10.114   There can be assignment by the company without the consent of either the client or the firm but notice has to be given by the company to both the client and the firm (but note there is no procedure in clause 12 for the service of notices on the client). Indeed, the assignment envisaged in the clause is 'by way of absolute legal assignment'. The effect of that provision is that that assignment has to comply with the formalities of section 136 of the Law of Property Act 1925 (see paragraph 4.7) and, in particular, there has to be express notice in writing to the debtor (the firm) by the assignor (the company) before a valid legal assignment can come into effect. The assignment envisaged in the clause is an *absolute* legal assignment. This involves an assignment that is not qualified by conditions, is not by way of charge and is in respect of the whole thing in action (see paragraphs 4.8 to 4.12). Once that absolute legal assignment has been effected, with all its formalities, the company (the assignor) has no right to sue on the warranty (see paragraph 4.15). There is no prohibition on further assignment by the assignee.

10.115 The clause does not appear to prohibit assignment in some other form; for example, the clause does not say that assignment can only be by the method set out in the clause, or that there shall be no right to assign unless it is effected in the manner provided in the clause. It does therefore seem likely that there can be an equitable assignment with or without notice to the client or the firm (see the discussion in paragraphs 4.17 and 4.21).

10.116 Clause 11S is the alternative version for warranties where the law of Scotland applies.

Clause 12

10.117 This clause provides for the formalities of the giving of notices but not in respect of the client. It is important to follow the requirements in this clause when giving notices under the warranty. Notices can be given by hand or sent by special delivery or recorded delivery. In respect of notices sent by special delivery or recorded delivery (but not by hand), the notice is deemed to have been received 48 hours after being posted. The effect of this deeming provision is that a posted notice takes effect even though the addressee may not actually have received it. There is no provision for service by fax.

10.118 It is surprising that there is no provision for the service of notices on the client in this clause. Whilst this is consistent with clauses 5, 6 and 7, which do not require notices to be served on the client, the operation of clause 11 does require service of notice on the client.

Clause 13

10.119 This clause provides a long stop date for liability under the warranty. A period of years has to be completed; time runs from practical completion of the premises under the building contract. Note that 'Premises' are not defined in CoWa/F (although they are in CoWa/P&T). It would be useful to amend the warranty to define premises. It is also important to check that the particular building contract in use on this project uses the phrase 'practical completion' and not some other term. If it is not used, then the warranty should be amended accordingly.

General

10.120 There is no provision negating the effect of the Contracts (Rights of Third Parties) Act 1999, because the warranty is a 1992 Edition, many years before the Act came into force. Accordingly, if it is desired to prevent third party rights under that Act, a purpose-drafted clause will have to be inserted into the warranty. A suitable draft might be as follows:

> 'Nothing in this Contract shall, or is intended to, create rights and/or benefits by reason of the Contracts (Rights of Third Parties) Act 1999 or otherwise in favour of any person who is not a party to this contract (a "third party") and no term or provision of this contract shall be, or is intended by either party to be, enforceable by any third party by reason of the Contracts (Rights of Third Parties) Act 1999 or otherwise.'

10.121   There are alternative attestation provisions for English law and the law of Scotland. Under the English law provisions for attestation, there are further alternatives for where the warranty is simply to be signed by the parties (under hand) or to be executed as a deed. In the event of the former, the limitation period is six years from the date of accrual of the cause of action whereas in relation to the latter it is 12 years from the date of accrual of the cause of action (see Chapter 6 for a discussion).

10.122   In addition to those matters, the issue of consideration will arise (see paragraphs 10.81 and 10.82). In the 'General advice' issued with the form, it is said at paragraph 6 that the warranty should not be signed under seal when it is collateral to an appointment that is under hand. Whilst that approach is sensible for consistency it does not necessarily need to be the case. A company providing finance may well insist on a warranty under seal at a time when the firm has already been appointed under a contract under hand. In this event, careful consideration should be given by the firm and the company as to the effective limitation period that they will obtain. Given the potential difficulties in relation to consideration vis-à-vis the client (see paragraphs 10.81 and 10.82), the parties may prefer, in any event, to execute the warranty under seal notwithstanding that the company may not thereby obtain the benefit of a longer limitation period.

10.123   In paragraphs 7 and 8 of the 'General advice' issued with the form, some points are made in relation to the professional indemnity aspects of the use of this form. Firstly, there is a reminder in paragraph 7 that whether or not a claim arising under the warranty will be met by the policy 'will depend upon the terms and conditions of the policy in force at the time when a claim is made'. In paragraphs 8 and 9 it is said that consultants with a *current* policy taken out under the RIBA, RICSIS, ACE or RIASIS schemes will not have a claim refused simply on the basis that it is brought under the terms of a collateral warranty provided it is in the terms of this standard form, unamended. Taking these two propositions in paragraph 7 and 8 together, the RIBA, RICSIS, ACE and RIASIS schemes do not appear to be accepting that a claim made at a date beyond the period of the *current* year of insurance will necessarily be met under the policy in force at that time – that will depend on the terms and conditions then in force. In view of the fact that the 'General advice' does not clearly state what is meant by the words 'consultants with a current policy', every consultant would be well advised to check with his then insurer prior to entering into this warranty whether or not they can do so without express permission from the insurer. Insurers' policies on these issues may change. Consultants who are insured with other insurers should adopt that procedure in any event.

10.124 The 'General advice' and the 'Commentary on clauses' which are printed at the front of the pad of forms with the warranty are set out in Appendix 4 to this book. Those short form notes do not affect the proper legal interpretation of the warranty and are, necessarily, in cryptic form.

Commentary on MCWa/F: The JCT Form of Agreement for Collateral Warranty for use where a warranty is to be given to a company providing finance (funder) for the proposed building works by a contractor (2001 Edition)

10.125 There is a commentary on the 'Enabling Clause' for use with this standard form starting at paragraph 7.55.

Clause 1

10.126 This clause contains a warranty by the contractor that he has complied, and will continue to comply, with the building contract. It then goes on to seek to limit the financial consequences of breach of that warranty.

10.127 Firstly, it says that the liability to the funder 'for costs under this Agreement shall be limited to the proportion of the Funder's losses' by reference to a net contribution provision.

10.128 The use of the words 'costs under this Agreement' and 'losses' need consideration. The expression 'costs under this Agreement' is difficult to understand. It is drafting that has the connotations of money payable *under* an express provision of an agreement but the preceding words say that the 'costs' are *in consequence of a breach* of warranty.

10.129 What is meant by 'costs'? There are no costs mentioned in the agreement. The more usual terminology would be to use words like 'damages for breach of the warranty in this Agreement'. So it is arguable that the draftsman did not mean 'costs' to have the same meaning as 'damages for breach of contract', otherwise he would have used the word 'damages'. 'Costs' has a particular meaning in litigation and arbitration but, presumably, the draftsman did not intend that meaning here in the context of the warranty. Did he mean, by the use of the word 'costs' to refer to 'direct loss', such as that recoverable under the first limb of *Hadley* v. *Baxendale* and implicitly, therefore, seeking to exclude second limb damages? If so, that is a legal construction that is very far from clear-cut.

10.130 'Losses' raises similar issues but may be wider in scope than 'costs'. For example, 'losses' might well include lost profit and/or lost interest.

10.131 It seems likely that funders will want to amend clause 1 of the warranty to deal with the potential uncertainties created by the drafting, which are referred to above.

10.132 The clause then goes on to limit the otherwise recoverable 'losses' by reference to a net contribution provision. There are spaces for completion in respect of other consultants and of sub-contractors, to be completed by discipline of consultants and trade/speciality of sub-contractors, not the names of either. The intention is that the drafting creates an assumption that all those other consultants and sub-contractors have not only also given undertakings to the fund but have paid their contribution to the fund. Leaving aside whether such provisions work as a matter of law (see paragraphs 9.87 to 9.91), the way this provision is intended to operate is likely to lead to further uncertainty because of the use of descriptions of disciplines and trade/specialities, which are inevitably not the same as the names of companies/firms; there is the possibility of a mismatch between those descriptions and the names of companies/firms.

10.133 Clause 1(b) makes it clear that in proceedings by the funder, the contractor can rely on the same defences that he would have if he had been sued under the building contract. Clause 1(c) expressly provides that the obligations of the contractor are not released or diminished by the funder appointing a person to carry out an independent enquiry.

10.134 This clause is in contradistinction to that in the JCT purchaser/tenant warranty (MCWa/P&T). There, the liability of the contractor is in respect of '...the reasonable costs of repair renewal and or reinstatement of any part or parts of the Works to the extent that the Purchaser/Tenant incurs such costs and/or the Purchaser/the Tenant is or becomes liable either directly or by way of financial contribution for such costs' (see clause 1 in Appendix 7). Then there is a provision limiting the contractor's liability for other losses (either in respect of each breach, or for all other liability) also in the same clause 1.

Clause 2

10.135 This is the excluded materials provision. There is no list set out. Rather, the exclusion is in respect of the use of materials other than '... in accordance with the guidelines contained in the edition of the publication *Good Practice in the Selection of Construction Materials* (Ove Arup & Partners) current at the date of the Building Contract'. This requires the studying of another document to ascertain the duties arising under the warranty. This provision will not be acceptable to a large number of funders because in many circumstances, it will require the exercise of judgement to ascertain whether particular materials do, or do not, fall foul of the requirement. That is the kind of uncertainty that funders are keen to avoid. For a further discussion of this provision, see 9.53 above.

Clauses 3 to 7

10.136 These provisions deal with the right of the funder, but the funder has no obligation to step into the shoes of the employer, in the following circumstances:

(1) Termination of the finance agreement (between the funder and the employer) by the funder or

(2) The contractor exercising a right of determination of the building contract or seeking to treat it as repudiated by the employer.

10.137   The standard form requires the contractor to copy notices to the employer under the building contract, to also be given to the funder. Furthermore, a minimum of seven days notice from the contractor to the funder has to be given (see clause 6(3) for the detail). There is provision in the standard form for that period of seven days to be altered. Most funders will require at least 21 days for that provision. After all, making a decision to step into a building contract is not one that a funder can make unless and until it has considered all its options and that will include the overall viability of the project at the time the issue arises for consideration.

10.138   Clause 7 provides that if the funder steps in, he is liable for payments to the contractor under the building contract, including any 'sums outstanding' at the date of the notice of step-in. If the funder instructs an 'appointee', then this same clause provides that the funder guarantees those same payments by the appointee.

Clause 8

10.139   Copyright in 'drawings, reports, models, specifications, plans, schedules, bills of quantities, calculations and other similar documents' remains vested in the contractor if the step-in rights are exercised by the funder. The funder (or its appointee) is granted a non-exclusive licence to copy and use those documents for the purposes set out in the clause, which include construction of the works. This provision should be reviewed on each project to see whether it needs amendment by deletion of the whole, or part or additions to cover project specific matters. For example, it does not expressly extend to CAD drawings or to programmes (whether electronic or paper-based).

Clause 9

10.140   This requires the contractor to maintain professional indemnity insurance. There are spaces to complete the basis of the insurance as appropriate (for any one occurrence or series of occurrences arising out of one event, or in the aggregate) and the limit of indemnity, together with the period for which insurance is to be maintained. There is also provision to state a different limit of indemnity in respect of claims arising out of or in connection with pollution and contamination. As to that, contractors should note that this clause only deals with insurance; it does not limit their liability under the warranty given in clause 1 to the sum stated in this clause. If contractors wish to have such a limit to their liability then they have to negotiate additional provisions to the standard form. That needs to be done before the building contract is entered into, otherwise the enabling clause (see paragraph 7.55) will mean that they may be bound to give a warranty in an unamended form.

10.141   The clause provides for the period of the insurance to run from sectional completion dates, if the building contract provides for sectional completion. It also provides that the funder is entitled to see documentary evidence that the insurance is in place.

Clause 11

10.142   This is the assignment provision and is drafted so that the warranty can be assigned twice only. The drafting may avoid the pitfall set out in paragraph 9.84, but see also paragraph 4.47.

Clause 13

10.143   This contains a contractual limitation on the period for the bringing of proceedings in relation to the warranty. It is a period (to be inserted in the space provided) after practical completion (or, if in sections, each section). This has the potential to be a cut off date earlier than that arising under the law of limitation of action (see Chapter 6)

Clause 14

10.144   This clause deals with delay. It provides that the contractor has no liability to the funder for delay unless the funder steps in under clause 5 or clause 6. The words are wide enough to cover both liquidated damages for delay and unliquidated damages for delay. Once the funder has stepped in, the contractor has no obligation to pay again in respect of liquidated damages where such damages have been paid to, or deducted by, the employer.

Clause 15

10.145   Here, the warranty purports to say that no third party has any rights that it might otherwise have had under the Contracts (Rights against Third Parties) Act 1999, although the Act is not specifically mentioned. It might have been preferable to do so. More importantly, the drafting may well not achieve its aim. Section 1(1)(b) of the Act provides that a third party may enforce a term of a contract to which he is a stranger *if the term purports to confer a benefit on him* (see paragraphs 3.20 to 3.27). The clause in this warranty does not expressly seek to exclude third parties from relying on the right created by section 1(1)(b) and it may well not do so. It will be likely to be effective to exclude any possible argument that a third party may enforce a term of the contract by reason of the contract expressly so providing (section 1(1)(a)). However, that is not generally where a problem arises in excluding the effect of the Act; it is rather on the 'conferring a benefit' proposition under section 1(1)(b).

Clause 16

10.146   The warranty is governed by English law. Accordingly, it should not be used in other jurisdictions, such as Scotland. In Scotland, there is a form for use under Scots law (see Appendix 8).

Chapter 11
Other Solutions: Present and Future

11.1 The difficulties created by the widespread use of collateral warranties are legion, and the length of this book is testimony to that proposition. Those difficulties have prompted many people to ponder whether there could be a better way to regulate these matters, with or without collateral warranties. What follows here is a discussion of some of those possibilities; some of these solutions are available now, and others are contemplating the future.

Possible solutions – the present

The Contracts (Rights of Third Parties) Act 1999

11.2 The Contracts (Rights of Third Parties) Act 1999 creates the possibility in England, Wales and Northern Ireland (but not Scotland) that a third party can enforce a term or terms in a contract to which he is not a party. This subject is of such importance that the whole of Chapter 3 is devoted to it.

11.3 Some of the consequences of the Act in relation to collateral warranties are dealt with in paragraphs 3.49 to 3.60, where the response to the Act of draftsmen is discussed, together with the possibilities of excluding third party rights, letting the Act apply by default and making express use of the Act.

11.4 It is the case that at the date of publication of this book, widespread use is not being made of the Act in the field of collateral warranties; in fact, the majority of collateral warranties and construction contracts have provisions expressly excluding third party rights that would otherwise be created by the Act.

Commercial leases

11.5 The terms of commercial leases usually include full repairing covenants on the part of tenants (see paragraphs 7.15 to 7.26). Where a developer/landlord is willing, and they sometimes but not often are, a great deal could be done to remove the need for collateral warranties by having terms in the lease which put the risk of repair in respect of latent defects on to the landlord, rather than the tenant. Some of these methods are set out below.

Excluding tenant's liability to remedy latent defects

11.6   This option is that the tenant is subjected to a full repairing obligation in the usual way, but that there is excluded from that obligation any liability on the tenant to remedy latent defects. Such a provision might be:

> 'Provided that nothing in this Lease shall be construed as obliging the Tenant to remedy any Defect of whose existence the Tenant has within the first ... years of the Term notified the Landlord or any want of repair which is attributable to such Defect and which manifests itself [within such period *or* at any time during the Term].'

11.7   Subject to the definition of 'Defect' which is dealt with below, this provision would put the liability for the repair of latent defects and any want of repair caused by the latent defect on to the landlord. Tenants looking at such a provision may well wish to exclude from their liability to service charges, the cost of remedying such latent defects.

Landlord to remedy latent defects

11.8   Another option is to have a provision in the lease that puts the obligation and the cost of remedying latent defects on to the landlord, rather than the tenant. Such an obligation might be in the following form:

> 'The Landlord shall at its own expense remedy any Defect of whose existence the Tenant shall within the first ... years of the Term have notified the Landlord and any want of repair which is attributable to any such Defect and which manifests itself [within such period *or* at any time during the Term].'

11.9   Again, subject to the definition of 'Defect', this provision puts the liability for the cost of repair in respect of latent defects on to the party, the landlord, who will usually be in the best position in law to make recovery from the parties responsible for the latent defects under the principal contracts between the landlord and the construction professions/the contractor/sub-contractors.

Payment to tenant of cost of repairs

11.10   A further possibility is to provide in the lease that the tenant is under a full repairing obligation but that in respect of the cost of repairing latent defects and any consequential repairs, the landlord has an obligation to reimburse the tenant. Such a provision would substantially relieve the tenant's burden, remove the need for collateral warranties, and leave the landlord in a position where he could pursue the parties with whom he is in contract, in respect of the loss that he has incurred under the lease, by reason of the latent defects. Such a clause might be:

> 'If the Tenant shall in compliance with its obligations under [the repairing covenant] of this Lease carry out any works to remedy any Defect of whose existence the Tenant shall within the first ... years of the Term have notified the Landlord or to remedy any want of repair which is attributable to any such Defect and which manifests itself [within such period *or* at any time during the Term] then provided that the Tenant prior to carrying out such works shall obtain the approval of the Landlord to the nature and extent of the works [such approval not to be unreasonably withheld [or delayed]] and complete such works to the [reasonable] satisfaction of the Landlord the Landlord shall pay or repay to the Tenant the costs and expenses of and incidental to such works following production to the Landlord of all relevant receipts and invoices or other reasonable evidence of such costs and expenses.'

Definition of 'defect'

11.11   In all the above provisions, to make the clauses workable, there has to be a careful and clear definition of 'defect'. It will be necessary to provide for the definition to include negligent design, workmanship, materials and plant not in accordance with the building contract, and negligent supervision of the construction of the building. Clearly, the tenant will wish to have as wide definitions as possible in those respects. On the other hand, the landlord is likely to want to exclude any defect that was patent, as opposed to latent, at the date of the lease. This could be achieved by excluding from the definition of 'defect' any defect that was visible or ought reasonably to have been visible to a competent surveyor at a time immediately before the date of the lease.

Supplemental deed

11.12   By means of a deed supplemental to the lease, the landlord can be obliged to take all reasonable steps, including proceedings, to enforce against third parties, (e.g. the architect, the engineer and/or the contractor), any rights which the landlord has in respect of the latent defects. In such an agreement, it is necessary to deal with the proceeds of such litigation (i.e. are they monies held in trust for the tenant by the landlord), the definition of defects and who is to pay the costs of any proceedings that may be necessary. In order that a tenant could feel reasonably secure in entering into such an arrangement, it is important that his legal advisers should investigate before entering into the lease what rights the landlord has against third parties; this may involve inspection of the building contract and the conditions of engagement of the professional team, as well as, in appropriate cases, collateral warranty agreements between designing sub-contractors and the landlord, and any other rights the landlord may have in sub-contracts.

Purchasers

11.13   The problems of *caveat emptor* (buyer beware) are discussed in paragraph 7.36; collateral warranties can often be avoided vis-à-vis a purchaser, and the method is determined by whether or not the purchase is completed before or after all the obligations (save in respect of latent defects) have been fulfilled under the principal contracts with the building contractor and the professional team.

11.14   If the purchase is to be completed before completion, then the appropriate way to give the purchaser the rights that he needs is for there to be a novation agreement in respect of each and every principal contract; draftsmen in preparing such agreements should have particular regard to the state of outstanding fees vis-à-vis the professional team and, in relation to the contractor, interim certificates/uncertified work in progress and retention monies. Where the purchase is completed at a later date (say after the release of retention to the main contractor and the payment of all fees due to the professional team), then the purchaser can be best protected by an absolute legal assignment of the benefits of the developer under each and every principal contract.

11.15   In order that the developer and purchaser can have the right to proceed by either novation or assignment, the terms of the principal contract may require amendment; for example, where there is to be a novation, it would be appropriate for the principal contract to contain a provision requiring the other party at the behest of the developer to enter into a novation agreement in the form of a draft attached; where there is to be an assignment, the principal contract should provide that that assignment can be effected by the developer without the consent of the other party to the principal contracts, or with their consent which consent is not to be unreasonably withheld.

11.16   However, where the purchase is agreed during construction but completion of the sale is to take place on practical completion of the building works, collateral warranties are necessary.

Insurance

11.17   The insurance industry has responded to some of the difficulties set out in this book. In particular, ten year non-cancellable insurance on the building (subject to various restrictions and extensions) is now more widely available than it used to be. The NEDO report *Building Users' Insurance Against Latent Defects* has stimulated discussion and action in this area.

BUILD

11.18   In 1988, a report was produced by the Construction Industry Sector Group of the National Economic Development Council. The report is called *Building Users' Insurance Against Latent Defects* known by the acronym BUILD. The committee that produced the report consisted of many distinguished people, (including architects, engineers, contractors, property developers, insurers and representatives of government), under the chairmanship of Professor Donald Bishop. The report contains an excellent analysis of the risks of building development to the parties involved in it. Among the many points that arise from this report is the recommendation that there should be 'BUILD insurance'. The essence of such insurance is that it should provide protection for a period of ten years, from the date of practical completion, on the basis that the insurance is a non-cancellable material damage policy against specified latent defects and damage. It is suggested that the cover initially would be limited to the structure, including foundations, the weather shield envelope and, optionally, loss of rent.

11.19   The policy would be taken out by the developer at a very preliminary stage, at least before any work is started on the site. The policy would be transferable to successive owners and to tenants of the whole building; in a situation where the whole building is not let to one tenant, the report envisages that such tenants would be indemnified by the landlord on a back to back basis with the terms of the BUILD policy held by the landlord. There would be a single premium to cover insurance and the necessary risk assessment and verification of the design and construction by independent consultants appointed by the insurer. There should be provision for inflation in building costs in respect of the cover, and realistic deductibles (excess). A further recommendation of the report was that insurers should waive their rights of subrogation without which, insurers would not be in a position to seek to recover their outlay from the party liable to the developer: for example, the architect, the engineer and/ or the contractor. At the time of publication of the report, this aspect was of considerable concern to insurers because it would create, in effect, a no fault insurance scheme.

11.20   At the time of the report, there were three underwriters in the market: Allianz, Norman and SCOR. They were joined by Sun Alliance and Commercial Union in 1989. The indication in BUILD of the premium, including the verification costs, was that they would be of the order of 1.3% to 1.7% of the rebuilding cost. Although a small percentage, the cost in money terms can look substantial; for example, on a £10 million rebuilding cost, the premium would be of the order of £150,000 if the rate were 1.5%. In addition, there are the fees and expenses of independent design checkers to be paid for. Some developers consider that level of premium, and other fees, a high cost to pay when they are in a position to obtain tenants who are prepared to take the risk on full repairing covenants, with collateral warranties, all free of cost to the developer. On the other hand, if the developer comes to sell the building, the fact that there is such insurance may enable a better price to be achieved, or alternatively, a sale to be achieved which might not be achieved in the absence of the cover. Furthermore, in times of over-supply of commercial buildings in the market place, it is conceivable that tenants may be more attracted to buildings with such cover than buildings without the cover.

Latent defects insurance

11.21   There are several insurers writing building defects type insurance; it simply is not possible to examine all the policy options available from all the insurers and underwriters in this field of insurance; however, it may be helpful to look in a little detail at the sorts of considerations that arise.

11.22   There is now much more flexibility in the insurance market as to the range of available cover than there was back in 1988/89. For example, traditional cover will extend to the structure and weatherproofing but that can be extended to include non-structural elements and mechanical and electrical services. Subject to particular project risks, structural and weatherproofing cover might cost somewhere between 0.65% and 1.0% of contract value. To add in non-structural elements and mechanical and electrical services (such as heating, ventilating, air-conditioning, water systems, lifts, escalators, electrical distribution systems, building management systems and so on) will take that figure up to 1% to 2% of contract value. If the cover is for loss of rent only, then the premiums are likely to be lower. These levels of premium are much less than was forecast in the BUILD report (see paragraph 11.20).

11.23   The possibility now exists for payment of the premium in annual instalments during the life of the cover, which life can now be twelve years rather than ten.

11.24   In addition to the premium, technical auditors will have to be paid for by the insured. Those technical auditors are there to check the design for insurers. There is, inevitably, an element of duplication of fees here in the sense that the building owner is paying professionals for the design and then paying other professionals to vet the design for insurers. However, it can be argued that this audit process is good for risk management of the design, workmanship, installation, choice of materials and testing.

11.25   Waiver of subrogation rights against architects, engineers and contractors is available but at increased premiums. At present, there is no relationship between this kind of waiver of subrogation and reduced professional indemnity premiums for the consultants. If there were, then such a reduction in professional indemnity premiums might assist to reduce fees and thus help to fund the latent defects insurance. It seems this is unlikely to happen at present.

11.26   Unfortunately, the use of latent defects insurance does not mean the end of collateral warranties. Most lawyers take the view when advising developers and tenants alike, that collateral warranties are still needed to plug any gaps that there may be in the extent of the cover provided by latent defects insurers.

Possible solutions – the future

Standard forms of collateral warranty

11.27   There can be no doubt that the agreement of widely accepted standard forms of collateral warranty would be of enormous benefit to tenants, property developers, funds, architects, engineers, contractors and sub-contractors. However, the wide diversity of forms that are needed to cater not only for the different parties, but also for the different methods of procurement make this a very large task indeed. It may be beneficial for there to be standard clauses agreed which can then be incorporated into the different forms of agreement in relation to the different methods of procurement. On the other hand, the competing vested interests of the parties in trying to agree such terms make the task exceedingly difficult. It would be very useful, however, if there could be some success in the agreement of standard forms that are universally acceptable, but the intervening years between the first and second editions of this book do not give rise to much hope of that happening.

11.28   The efforts of the JCT, BPF, the RIAS, the RIBA, ACE and RICS are much to be welcomed in this field of endeavour.

Developments in insurance

11.29   There can be no doubt that the insurance market is responding to the difficulties created by the changes in the law of tort in relation to latent building defects and owners, purchasers and tenants. Premium levels on ten year non-cancellable building insurance appear to be becoming more competitive and the optional cover available is becoming wider; the fact that it is now sometimes possible, subject to insurers' approval in each case, to obtain insurance of this type on completed buildings is a helpful development.

11.30   Waiver of subrogation rights against third parties is now sometimes available on payment of an increased premium. That development is particularly to be welcomed. If it became more widespread, then the prospect of lengthy and costly multi-party litigation in relation to building defects would be substantially reduced.

European Community proposals

11.31   In February of 1990, Claude Mathurin, Ingènieur General des Ponts et Chaussèes, delivered his long-awaited final report to the Commission of

European Communities in Brussels: *Study of responsibilities, guarantees and insurance in the construction industry with a view to harmonisation at Community level.* The report follows a resolution adopted by the European Parliament on 12 October 1988 calling for the standardisation of contracts and controls in the construction industry and the harmonisation of responsibilities, and of the standards governing after-sales guarantees on housing. The thinking behind this resolution was in part that if a developer in country A wished to retain an architect in country B and an engineer in country C with a view to a development in country D, to be constructed by a contractor from country E, then the internal market in the EEC should permit that to happen, by a rationalisation and a harmonisation of procedures. It is easy to appreciate, therefore, the monumental task that faced Claude Mathurin. Even the start of his consideration of these issues would have been daunting to most people: what is the relevant law in each of the countries of the European Union in relation to these issues?

11.32   The final report was examined in the first edition of this book. However, Europe has made little progress in pursuing this area of harmonisation such that it seems unlikely that any action will be taken in the foreseeable future that will be of benefit in the field of collateral warranties.

Statutory provisions

Amendments to the Defective Premises Act 1972

11.33   The name of this Act is a little misleading; it applies only to the provision of dwellings. The Act is considered in Chapter 5.

11.34   The Act contains its own limitation period in respect of the duty, namely six years from the time when the dwelling was completed, but if further work is done to rectify the work that has already been done, then it is six years from the date when that further work was finished.

11.35   The Defective Premises Act could be reworked to extend its scope to construction work in general. Such drafting would have to take into account at least the following points:

- Who is to owe the statutory duty? This should be wide enough to encompass architects, engineers, contractors and sub-contractors.
- In respect of what type of buildings is the duty to be owed? Policy considerations might restrict the construction work definition. For example, what would be the position on civil engineering projects and government owned buildings?
- To whom is the duty to be owed? As in the present Act, there is no reason why the duty should not be owed to the person who ordered the construction work *and* every person who acquires an interest (legal or equitable) in the property.
- What should the nature of the duty be? It does seem that fit for

habitation would probably not be the best type of duty to be owed in relation to building works generally for the simple reason that not all building works are intended to be fit for habitation; although not all building works are intended for occupation, it may be that that could be a suitable substitution for habitation in cases where there is to be occupation. In any event, the duty should extend to carrying out the work in a professional manner with materials of good quality and in a good and workmanlike manner.

- There may well have to be limitations on the duty, such as the 'instructions' exception in section 1(2) of the present Act.
- What should the limitation period be? It does seem that the most likely contender would be ten years from practical completion, or from the date of completion of work to remedy defective work, whichever is the later. This would be consistent with the recommendations of the DTI *Professional Liability Report* and with the ten-year non-cancellable building insurance policies (see paragraphs 11.17 to 11.26).

11.36   It is an anomaly of our present law that certain types of duty are created by statute in respect of the provision of dwellings but not in respect of other types of construction; that anomaly is even more profound since the restriction of tortious duties in *D. & F. Estates Limited and Others v. The Church Commissioners for England and Others* and in *Murphy v. Brentwood District Council*.

11.37   However, this route to reform is a highly unlikely one now that the Contracts (Rights of Third Parties) Act 1999 is in place. It can be argued that that Act creates the rights that are needed and that, therefore, no further legislation is necessary in relation to the Defective Premises Act.

11.38   It cannot be in the public interest for there to be a continuation of the present explosive use of collateral warranties, which can only, in the long term, create difficulties of legal interpretation, uncertainty and extensive legal costs for those involved in the consequential litigation on the warranties that have been and will be signed, but it presently looks as if their use will continue.

Appendix 1
Housing Grants, Construction and Regeneration Act 1996

(Part II, Sections 104 to 117 only)

PART II

CONSTRUCTION CONTRACTS

Introductory provisions

104.–(1) In this Part a "construction contract" means an agreement with a person for any of the following– *(margin note: Construction contracts.)*

  (a) the carrying out of construction operations;

  (b) arranging for the carrying out of construction operations by others, whether under sub-contract to him or otherwise;

  (c) providing his own labour, or the labour of others, for the carrying out of construction operations.

(2) References in this Part to a construction contract include an agreement–

  (a) to do architectural, design, or surveying work, or

  (b) to provide advice on building, engineering, interior or exterior decoration or on the laying-out of landscape,

in relation to construction operations.

(3) References in this Part to a construction contract do not include a contract of employment (within the meaning of the Employment Rights Act 1996).

(4) The Secretary of State may by order add to, amend or repeal any of the provisions of subsection (1), (2) or (3) as to the agreements which are construction contracts for the purposes of this Part or are to be taken or not to be taken as included in references to such contracts.

No such order shall be made unless a draft of it has been laid before and approved by a resolution of each House of Parliament.

(5) Where an agreement relates to construction operations and other matters, this Part applies to it only so far as it relates to construction operations.

An agreement relates to construction operations so far as it makes provision of any kind within subsection (1) or (2).

(6) This Part applies only to construction contracts which–

  (a) are entered into after the commencement of this Part, and

  (b) relate to the carrying out of construction operations in England, Wales or Scotland.

(7) This Part applies whether or not the law of England and Wales or Scotland is otherwise the applicable law in relation to the contract.

105.–(1) In this Part "construction operations" means, subject as follows, operations of any of the following descriptions–

(a) construction, alteration, repair, maintenance, extension, demolition or dismantling of buildings, or structures forming, or to form, part of the land (whether permanent or not).

(b) construction, alteration, repair, maintenance, extension, demolition or dismantling of any works forming, or to form, part of the land, including (without prejudice to the foregoing) walls, roadworks, power-lines, telecommunication apparatus, aircraft runways, docks and harbours, railways, inland waterways, pipe-lines, reservoirs, water-mains, wells, sewers, industrial plant and installations for purposes of land drainage, coast protection or defence;

(c) installation in any building or structure of fittings forming part of the land, including (without prejudice to the foregoing) systems of heating, lighting, air-conditioning, ventilation, power supply, drainage, sanitation, water supply or fire protection, or security or communications systems;

(d) external or internal cleaning of buildings and structures, so far as carried out in the course of their construction, alteration, repair, extension or restoration;

(e) operations which form an integral part of, or are preparatory to, or are for rendering complete, such operations as are previously described in this subsection, including site clearance, earthmoving, excavation, tunnelling and boring, laying of foundations, erection, maintenance or dismantling of scaffolding, site restoration, landscaping and the provision of roadways and other access works;

(f) painting or decorating the internal or external surfaces of any building or structure.

(2) The following operations are not construction operations within the meaning of this Part–

(a) drilling for, or extraction of, oil or natural gas;

(b) extraction (whether by underground or surface working) of minerals; tunnelling or boring, or construction of underground works, for this purpose;

(c) assembly, installation or demolition of plant or machinery, or erection or demolition of steelwork for the purposes of supporting or providing access to plant or machinery, on a site where the primary activity is–

(i) nuclear processing, power generation, or water or effluent treatment, or

(ii) the production, transmission, processing or bulk storage (other than warehousing) of chemicals, pharmaceuticals, oil, gas, steel or food and drink;

(d) manufacture or delivery to site of–

(i) building or engineering components or equipment,

    (ii) materials, plant or machinery, or

    (iii) components for systems of heating, lighting, air-conditioning, ventilation, power supply, drainage, sanitation, water supply or fire protection, or for security or communications systems,

    except under a contract which also provides for their installation;

  (e) the making, installation and repair of artistic works, being sculptures, murals and other works which are wholly artistic in nature.

(3) The Secretary of State may by order add to, amend or repeal any of the provisions of subsection (1) or (2) as to the operations and work to be treated as construction operations for the purposes of this Part.

(4) No such order shall be made unless a draft of it has been laid before and approved by a resolution of each House of Parliament.

106.–(1) This Part does not apply–

  (a) to a construction contract with a residential occupier (see below), or

  (b) to any other description of construction contract excluded from the operation of this Part by order of the Secretary of State.

Provisions not applicable to contract with residential occupier.

(2) A construction contract with a residential occupier means a construction contract which principally relates to operations on a dwelling which one of the parties to the contract occupies, or intends to occupy, as his residence.

In this subsection "dwelling" means a dwelling-house or a flat; and for this purpose–

  "dwelling-house" does not include a building containing a flat; and

  "flat" means separate and self-contained premises constructed or adapted for use for residential purposes and forming part of a building from some other part of which the premises are divided horizontally.

(3) The Secretary of State may by order amend subsection (2).

(4) No order under this section shall be made unless a draft of it has been laid before and approved by a resolution of each House of Parliament.

107.–(1) The provisions of this Part apply only where the construction contract is in writing, and any other agreement between the parties as to any matter is effective for the purposes of this Part only if in writing.

Provisions applicable only to agreements in writing.

The expressions "agreement", "agree" and "agreed" shall be construed accordingly.

(2) There is an agreement in writing–

  (a) if the agreement is made in writing (whether or not it is signed by the parties),

  (b) if the agreement is made by exchange of communications in writing, or

  (c) if the agreement is evidenced in writing.

(3) Where parties agree otherwise than in writing by reference to terms which are in writing, they make an agreement in writing.

(4) An agreement is evidenced in writing if an agreement made otherwise than in writing is recorded by one of the parties, or by a third party, with the authority of the parties to the agreement.

(5) An exchange of written submissions in adjudication proceedings, or in arbitral or legal proceedings in which the existence of an agreement otherwise than in writing is alleged by one party against another party and not denied by the other party in his response constitutes as between those parties an agreement in writing to the effect alleged.

(6) References in this Part to anything being written or in writing include its being recorded by any means.

Adjudication

Right to refer disputes to adjudication.

108.–(1) A party to a construction contract has the right to refer a dispute arising under the contract for adjudication under a procedure complying with this section.

For this purpose "dispute" includes any difference.

(2) The contract shall–
   (a) enable a party to give notice at any time of his intention to refer a dispute to adjudication;
   (b) provide a timetable with the object of securing the appointment of the adjudicator and referral of the dispute to him within 7 days of such notice;
   (c) require the adjudicator to reach a decision within 28 days of referral or such longer period as is agreed by the parties after the dispute has been referred;
   (d) allow the adjudicator to extend the period of 28 days by up to 14 days, with the consent of the party by whom the dispute was referred;
   (e) impose a duty on the adjudicator to act impartially; and
   (f) enable the adjudicator to take the initiative in ascertaining the facts and the law.

(3) The contract shall provide that the decision of the adjudicator is binding until the dispute is finally determined by legal proceedings, by arbitration (if the contract provides for arbitration or the parties otherwise agree to arbitration) or by agreement.

The parties may agree to accept the decision of the adjudicator as finally determining the dispute.

(4) The contract shall also provide that the adjudicator is not liable for anything done or omitted in the discharge or purported discharge of his functions as adjudicator unless the act or omission is in bad faith, and that any employee or agent of the adjudicator is similarly protected from liability.

(5) If the contract does not comply with the requirements of subsections (1) to (4), the adjudication provisions of the Scheme for Construction Contracts apply.

(6) For England and Wales, the Scheme may apply the provisions of the Arbitration Act 1996 with such adaptations and modifications as appear to the Minister making the scheme to be appropriate.

For Scotland, the Scheme may include provision conferring powers on courts in relation to adjudication and provision relating to the enforcement of the adjudicator's decision.

Payment

109.–(1) A party to a construction contract is entitled to payment by instalments, stage payments or other periodic payments for any work under the contract unless–

Entitlement to
stage payments.

- (a) it is specified in the contract that the duration of the work is to be less than 45 days, or
- (b) it is agreed between the parties that the duration of the work is estimated to be less than 45 days.

(2) The parties are free to agree the amounts of the payments and the intervals at which, or circumstances in which, they become due.

(3) In the absence of such agreement, the relevant provisions of the Scheme for Construction Contracts apply.

(4) References in the following sections to a payment under the contract include a payment by virtue of this section.

110.–(1) Every construction contract shall–

Dates for
payment.

- (a) provide an adequate mechanism for determining what payments become due under the contract, and when, and
- (b) provide for a final date for payment in relation to any sum which becomes due.

The parties are free to agree how long the period is to be between the date on which a sum becomes due and the final date for payment.

(2) Every construction contract shall provide for the giving of notice by a party not later than five days after the date on which a payment becomes due from him under the contract, or would have become due if–

- (a) the other party had carried out his obligations under the contract, and
- (b) no set-off or abatement was permitted by reference to any sum claimed to be due under one or more other contracts,

specifying the amount (if any) of the payment made or proposed to be made, and the basis on which that amount was calculated.

(3) If or to the extent that a contract does not contain such provision as is mentioned in subsection (1) or (2), the relevant provisions of the Scheme for Construction Contracts apply.

111.–(1) A party to a construction contract may not withhold payment after the final date for payment of a sum due under the contract unless he has given an effective notice of intention to withhold payment.

Notice of
intention to
withhold
payment.

The notice mentioned in section 110(2) may suffice as a notice of intention to withhold payment if it complies with the requirements of this section.

(2) To be effective such a notice must specify–
    (a) the amount proposed to be withheld and the ground for with-
       holding payment, or
    (b) if there is more than one ground, each ground and the amount
       attributable to it,
and must be given not later than the prescribed period before the final
date for payment.

(3) The parties are free to agree what that prescribed period is to be.
   In the absence of such agreement, the period shall be that provided by
the Scheme for Construction Contracts.

(4) Where an effective notice of intention to withhold payment is
given, but on the matter being referred to adjudication it is decided that
the whole or part of the amount should be paid, the decision shall be
construed as requiring payment not later than–
    (a) seven days from the date of the decision, or
    (b) the date which apart from the notice would have been the final
       date for payment,
whichever is the later.

Right to suspend performance for non-payment.

112.–(1) Where a sum due under a construction contract is not
paid in full by the final date for payment and no effective notice
to withhold payment has been given, the person to whom the sum
is due has the right (without prejudice to any other right or remedy)
to suspend performance of his obligations under the contract to the party
by whom payment ought to have been made ("the party in default").

(2) The right may not be exercised without first giving to the party in
default at least seven days' notice of intention to suspend performance,
stating the ground or grounds on which it is intended to suspend
performance.

(3) The right to suspend performance ceases when the party in default
makes payment in full of the amount due.

(4) Any period during which performance is suspended in pursuance
of the right conferred by this section shall be disregarded in computing
for the purposes of any contractual time limit the time taken, by the party
exercising the right or by a third party, to complete any work directly or
indirectly affected by the exercise of the right.
   Where the contractual time limit is set by reference to a date rather
than a period, the date shall be adjusted accordingly.

Prohibition of conditional payment provisions.

113.–(1) A provision making payment under a construction contract
conditional on the payer receiving payment from a third person is ineffective, unless that third person, or any other person payment by whom
is under the contract (directly or indirectly) a condition of payment by
that third person, is insolvent.

(2) For the purposes of this section a company becomes insolvent–
    (a) on the making of an administration order against it under Part II
      of the Insolvency Act 1986,
    (b) on the appointment of an administrative receiver or a receiver or

manager of its property under Chapter I of Part III of that Act, or the appointment of a receiver under Chapter II of that Part,

    (c) on the passing of a resolution for voluntary winding-up without a declaration of solvency under section 89 of that Act, or

    (d) on the making of a winding-up order under Part IV or V of that Act.

(3) For the purposes of the section a partnership becomes insolvent–

    (a) on the making of a winding-up order against it under any provision of the Insolvency Act 1986 as applied by an order under section 420 of that Act, or

    (b) when sequestration is awarded on the estate of the partnership under section 12 of the Bankruptcy (Scotland) Act 1985 or the partnership grants a trust deed for its creditors.

(4) For the purposes of this section an individual becomes insolvent–

    (a) on the making of a bankruptcy order against him under Part IX of the Insolvency Act 1986, or

    (b) on the sequestration of his estate under the Bankruptcy (Scotland) Act 1985 or when he grants a trust deed for his creditors.

(5) A company, partnership or individual shall also be treated as insolvent on the occurrence of any event corresponding to those specified in subsection (2), (3) or (4) under the law of Northern Ireland or of a country outside the United Kingdom.

(6) Where a provision is rendered ineffective by subsection (1), the parties are free to agree other terms for payment.

In the absence of such agreement, the relevant provisions of the Scheme for Construction Contracts apply.

Supplementary provisions

114.–(1) The Minister shall by regulations make a scheme ("the Scheme for Construction Contracts") containing provision about the matters referred to in the preceding provisions of this Part.

The Scheme for Construction Contracts.

(2) Before making any regulations under this section the Minister shall consult such persons as he thinks fit.

(3) In this section "the Minister" means–

    (a) for England and Wales, the Secretary of State, and

    (b) for Scotland, the Lord Advocate.

(4) Where any provisions of the Scheme for Construction Contracts apply by virtue of this Part in default of contractual provision agreed by the parties, they have effect as implied terms of the contract concerned.

(5) Regulations under this section shall not be made unless a draft of them has been approved by resolution of each House of Parliament.

115.–(1) The parties are free to agree on the manner of service of any notice or other document required or authorised to be served in pursuance of the construction contract or for any of the purposes of this Part.

Service of notices, &c.

(2) If or to the extent that there is no such agreement the following provisions apply.

(3) A notice or other document may be served on a person by any effective means.

(4) If a notice or other document is addressed, pre-paid and delivered by post–

    (a) to the addressee's last known principal residence or, if he is or has been carrying on a trade, profession or business, his last known principal business address, or

    (b) where the addressee is a body corporate, to the body's registered or principal office,

it shall be treated as effectively served.

(5) This section does not apply to the service of documents for the purposes of legal proceedings, for which provision is made by rules of court.

(6) References in this Part to a notice or other document include any form of communication in writing and references to service shall be construed accordingly.

Reckoning periods of time. 116.–(1) For the purposes of this Part periods of time shall be reckoned as follows.

(2) Where an act is required to be done within a specified period after or from a specified date, the period begins immediately after that date.

(3) Where the period would include Christmas Day, Good Friday or a day which under the Banking and Financial Dealings Act 1971 is a bank holiday in England and Wales or, as the case may be, in Scotland, that day shall be excluded.

Crown application. 117.–(1) This Part applies to a construction contract entered into by or on behalf of the Crown otherwise than by or on behalf of Her Majesty in her private capacity.

(2) This Part applies to a construction contract entered into on behalf of the Duchy of Cornwall notwithstanding any Crown interest.

(3) Where a construction contract is entered into by or on behalf of Her Majesty in right of the Duchy of Lancaster, Her Majesty shall be represented, for the purposes of any adjudication or other proceedings arising out of the contract by virtue of this Part, by the Chancellor of the Duchy or such person as he may appoint.

(4) Where a construction contract is entered into on behalf of the Duchy of Cornwall, the Duke of Cornwall or the possessor for the time being of the Duchy shall be represented, for the purposes of any adjudication or other proceedings arising out of the contract by virtue of this Part, by such person as he may appoint.

Appendix 2

The Law Commission Report No. 242: Privity of Contract: Contracts for the benefit of Third Parties

Published in 1996
(Extract in relation to construction)

The Third Party Rule Causes Difficulties in Commercial Life

3.9   Lest it be erroneously thought that the third party rule nowadays causes no real difficulties in commercial life, or that the case for reform is purely theoretical rather than practical, we have chosen two types of contracts – construction contracts and insurance contracts – to illustrate some of the difficulties caused by the rule.

(1) Construction Contracts

3.10   Both simple construction contracts involving only an employer and a builder, and complex construction contracts involving several main contractors, many subcontractors and design professionals are affected by the third party rule.

3.11   Simple construction contracts illustrate the difficulties which can arise when one contracting party agrees to pay for work to be done by another contracting party which will benefit a third party to the contract. Say, for example, a client contracts with a builder for work to be done on the home of an elderly relative. If the work is done defectively, it is only the client who has a contractual right to sue the builder for its failure to deliver the promised performance. On traditional principles, and subject to the decisions in *Linden Gardens Trust* v. *Lenesta Sludge Disposals Ltd* and *Darlington BC* v. *Wiltshier Northern Ltd,* the client can often only recover nominal damages, since he or she will have suffered no direct financial loss as a result of the builder's failure to perform. The elderly relative could not himself sue for breach of contract, and the tort of negligence does not normally allow the recovery of pure economic loss. Therefore the elderly relative could not recover the cost of repairs in the tort of negligence and, if forced to move to alternative accommodation while the repairs were being carried out, could not recover consequent loss and expense either.

3.12   In complex construction projects, there will be a web of agreements between the participants in the project, allocating responsibilities and liabilities between the client (and sometimes its financiers), the main contractor, specialist subcontractors and consultants (architects, engineers and surveyors). Most significant construction projects in the UK are carried out under one of three major contractual procurement routes, and so the documentation used is very often highly standardised.

3.13   The third party rule means that only the parties within each contractual relationship can sue each other. The unfortunate result is that one cannot in the 'main' contracts simply extend the benefit of the architect's and engineer's duties of care and skill, and the contractor's duties to build according to the specifications, to subsequent purchasers and tenants of the development, or to funding institutions who might suffer loss as a result of the defective execution of the works. This cannot be achieved under the present third party rule without either joining the third party in question into the contract which contains these obligations, which in the case of a subsequent purchaser or tenant is impractical, since their identity may be unknown at the commencement of the works, or even for a long time afterwards, or executing a separate document – a 'collateral warranty' – extending the benefit of the duties in question. Were it not for these collateral warranties, the third party rule would prevent contractual actions by subsequent owners of completed buildings against the architect, engineer, main contractor or sub-contractor whose defective performance may have caused the loss or damage to them.

3.14   In an effort to overcome the privity deficiency of the law of contract, attempts were earlier made by subsequent owners of defective premises to sue in tort, and the expansion of the categories of negligence following *Hedley Byrne & Co* v. *Heller & Partners Ltd,* and particularly *Anns* v. *Merton London Borough Council,* initially resulted in such claims being successful. However, the law is now set against the recovery in negligence of economic loss caused by defective construction. In *D & F Estates Ltd* v. *Church Commissioners for England,* it was held that a builder was not liable in tort to a subsequent purchaser in respect of the cost of repair of defects to a building. The House of Lords then overruled *Anns* v. *Merton LBC* in *Murphy* v. *Brentwood District Council,* in holding that a local authority, which negligently failed to ensure that the builder complied with relevant by-laws and Building Regulations, owed no duty of care in tort as regards defects in the building causing pure economic loss and, in a decision handed down on the same day, confirmed the approach taken in the *D. & F. Estates Ltd* case in relation to builders. As a result of these cases, a subsequent owner or purchaser now has little protection in tort.

3.15   A typical collateral warranty given by an architect, engineer or main contractor excludes consequential economic loss and limits the defendant's liability, having regard to other claims of the warrantee, to a just and equitable proportion of the third party's loss. A typical warranty in favour of a finance house will also contain provisions permitting the finance house to take over the benefit of the contractor's or architect's or engineer's appointment contracts on condition of payment of liabilities, if the main finance contract is determined for any breach on the part of the employer, so that the finance house could ensure the continuance of work on the development notwithstanding some breach of the loan agreement by which the original employer was financed. There will also be provisions permitting the finance house, purchaser or tenant a licence to copy and use for specified purposes any designs or documents that are the property of the contractor or architect or engineer, and a clause undertaking that the contractor or architect or engineer will maintain professional indemnity insurance in a specified sum for a specified period. Finally, the warranty will normally permit assignment by the finance house, purchaser or tenant without any consent of the warrantor being required. These collateral warranties are generally supported by separate nominal considerations or are made under deed and thus are not tied to consideration in the main contract.

3.16 It is important to add that, where the benefits of the obligations undertaken by the warranty are assigned to sub-financiers or further purchasers or tenants down the line, this can give rise to further difficulties arising from the law of assignment. In particular, there is the difficulty as to whether an assignee can recover full damages, which was in issue in *Linden Gardens Trust* v. *Lenesta Sludge Disposals Ltd* and *Darlington Borough Council* v. *Wiltshier Northern Ltd,* and which led to the application, and extension, of an exemption to the normal rule on quantification of damages.

3.17 Our proposed reforms would enable contracting parties to avoid the need for collateral warranties by simply laying down third party rights in the main contract. Moreover, our proposed reforms would enable the contracting parties to mirror the terms in existing collateral warranties. Although this involves 'jumping ahead' to some details of our proposed reforms, it is worth explaining this latter point in some detail. Applying our conditional benefit approach discussed in Part X below, there is no reason why the architects', engineers' and contractors' liability to a third party could not be limited, as it presently is under collateral warranty agreements, so as to exclude consequential loss and so as to be limited to a specified share or a just and equitable share of the third party's loss. As regards defences, a claim by a third party under our proposed legislation, as we examine in Part X below, will be subject to defences and set-offs arising from, or in connection with, the contract and relevant to the particular contractual provision being enforced by the third party and which would have been available against the promisee. But this is a default rule only and the contracting parties can provide for a wider or narrower sphere of operation for defences and set-offs, if they so wish. So the present position under collateral warranties, whereby the claim is subject to defences arising under the main contract, is, or can be, replicated. What about the variation of the contract by the original contracting parties? A collateral warranty, once executed, may not be varied without the consent of the benefited third party purchaser, tenant or finance house. Our proposals are, on the face of it, more flexible in that the contracting parties can vary the contract without the third party's consent until the third party has relied on the contract or has accepted it. In practice, however, this ability to vary is not likely to be of any great advantage to the contracting parties because, assuming that the promisor could reasonably be expected to have foreseen that the third party would rely on the contract, they would be certain that it was safe to vary or cancel the original contract if they first communicated with the third party to ensure that there has been no reliance. It should be stressed that when we refer to variation, we are referring to variation of the contract. The work in building contracts is commonly subject to variation and, if so, would obviously continue to be variable irrespective of the third party's reliance or acceptance.

3.18 So, in our view, our proposals would enable the contracting parties to replicate the advantages of collateral warranties without the inconvenience of actually drafting and entering into separate contracts. Moreover, our proposals may carry a limited degree of extra flexibility as regards variation.

3.19 A further advantage of our legislative reform, as against collateral warranty agreements, is that it would not be necessary to assign the benefit of a provision extending the contractor's or architect's duty of care to sub-financiers and other purchasers and tenants down the line, since these persons could simply be named as potential beneficiaries of the clause by class. Thus, the difficulties caused by quantification of damages in claims arising under assigned collateral warranty agreements would be entirely removed.

3.20   The main contract between the client and main contractor may also contain exclusion clauses limiting the liability of the main contractor for certain types of defective performance. The main contractor may have entered into these clauses on its own behalf and on behalf of sub-contractors, in an effort to enable sub-contractors to take advantage, in actions against them in the tort for negligence, of the limitations and exemptions contained in such clauses. As we have seen, the exception to the third party rule, developed in *New Zealand Shipping Ltd* v. *A. M. Satterthwaite & Co Ltd,* may not necessarily work in this context albeit that the courts have sometimes allowed third parties to take advantage of the exclusion clause by regarding it as negativing the duty of care that would otherwise have arisen. A reform of the privity rule would permit contractors and clients straightforwardly and uncontroversially to extend the benefit of exclusion clauses in their contract to employees, sub-contractors and others.

3.21   Further problems may arise as regards payment obligations. At present, when a main contractor fails to pay a sub-contractor for work performed, the sub-contractor will have no right to sue the client directly for payment, although the client will be entitled to take the benefit of the sub-contractor's work. It may be that the participants in a construction project might wish to provide for payment direct by the client to the sub-contractor for work performed under a sub-contract, and to give the sub-contractor corresponding right to sue the client for the agreed sum once the work is performed. Even if such a right were expressly provided for, the present third party rule would prevent such an express term from being enforceable by the sub-contractor against the client unless the sub-contractor and client are in a contractual relationship.

3.22   Employers may make arrangements with contractors which are designed by both parties to benefit neighbours with regard to issues such as noise, access and working hours. The intended recipients of these benefits may be protected via the tort of nuisance but reform of the third party rule would enable the parties to give the neighbours the right to enforce the contract which would give them additional, and often significantly better, protection that in tort.

3.23   Our attention has also recently been drawn to the difficulties caused by the third party rule in the offshore oil and gas industry, which provide an excellent example of the anomalies and inconsistencies generated by the rule in practice. We understand that for many years, major oil companies and their advisers have attempted to minimise litigation arising from drilling contracts in the North Sea. This has largely been achieved by the use of cross indemnities between oil companies and contractors, which to be effective, must not only benefit the parties to the contracts in question but also all other companies in their respective groups, their employees, agents, sub-contractors and co-licensees. This is because, for example, it will often be unclear at the outset of a project which member of a client company's group will operate a platform and will thus be caused loss by any failings on the part of the contractor. An indemnity should therefore ideally benefit all companies likely to be affected. It is generally impractical for more than a handful of the beneficiaries of the indemnities given to be made parties to the contract. Moreover, careful drafting of the indemnity is necessary to ensure that those made parties to the indemnity can recover losses actually sustained by other group companies and beneficiaries. At present therefore, the third party rule is circumvented by making the parties to the contract agents or trustees for the other beneficiaries. Some devices used have become even more complex than those commonly employed in the construction industry, with webs of mutual cross-indemnities, back to back indemnity agreements, and incorporation of all main contract provisions into sub-contracts. We understand that there is concern among legal advisers as to the validity of these circumventions of privity. Our proposed reform will permit contractors and employers straightforwardly to extend the benefit of indemnity and exemption clauses contained in a contract to other companies in a group, employees, sub-contractors and others...

Appendix 3
Contracts (Rights of Third Parties) Act 1999

Right of third party to enforce contractual term.

1.–(1) Subject to the provisions of this Act, a person who is not a party to a contract (a "third party") may in his own right enforce a term of the contract if–

  (a) the contract expressly provides that he may, or
  (b) subject to subsection (2), the term purports to confer a benefit on him.

(2) Subsection (1)(b) does not apply if on a proper construction of the contract it appears that the parties did not intend the term to be enforceable by the third party.

(3) The third party must be expressly identified in the contract by name, as a member of a class or as answering a particular description but need not be in existence when the contract is entered into.

(4) This section does not confer a right on a third party to enforce a term of a contract otherwise than subject to and in accordance with any other relevant terms of the contract.

(5) For the purpose of exercising his right to enforce a term of the contract, there shall be available to the third party any remedy that would have been available to him in an action for breach of contract if he had been a party to the contract (and the rules relating to damages, injunctions, specific performance and other relief shall apply accordingly).

(6) Where a term of a contract excludes or limits liability in relation to any matter references in this Act to the third party enforcing the term shall be construed as references to his availing himself of the exclusion or limitation.

(7) In this Act, in relation to a term of a contract which is enforceable by a third party –
  "the promisor" means the party to the contract against whom the term is enforceable by the third party, and
  "the promisee" means the party to the contract by whom the term is enforceable against the promisor.

Variation and rescission of contract.

2.–(1) Subject to the provisions of this section, where a third party has a right under section 1 to enforce a term of the contract, the parties to the contract may not, by agreement, rescind the contract, or vary it in such a way as to extinguish or alter his entitlement under that right, without his consent if–

(a) the third party has communicated his assent to the term to the promisor,

(b) the promisor is aware that the third party has relied on the term, or

(c) the promisor can reasonably be expected to have foreseen that the third party would rely on the term and the third party has in fact relied on it.

(2) The assent referred to in subsection (1)(a)–

(a) may be by words or conduct, and

(b) if sent to the promisor by post or other means, shall not be regarded as communicated to the promisor until received by him.

(3) Subsection (1) is subject to any express term of the contract under which–

(a) the parties to the contract may by agreement rescind or vary the contract without the consent of the third party, or

(b) the consent of the third party is required in circumstances specified in the contract instead of those set out in subsection (1)(a) to (c).

(4) Where the consent of a third party is required under subsection (1) or (3), the court or arbitral tribunal may, on the application of the parties to the contract, dispense with his consent if satisfied–

(a) that his consent cannot be obtained because his whereabouts cannot reasonably be ascertained, or

(b) that he is mentally incapable of giving his consent.

(5) The court or arbitral tribunal may, on the application of the parties to a contract, dispense with any consent that may be required under subsection (1)(c) if satisfied that it cannot reasonably be ascertained whether or not the third party has in fact relied on the term.

(6) If the court or arbitral tribunal dispenses with a third party's consent, it may impose such conditions as it thinks fit, including a condition requiring the payment of compensation to the third party.

(7) The jurisdiction conferred on the court by subsections (4) to (6) is exercisable by both the High Court and a county court.

3.–(1) Subsections (2) to (5) apply where, in reliance on section 1, proceedings for the enforcement of a term of a contract are brought by a third party. *Defences etc. available to promisor.*

(2) The promisor shall have available to him by way of defence or set-off any matter that–

(a) arises from or in connection with the contract and is relevant to the term, and

(b) would have been available to him by way of defence or set-off if the proceedings had been brought by the promisee.

(3) The promisor shall also have available to him by way of defence or set-off any matter if–

(a) an express term of the contract provides for it to be available to him in proceedings brought by the third party, and

    (b) it would have been available to him by way of defence or set-off if the proceedings had been brought by the promisee.

(4) The promisor shall also have available to him–
    (a) by way of defence or set-off any matter, and
    (b) by way of counterclaim any matter not arising from the contract,
that would have been available to him by way of defence or set-off or, as the case may be, by way of counterclaim against the third party if the third party had been a party to the contract.

(5) Subsections (2) and (4) are subject to any express term of the contract as to the matters that are not to be available to the promisor by way of defence, set-off or counterclaim.

(6) Where in any proceedings brought against him a third party seeks in reliance on section 1 to enforce a term of a contract (including, in particular, a term purporting to exclude or limit liability), he may not do so if he could not have done so (whether by reason of any particular circumstances relating to him or otherwise) had he been a party to the contract.

Enforcement of contract by promisee.   4. Section 1 does not affect any right of the promisee to enforce any term of the contract.

Protection of promisor from double liability.   5. Where under section 1 a term of a contract is enforceable by a third party, and the promisee has recovered from the promisor a sum in respect of–
    (a) the third party's loss in respect of the term, or
    (b) the expense to the promisee of making good to the third party the default of the promisor,
then, in any proceedings brought in reliance on that section by the third party, the court or arbitral tribunal shall reduce any award to the third party to such extent as it thinks appropriate to take account of the sum recovered by the promisee.

Exceptions.   6.–(1) Section 1 confers no rights on a third party in the case of a contract on a bill of exchange, promissory note or other negotiable instrument.

(2) Section 1 confers no rights on a third party in the case of any contract binding on a company and its members under section 14 of the Companies Act 1985.

(3) Section 1 confers no right on a third party to enforce–
    (a) any term of a contract of employment against an employee,
    (b) any term of a worker's contract against a worker (including a home worker), or
    (c) any term of a relevant contract against an agency worker.

(4) In subsection (3)–
    (a) "contract of employment", "employee", "worker's contract", and "worker" have the meaning given by section 54 of the National Minimum Wage Act 1998,
    (b) "home worker" has the meaning given by section 35(2) of that Act,

    (c) "agency worker" has the same meaning as in section 34(1) of that Act, and

    (d) 'relevant contract' means a contract entered into, in a case where section 34 of that Act applies, by the agency worker as respects work falling within subsection (1)(a) of that section.

(5) Section 1 confers no rights on a third party in the case of–

    (a) a contract for the carriage of goods by sea, or

    (b) a contract for the carriage of goods by rail or road, or for the carriage of cargo by air, which is subject to the rules of the appropriate international transport convention,

except that a third party may in reliance on that section avail himself of an exclusion or limitation of liability in such a contract.

(6) In subsection (5) "contract for the carriage of goods by sea" means a contract of carriage–

    (a) contained in or evidenced by a bill of lading, sea waybill or a corresponding electronic transaction, or

    (b) under or for the purposes of which there is given an undertaking which is contained in a ship's delivery order or a corresponding electronic transaction.

(7) For the purposes of subsection (6)–

    (a) "bill of lading", "sea waybill" and "ship's delivery order" have the same meaning as in the Carriage of Goods by Sea Act 1992, and

    (b) a corresponding electronic transaction is a transaction within section 1(5) of that Act which corresponds to the issue, indorsement, delivery or transfer of a bill of lading, sea waybill or ship's delivery order.

(8) In subsection (5) "the appropriate international transport convention" means–

    (a) in relation to a contract for the carriage of goods by rail, the Convention which has the force of law in the United Kingdom under section 1 of the International Transport Conventions Act 1983,

    (b) in relation to a contract for the carriage of goods by road, the Convention which has the force of law in the United Kingdom under section 1 of the Carriage of Goods by Road Act 1965, and

    (c) in relation to a contract for the carriage of cargo by air–

        (i) the Convention which has the force of law in the United Kingdom under section 1 of the Carriage by Air Act 1961, or

        (ii) the Convention which has the force of law under section 1 of the Carriage by Air (Supplementary Provisions) Act 1962, or

        (iii) either of the amended Conventions set out in Part B of Schedule 2 or 3 to the Carriage by Air Acts (Application of Provisions) Order 1967.

7.–(1) Section 1 does not affect any right or remedy of a third party that exists or is available apart from this Act.

(2) Section 2(2) of the Unfair Contract Terms Act 1977 (restriction on

Supplementary provisions relating to third party.

exclusion etc. of liability for negligence) shall not apply where the negligence consists of the breach of an obligation arising from a term of a contract and the person seeking to enforce it is a third party acting in reliance on section 1.

(3) In sections 5 and 8 of the Limitation Act 1980 the references to an action founded on a simple contract and an action upon a specialty shall respectively include references to an action brought in reliance on section 1 relating to a simple contract and an action brought in reliance on that section relating to a specialty.

(4) A third party shall not, by virtue of section 1(5) or 3(4) or (6), be treated as a party to the contract for the purposes of any other Act (or any instrument made under any other Act).

Arbitration provisions.

8.–(1) Where–

(a) a right under section 1 to enforce a term ("the substantive term") is subject to a term providing for the submission of disputes to arbitration ("the arbitration agreement"), and

(b) the arbitration agreement is an agreement in writing for the purposes of Part I of the Arbitration Act 1996,

the third party shall be treated for the purposes of that Act as a party to the arbitration agreement as regards disputes between himself and the promisor relating to the enforcement of the substantive term by the third party.

(2) Where–

(a) a third party has a right under section 1 to enforce a term providing for one or more descriptions of dispute between the third party and the promisor to be submitted to arbitration ("the arbitration agreement"),

(b) the arbitration agreement is an agreement in writing for the purposes of Part I of the Arbitration Act 1996, and

(c) the third party does not fall to be treated under subsection (1) as a party to the arbitration agreement,

the third party shall, if he exercises the right, be treated for the purposes of that Act as a party to the arbitration agreement in relation to the matter with respect to which the right is exercised, and be treated as having been so immediately before the exercise of the right.

Northern Ireland.

9.–(1) In its application to Northern Ireland, this Act has effect with the modifications specified in subsections (2) and (3).

(2) In section 6(2), for "section 14 of the Companies Act 1985" there is substituted "Article 25 of the Companies (Northern Ireland) Order 1986".

(3) In section 7, for subsection (3) there is substituted–

"(3) In Articles 4(a) and 15 of the Limitation (Northern Ireland) Order 1989, the references to an action founded on a simple contract and an action upon an instrument under seal shall respectively include references to an action brought in reliance on section 1 relating to a simple contract and an action brought in reliance on that section relating to a contract under seal''.

(4) In the Law Reform (Husband and Wife) (Northern Ireland) Act 1964, the following provisions are hereby repealed–

(a) section 5, and
(b) in section 6, in subsection (1)(a), the words ''in the case of section 4'' and ''and in the case of section 5 the contracting party'' and, in subsection (3), the words ''or section 5''.

10.–(1) This Act may be cited as the Contracts (Rights of Third Parties) Act 1999.

Short title, commencement and extent.

(2) This Act comes into force on the day on which it is passed but, subject to subsection (3), does not apply in relation to a contract entered into before the end of the period of six months beginning with that day.

(3) The restriction in subsection (2) does not apply in relation to a contract which–

(a) is entered into on or after the day on which this Act is passed, and
(b) expressly provides for the application of this Act.

(4) This Act extends as follows–

(a) section 9 extends to Northern Ireland only;
(b) the remaining provisions extend to England and Wales and Northern Ireland only.

Appendix 4
CoWa/F

The BPF, ACE, RIAS, RIBA, RICS, Form of Agreement for Collateral Warranty for use where a warranty is to be given to a company providing finance for a proposed development (Third Edition 1992)

This document is the copyright of The British Property Federation, The Association of Consulting Engineers, The Royal Incorporation of Architects in Scotland, The Royal Institute of British Architects and the Royal Institution of Chartered Surveyors

Form of Agreement for

Collateral Warranty
for funding institutions

CoWa/F

The forms in this pad are for use where a warranty is to be given to a company providing finance for a proposed development. They must not in any circumstances be provided in favour of prospective purchasers or tenants.

General advice

1. The term "collateral agreement", "duty of care letter" or "collateral warranty" is often used without due regard to the strict legal meaning of the phrase. It is used here for agreements with a funding institution putting up money for construction and development.

2. The purpose of the Agreement is to bind the party giving the warranty in contract where no contract otherwise exist. This can have implications in terms of professional liability and could cause exposure to claims which might otherwise not have existed under Common Law.

3. The information and guidance contained in this note is designed to assist consultants faced with a request that collateral agreements be entered into.

4. The use of the word "collateral" is not accidental. It is intended to refer to an agreement that is an adjunct to another or principal agreement, namely the conditions of appointment of the consultant. It is imperative therefore that before collateral warranties are executed the consultant's terms and conditions of appointment have been agreed between the client and the consultant and set down in writing.

5. Under English Law the terms and conditions of the consultant's appointment may be "under hand" or executed as a Deed. In the latter case the length of time that claims may be brought under the Agreement is extended from six years to twelve years.

6. Under English Law this Form of Agreement for Collateral Warranty is designed for use under hand or to be executed as a Deed. It should not be signed as a Deed when it is collateral to an appointment which is under hand.

7. The acceptance of a claim under the consultant's professional indemnity policy, brought under the terms of a collateral warranty, will depend upon the terms and conditions of the policy in force at the time when a claim is made.

8. Consultants with a current indemnity insurance policy taken out under the RIBA, RICSIS, ACE or RIASIS schemes will not have a claim refused simply on the basis that it is brought under the terms of a collateral warranty provided that warranty is in this form. In other respects the claim will be treated in accordance with policy terms and conditions in the normal way. **Consultants insured under different policies** must seek the advice of their brokers or insurers.

9. **Amendment to the clauses should be resisted.** Insurers' approval as mentioned above is in respect of the unamended clauses only.

Commentary on Clauses

Recitals A, B and C are self-explanatory and need completion. The Consultant is described in the form as "The Firm". The following notes are to assist in understanding the use of the document:

Clause 1
This confirms the duty of care that will be owed to the Company. The words in square brackets enable the clause to reflect exactly the provisions contained within the terms and conditions of the Appointment.

Paragraphs (a) and (b) qualify and limit in two ways the Firm's liability in the event of a breach of the duty of care.

1 (a) By this provision the Firm's potential liability is limited. The intention is that the effect of "several" liability at Common Law is negated. When the Firm agrees - probably at the time of appointment - to sign a warranty at a future date, the list should include the names, if known, or otherwise the description or profession, of those responsible for the design of the relevant parts of the Development and the general contractor. When the warranty is signed, the list should be completed with the names of those previously referred to by description or profession.

1 (b) By this clause, the Company is bound by any limitations on liability that may exist in the conditions of the Appointment. Furthermore, the consultant has the same rights of defence that would have been available had the relevant claim been made by the Client under the Appointment.

Clause 2
As a consultant it is not possible to give assurances beyond those to the effect that materials as listed have not been nor will be specified. Concealed use of such materials by a contractor could possibly occur, hence the very careful restriction in terms of this particular warranty. Further materials may be added.

Clause 4
This obliges the consultant to ensure that all fees due and owing including VAT at the time the warranty is entered into have been paid.

Clause 5
This entitles the funding organisation to take over the consultant's appointment from the client on terms that all fees outstanding will be discharged by the funding authority (see Clause 7).

Clause 6
This affects the consultant's right to determine the appointment with the client in the sense that the funding authority will be given the opportunity of taking over the appointment, again subject to the payment of all fees which is the purpose of **Clause 7**.

Clause 8
Reasonable use by the Company of drawings and associated documents is necessary in most cases. By this clause, the Company is given the rights that might be reasonably expected but it does not allow the reproduction of the designs for any purpose outside the scope of the Development.

Clause 9
This confirms that professional indemnity insurance will be maintained in so far as it is reasonably possible to do so. Professional indemnity insurance is on the basis of annual contracts and the terms and conditions of a policy may change from renewal to renewal.

Clause 11
This clause indicates the right of assignment by the funding institution.

Clause 11S
This is applicable in Scotland in relation to assignations.

Clause 12
This identifies the method of giving Notice under Clauses 5, 6, 11 & 11S

Clause 13
This needs completion. The clause makes clear that any liability that the Firm has by virtue of this Warranty ceases on the expiry of the stated period of years after practical completion of the Premises. (Note: the practical completion of the Development may be later).

Under English law the period should not exceed 6 years for agreements under hand, nor 12 years for those executed as a Deed.

In Scotland, the Prescription and Limitations (Scotland) Act 1973 prescribes a 5 year period.

Clause 14 and Attestation below
The appropriate method of execution by the Firm, the Client and the Company should be checked carefully.

Clause 14S and Testing Clause below
This assumes the Firm is a partnership and the Client and the Company are Limited Companies. Otherwise legal advice should be taken.

N.B. The above advice and commentary is not intended to affect the interpretation of this Collateral Warranty. It is based on the terms of insurance current at the date of publication. All parties to the Agreement should ensure the terms of insurance have not changed.

Published by
The British Property Federation Limited
35 Catherine Place, London SW1E 6DY Telephone: 0171-828 0111

© The British Property Federation, The Association of Consulting Engineers, The Royal Incorporation of Architects in Scotland, The Royal Institute of British Architects and The Royal Institution of Chartered Surveyors. 1992.

ISBN 0 900101 08 6

Warranty Agreement   CoWa/F

Note

This form is to be used where the warranty is to be given to a company providing finance for the proposed development. Where that company is acting as an agent for a syndicate of banks, a recital should be added to refer to this as appropriate.

THIS AGREEMENT

(In Scotland, leave blank. For applicable date see Testing Clause on page 5)

is made the .... day of .... 19 ....

BETWEEN:-

(insert name of the Consultant)

(1) ....

of/whose registered office is situated at ....

.... ("the Firm");

(insert name of the Firm's Client)

(2) ....

whose registered office is situated at ....

.... ("the Client"); and

(insert name of the financier)

(3) ....

whose registered office is situated at ....

("the Company" which term shall include all permitted assignees under this agreement).

WHEREAS:-

A. The Company has entered into an agreement ("the Finance Agreement") with the Client for the provision of certain finance in connection with the carrying out of

(insert description of the works)

....

....

(insert address of the development)

at ....

....

.... ("the Development").

(insert date of appointment)
(delete/complete as appropriate)

B. By a contract ("the Appointment") dated ....
the Client has appointed the Firm as [architects/consulting structural engineers/consulting building services engineers/   surveyors] in connection with the Development.

(insert name of building contractor or "a building contractor to be selected by the Client")

C. The Client has entered or may enter into a building contract ("the Building Contract") with

....

....

....

for the construction of the Development.

CoWa/F   3rd Edition
© BPF, ACE, RIAS, RIBA, RICS 1992

NOW IN CONSIDERATION OF THE PAYMENT OF ONE POUND (£1) BY THE COMPANY TO THE FIRM (RECEIPT OF WHICH THE FIRM ACKNOWLEDGES) IT IS HEREBY AGREED as follows:-

(delete "and care" or "care and diligence" to reflect terms of the Appointment)

1. The Firm warrants that it has exercised and will continue to exercise reasonable skill [and care] [care and diligence] in the performance of its duties to the Client under the Appointment. In the event of any breach of this warranty:

    (a) the Firm's liability for costs under this Agreement shall be limited to that proportion of the Company's losses which it would be just and equitable to require the Firm to pay having regard to the extent of the Firm's responsibility for the same and on the basis that ......

(insert the names of other intended warrantors)

    ....

    ....

    ....

    ....

    ....shall be deemed to have provided contractual undertakings on terms no less onerous than this Clause 1 to the Company in respect of the performance of their services in connection with the Development and shall be deemed to have paid to the Company such proportion which it would be just and equitable for them to pay having regard to the extent of their responsibility;

    (b) the Firm shall be entitled in any action or proceedings by the Company to rely on any limitation in the Appointment and to raise the equivalent rights in defence of liability as it would have against the Client under the Appointment;

(delete where the Firm is the quantity surveyor)

2. [Without prejudice to the generality of Clause 1, the Firm further warrants that it has exercised and will continue to exercise reasonable skill and care to see that, unless authorised by the Client in writing or, where such authorisation is given orally, confirmed by the Firm to the Client in writing, none of the following has been or will be specified by the Firm for use in the construction of those parts of the Development to which the Appointment relates:-

    (a) high alumina cement in structural elements;

    (b) wood wool slabs in permanent formwork to concrete;

    (c) calcium chloride in admixtures for use in reinforced concrete;

    (d) asbestos products;

    (e) naturally occurring aggregates for use in reinforced concrete which do not comply with British Standard 882: 1983 and/or naturally occurring aggregates for use in concrete which do not comply with British Standard 8110: 1985.

(further specific materials may be added by agreement)

    (f)

    ]

3. The Company has no authority to issue any direction or instruction to the Firm in relation to performance of the Firm's services under the Appointment unless and until the Company has given notice under Clauses 5 or 6.

CoWa/F   3rd Edition
          © BPF, ACE, RIAS, RIBA, RICS 1992

4. The Firm acknowledges that the Client has paid all fees and expenses properly due and owing to the Firm under the Appointment up to the date of this Agreement. The Company has no liability to the Firm in respect of fees and expenses under the Appointment unless and until the Company has given notice under Clauses 5 or 6.

5. The Firm agrees that, in the event of the termination of the Finance Agreement by the Company, the Firm will, if so required by notice in writing given by the Company and subject to Clause 7, accept the instructions of the Company or its appointee to the exclusion of the Client in respect of the Development upon the terms and conditions of the Appointment. The Client acknowledges that the Firm shall be entitled to rely on a notice given to the Firm by the Company under this Clause 5 as conclusive evidence for the purposes of this Agreement of the termination of the Finance Agreement by the Company.

6. The Firm further agrees that it will not without first giving the Company not less than twenty one days' notice in writing exercise any right it may have to terminate the Appointment or to treat the same as having been repudiated by the Client or to discontinue the performance of any services to be performed by the Firm pursuant thereto. Such right to terminate the Appointment with the Client or treat the same as having been repudiated or discontinue performance shall cease if, within such period of notice and subject to Clause 7, the Company shall give notice in writing to the Firm requiring the Firm to accept the instructions of the Company or its appointee to the exclusion of the Client in respect of the Development upon the terms and conditions of the Appointment.

7. It shall be a condition of any notice given by the Company under Clauses 5 or 6 that the Company or its appointee accepts liability for payment of the fees and expenses payable to the Firm under the Appointment and for performance of the Client's obligations including payment of any fees and expenses outstanding at the date of such notice. Upon the issue of any notice by the Company under Clauses 5 or 6, the Appointment shall continue in full force and effect as if no right of termination on the part of the Firm had arisen and the Firm shall be liable to the Company and its appointee under the Appointment in lieu of its liability to the Client. If any notice given by the Company under Clauses 5 or 6 requires the Firm to accept the instructions of the Company's appointee, the Company shall be liable to the Firm as guarantor for the payment of all sums from time to time due to the Firm from the Company's appointee.

8. The copyright in all drawings, reports, models, specifications, bills of quantities, calculations and other similar documents provided by the Firm in connection with the Development (together referred to in this Clause 8 as "the Documents") shall remain vested in the Firm but, subject to the Firm having received payment of any fees agreed as properly due under the Appointment, the Company and its appointee shall have a licence to copy and use the Documents and to reproduce the designs and content of them for any purpose related to the Premises including, but without limitation, the construction, completion, maintenance, letting, promotion, advertisement, reinstatement, refurbishment and repair of the Development. Such licence shall enable the Company and its appointee to copy and use the Documents for the extension of the Development but such use shall not include a licence to reproduce the designs contained in them for any extension of the Development. The Firm shall not be liable for any such use by the Company or its appointee of any of the Documents for any purpose other than that for which the same were prepared by or on behalf of the Firm.

9. The Firm shall maintain professional indemnity insurance in an amount of not less than
    (insert amount)
    pounds (£   )
    for any one occurrence or series of occurrences arising out of any one event for a period
    (insert period)
    of   years from the date of practical completion of the Development for the purposes of the Building Contract, provided always that such insurance is available at commercially reasonable rates. The Firm shall immediately inform the Company if such insurance ceases to be available at commercially reasonable rates in order that the Firm and the Company can discuss means of best protecting the respective positions of the Company and the Firm in respect of the Development in the absence of such insurance. As and when it is reasonably requested to do so by the Company or its appointee under the Clauses 5 or 6, the Firm shall produce for inspection documentary evidence that its professional indemnity insurance is being maintained.

CoWa/F   3rd Edition
© BPF, ACE, RIAS, RIBA, RICS 1992

10. The Client has agreed to be a party to this Agreement for the purposes of acknowledging that the Firm shall not be in breach of the Appointment by complying with the obligations imposed on it by Clauses 5 and 6.

(delete if under Scots law)
[11. This Agreement may be assigned by the Company by way of absolute legal assignment to another company providing finance or re-finance in connection with the carrying out of the Development without the consent of the Client or the Firm being required and such assignment shall be effective upon written notice thereof being given to the Client and to the Firm]

(delete if under English law)
[11S. *The Company shall be entitled to assign or transfer its rights under this Agreement to any other company providing finance or re-finance in connection with the carrying out of the Development without the consent of the Client or the Firm being required subject to written notice of such assignation being given to the Firm in accordance with Clause 12 hereof.*]

12. Any notice to be given by the Firm hereunder shall be deemed to be duly given if it is delivered by hand at or sent by registered post or recorded delivery to the Company at its registered office and any notice given by the Company hereunder shall be deemed to be duly given if it is addressed to "The Senior Partner"/"The Managing Director" and delivered by hand at or sent by registered post or recorded delivery to the above-mentioned address of the Firm or to the principal business address of the Firm for the time being and, in the case of any such notices, the same shall if sent by registered post or recorded delivery be deemed to have been received forty eight hours after being posted.

(complete as appropriate)
13. No action or proceedings for any breach of this Agreement shall be commenced against the Firm after the expiry of   years from the date of practical completion of the Premises under the Building Contract.

(delete if under Scots law)
[14. The construction validity and performance of this agreement shall be governed by English Law and the parties agree to submit to the non-exclusive jurisdiction of the English Courts.

(alternatives: delete as appropriate)
[AS WITNESS the hands of the parties the day and year first before written.

Signed by or on behalf of the Firm ....

(for Agreement executed under hand and NOT as a Deed)
in the presence of: ....

Signed by or on behalf of the Client ....

in the presence of: ....

Signed by or on behalf of the Company ....

in the presence of: ....]

(this must only apply if the Appointment is executed as a Deed)
[IN WITNESS WHEREOF this Agreement was executed as a Deed and delivered the day and year first before written.

by the Firm

....

....

....

by the Client

....

....

....

by the Company

....

....

....]]

CoWa/F   3rd Edition
© BPF, ACE, RIAS, RIBA, RICS 1992

14S. *This Agreement shall be construed and the rights of the parties and all matters arising hereunder shall be determined in all respects according to the Law of Scotland.*

IN WITNESS WHEREOF these presents are executed as follows:-

SIGNED by the above named Firm at ....

on the .... *day of* .... *Nineteen hundred and* ....

as follows:-

.... *(Firm's signature)*

Signature .... *Full Name* ....

Address ....

.... *Occupation* ....

Signature .... *Full Name* ....

Address ....

.... *Occupation* ....

SIGNED by the above named Client at ....

on the .... *day of* .... *Nineteen hundred and* ....

as follows:-

For and on behalf of the Client

.... *Director/Authorised Signatory*

.... *Director/Authorised Signatory*

SIGNED by the above named Company at ....

on the .... *day of* .... *Nineteen hundred and* ....

as follows:-

For and on behalf of the Company

.... *Director/Authorised Signatory*

.... *Director/Authorised Signatory*]

CoWa/F 3rd Edition
© BPF, ACE, RIAS, RIBA, RICS 1992

Appendix 5
CoWa/P&T

The BPF, ACE, RIAS, RIBA, RICS, Form of Agreement for Collateral Warranty for use where a warranty is to be given to a purchaser or tenant of premises in a commercial and/or industrial development (Second Edition 1993)

This document is the copyright of The British Property Federation, The Association of Consulting Engineers, The Royal Incorporation of Architects in Scotland, The Royal Institute of British Architects and the Royal Institution of Chartered Surveyors

Form of Agreement for

Collateral Warranty
for purchasers & tenants | CoWa/P&T

The forms in this pad are for use where a warranty is to be given to a purchaser or tenant of a whole building in a commercial and/or industrial development, or a part of such a building. It is essential that the number of warranties to be given to tenants in one building should sensibly be limited.

General advice

1. The term "collateral agreement", "duty of care letter" or "collateral warranty" is often used without due regard to the strict legal meaning of the phrase. It is used here for agreements with tenants or purchasers of the whole or part of a commercial and/or industrial development.

2. The purpose of the Agreement is to bind the party giving the warranty in contract where no contract would otherwise exist. This can have implications in terms of professional liability and could cause exposure to claims which might otherwise not have existed under Common Law.

3. The information and guidance contained in this note is designed to assist consultants faced with a request that collateral agreements be entered into.

4. The use of the word 'collateral' is not accidental. It is intended to refer to an agreement that is an adjunct to another or principal agreement, namely the conditions of appointment of the consultant. It is imperative therefore that before collateral warranties are executed the consultant's terms and conditions of appointment have been agreed between the client and the consultant and set down in writing.

5. Under English Law the terms and conditions of the consultant's appointment may be 'under hand' or executed as a Deed. In the latter case the length of time that claims may be brought under the Agreement is extended from six years to twelve years.

6. Under English Law this Form of Agreement for Collateral Warranty is designed for use under hand or to be executed as a Deed. It should not be signed as a Deed when it is collateral to an appointment which is under hand.

7. The acceptance of a claim under the consultant's professional indemnity policy, brought under the terms of a collateral warranty, will depend upon the terms and conditions of the policy in force at the time when a claim is made.

8. Consultants with a current indemnity insurance policy taken out under the RIBA, RICSIS, ACE or RIASIS schemes will not have a claim refused simply on the basis that it is brought under the terms of a collateral warranty provided that warranty is in this form. In other respects the claim will be treated in accordance with policy terms and conditions in the normal way. **Consultants insured under different policies** must seek the advice of their brokers or insurers.

9. **Amendment to the clauses should be resisted.** Insurers' approval as mentioned above is in respect of the unamended clauses only.

Commentary on Clauses

Recital A.

This needs completion.

When this warranty is to be given in favour of a purchaser or tenant of part of the Development, the following words in square brackets must be deleted.

["The Premises" are also referred to as "the Development" in this Agreement".]

Care must be taken in describing "the Premises" accurately.

When this warranty is to be given in favour of a purchaser or tenant of the entire development, the terms "the Premises" and "the Development" are synonymous.

The following words in square brackets must be deleted

[forming part of. ....

at. ....("the Development").]

Recitals B & C

These are self explanatory but need completion.

Clause 1

This confirms the duty of care that will be owed to the Purchaser/the Tenant. The words in square brackets enable the clause to reflect exactly the provisions contained within the terms and conditions of the Appointment.

Paragraphs (a),(b) and (c) qualify and limit in three ways the Firm's liability in the event of a breach of the duty of care.

1 (a) By this provision, the Firm is liable for the reasonable costs of repair renewal and or reinstatement of the Development insofar as the Purchaser/the Tenant has a financial obligation to pay or contribute to the cost of that repair. Other losses are expressly excluded.

1 (b) By this provision the Firm's potential liability is limited. The intention is that the effect of "several" liability at Common Law is negated. When the Firm agrees - probably at the time of appointment - to sign a warranty at a future date, the list should include the names, if known, or otherwise the description or profession, of those responsible for the design of the relevant parts of the Development and the general contractor. When the warranty is signed, the list should be completed with the names of those previously referred to by description or profession.

1 (c) By this clause, the Purchaser/ the Tenant is bound by any limitations on liability that may exist in the conditions of the Appointment. Furthermore, the consultant has the same rights of defence that would have been available had the relevant claim been made by the Client under the Appointment.

1 (d) This states the relationship between the Firm and any consultant employed by the Purchaser/the Tenant to survey the premises.

Clause 2

As a consultant it is not possible to give assurances beyond those to the effect that materials as listed have not been nor will be specified. Concealed use of such materials by a contractor could possibly occur, hence the very careful restriction in terms of this particular warranty. Further materials may be added.

N.B. The above advice and commentary is not intended to affect the interpretation of this Collateral Warranty. It is based on the terms of insurance current at the date of publication. All parties to the Agreement should ensure the terms of insurance have not changed.

Clause 3

This obliges the consultant to ensure that all fees due and owing including VAT at the time the warranty is entered into have been paid.

Clause 4

This is included to make it clear that the Purchaser/the Tenant has no power or authority to direct or instruct the Firm in its duties to the Client.

Clause 5

Reasonable use by the Purchaser/the Tenant of drawings and associated documents is necessary in most cases. By this clause, the Purchaser/ the Tenant is given the rights that might be reasonably expected but it does not allow the reproduction of the designs for any purpose outside the scope of the Development.

Clause 6

This confirms that professional indemnity insurance will be maintained in so far as it is reasonably possible to do so. Professional indemnity insurance is on the basis of annual contracts and the terms and conditions of a policy may change from renewal to renewal.

Clause 7

This allows the Purchaser/the Tenant to assign the benefit of this Warranty provided it is done by formal legal assignment and relates to the entire interest of the original Purchaser/Tenant. By this clause any right of assignment may be limited or extinguished. If it is to be extinguished the word "not" shall be inserted after "may" and all words after "the Purchaser/the Tenant" deleted. If it is agreed that there should be a limited number of assignments, the precise number should be inserted in the space between "assigned" and "by the Purchaser/the Tenant".

Clause 7S

This is applicable in Scotland in relation to assignations. Completion is as for Clause 7.

Clause 8

This identifies the method of giving Notice under Clause 7 & 7S.

Clause 9

This needs completion. The clause makes clear that any liability that the Firm has by virtue of this Warranty ceases on the expiry of the stated period of years after practical completion of the Premises. (Note: the practical completion of the Development may be later).

Under English law the period should not exceed 6 years for agreements under hand, nor 12 years for those executed as a Deed.

In Scotland, the Prescription and Limitations (Scotland) Act 1973 prescribes a 5 year period.

Clause 10 and Attestation below

The appropriate method of execution by the Firm and the Purchaser/the Tenant should be checked carefully.

Clause 10S and Testing Clause below

This assumes the Firm is a partnership and the Purchaser/the Tenant is a Limited Company. Otherwise legal advice should be taken.

Published by
The British Property Federation Limited
35 Catherine Place, London SW1E 6DY Telephone: 0171-828 0111

© The British Property Federation, The Association of Consulting Engineers, The Royal Incorporation of Architects in Scotland, The Royal Institute of British Architects and The Royal Institution of Chartered Surveyors. 1992.

ISBN 0 900101 08 7

Warranty Agreement  CoWa/P&T

(In Scotland, leave blank. For applicable date see Testing Clause on page 4)

THIS AGREEMENT

is made the ....day of 199

BETWEEN:-

(insert name of the Consultant)

(1) ....

of/whose registered office is situated at ....

....("the Firm"), and

(insert name of the Purchaser/the Tenant)

(2) ....

whose registered office is situated at ....

....

(delete as appropriate)

("the Purchaser"/"the Tenant" which term shall include all permitted assignees under this Agreement).

WHEREAS:-

(delete as appropriate)

A. The Purchaser/the Tenant has entered into an agreement to purchase/an agreement to lease/a lease with

....

.... ("the Client") relating to

(insert description of the premises)

....

....

....("the Premises")

(delete as appropriate)

[forming part of..

....

(insert description of the development)

....

(insert address of the development)

at ....

.... ("the Development").]

(delete as appropriate)

["The Premises" are also referred to as "the Development" in this Agreement.]

(insert date of appointment) (delete/complete as appropriate)

B. By a contract ("the Appointment") dated ....
the Client has appointed the Firm as [architects/consulting structural engineers/consulting building services engineers/   surveyors] in connection with the Development.

C. The Client has entered or may enter into a contract ("the Building Contract") with

(insert name of building contractor or "a building contractor to be selected by the Client")

....

....

....

for the construction of the Development.

CoWa/P&T   2nd Edition
© BPF, ACE, RIAS, RICS, RIBA 1993

265

NOW IN CONSIDERATION OF THE PAYMENT OF ONE POUND (£1) BY THE PURCHASER/ THE TENANT TO THE FIRM (RECEIPT OF WHICH THE FIRM ACKNOWLEDGES) IT IS HEREBY AGREED as follows:-

1. The Firm warrants that it has exercised and will continue to exercise reasonable skill [and care] [care and diligence] in the performance of its services to the Client under the Appointment. In the event of any breach of this warranty:

    (a) subject to paragraphs (b) and (c) of this clause, the Firm shall be liable for the reasonable costs of repair renewal and/or reinstatement of any part or parts of the Development to the extent that

    – the Purchaser/the Tenant incurs such costs and/or
    – the Purchaser/the Tenant is or becomes liable either directly or by way of financial contribution for such costs.

    The Firm shall not be liable for other losses incurred by the Purchaser/the Tenant.

    (b) the Firm's liability for costs under this Agreement shall be limited to that proportion of such costs which it would be just and equitable to require the Firm to pay having regard to the extent of the Firm's responsibility for the same and on the basis that ....

    ....

    ....

    ....

    ....

    ....shall be deemed to have provided contractual undertakings on terms no less onerous than this Clause 1 to the Purchaser/the Tenant in respect of the performance of their services in connection with the Development and shall be deemed to have paid to the Purchaser/the Tenant such proportion which it would be just and equitable for them to pay having regard to the extent of their responsibility;

    (c) the Firm shall be entitled in any action or proceedings by the Purchaser/the Tenant to rely on any limitation in the Appointment and to raise the equivalent rights in defence of liability as it would have against the Client under the Appointment;

    (d) the obligations of the Firm under or pursuant to this Clause 1 shall not be released or diminished by the appointment of any person by the Purchaser/the Tenant to carry out any independent enquiry into any relevant matter.

2. [Without prejudice to the generality of Clause 1, the Firm further warrants that it has exercised and will continue to exercise reasonable skill and care to see that, unless authorised by the Client in writing or, where such authorisation is given orally, confirmed by the Firm to the Client in writing, none of the following has been or will be specified by the Firm for use in the construction of those parts of the Development to which the Appointment relates:-

    (a) high alumina cement in structural elements;

    (b) wood wool slabs in permanent formwork to concrete;

    (c) calcium chloride in admixtures for use in reinforced concrete;

    (d) asbestos products;

    (e) naturally occurring aggregates for use in reinforced concrete which do not comply with British Standard 882: 1983 and/or naturally occurring aggregates for use in concrete which do not comply with British Standard 8110: 1985.

    (f)

    In the event of any breach of this warranty the provisions of Clauses 1a, b, c and d shall apply.]

CoWa/P&T   2nd Edition
© BPF, ACE, RIAS, RICS, RIBA 1993

266

3. The Firm acknowledges that the Client has paid all fees and expenses properly due and owing to the Firm under the Appointment up to the date of this Agreement.

4. The Purchaser/the Tenant has no authority to issue any direction or instruction to the Firm in relation to the Appointment.

5. The copyright in all drawings, reports, models, specifications, bills of quantities, calculations and other documents and information prepared by or on behalf of the Firm in connection with the Development (together referred to in this Clause 5 as "the Documents") shall remain vested in the Firm but, subject to the Firm having received payment of any fees agreed as properly due under the Appointment, the Purchaser/the Tenant and its appointee shall have a licence to copy and use the Documents and to reproduce the designs and content of them for any purpose related to the Premises including, but without limitation, the construction, completion, maintenance, letting, promotion, advertisement, reinstatement, refurbishment and repair of the Premises. Such licence shall enable the Purchaser/the Tenant and its appointee to copy and use the Documents for the extension of the Premises but such use shall not include a licence to reproduce the designs contained in them for any extension of the Premises. The Firm shall not be liable for any use by the Purchaser/the Tenant or its appointee of any of the Documents for any purpose other than that for which the same were prepared by or on behalf of the Firm.

(insert amount)

(insert period)

6. The Firm shall maintain professional indemnity insurance in an amount of not less than pounds (£   ) for any one occurrence or series of occurrences arising out of any one event for a period of   years from the date of practical completion of the Premises under the Building Contract, provided always that such insurance is available at commercially reasonable rates. The Firm shall immediately inform the Purchaser/the Tenant if such insurance ceases to be available at commercially reasonable rates in order that the Firm and the Purchaser/the Tenant can discuss means of best protecting the respective positions of the Purchaser/the Tenant and the Firm in the absence of such insurance. As and when it is reasonably requested to do so by the Purchaser/the Tenant or its appointee the Firm shall produce for inspection documentary evidence that its professional indemnity insurance is being maintained.

(insert number
of times)

(delete if under
Scots law)

7. This Agreement may   be assigned   by the Purchaser/the Tenant by way of absolute legal assignment to another person taking an assignment of the Purchaser's/the Tenant's interest in the Premises without the consent of the Client or the Firm being required and such assignment shall be effective upon written notice thereof being given to the Firm. No further assignment shall be permitted.]

(insert number of times)

(delete if under
English law)

7S. *The Purchaser/the Tenant shall   be entitled to assign or transfer his/their rights under this Agreement   to any other person acquiring the Purchaser's/the Tenant's interest in the whole of the Premises without the consent of the Firm subject to written notice of such assignation being given to the Firm in accordance with Clause 8 hereof. Nothing in this clause shall permit any party acquiring such right as assignee or transferee to enter into any further assignation or transfer to anyone acquiring subsequently an interest in the Premises from him.]*

8. Any notice to be given by the Firm hereunder shall be deemed to be duly given if it is delivered by hand at or sent by registered post or recorded delivery to the Purchaser/the Tenant at its registered office and any notice given by the Purchaser/the Tenant hereunder shall be deemed to be duly given if it is addressed to "The Senior Partner"/"The Managing Director" and delivered by hand at or sent by registered post or recorded delivery to the above-mentioned address of the Firm or to the principal business address of the Firm for the time being and, in the case of any such notices, the same shall if sent by registered post or recorded delivery be deemed to have been received forty eight hours after being posted.

(complete as
appropriate)

9. No action or proceedings for any breach of this Agreement shall be commenced against the Firm after the expiry of   years from the date of practical completion of the Premises under the Building Contract.

CoWa/P&T   2nd Edition
© BPF, ACE, RIAS, RICS, RIBA 1993

267

[10. The construction validity and performance of this Agreement shall be governed by English law and the parties agree to submit to the non-exclusive jurisdiction of the English Courts.

[**AS WITNESS** the hands of the parties the day and year first before written.

Signed by or on behalf of the Firm ....

  in the presence of: ....

Signed by or on behalf of the Purchaser/the Tenant ....

  in the presence of: ....]

[**IN WITNESS WHEREOF** this Agreement was executed as a Deed and delivered the day and year first before written.

by the Firm

....

....

....

....

by the Purchaser/the Tenant

....

....

....

....]]

10S. *This Agreement shall be construed and the rights of the parties and all matters arising hereunder shall be determined in all respects according to the Law of Scotland.*

IN WITNESS WHEREOF these presents are executed as follows:-

SIGNED by the above named Firm at ....

on the ....day of.........................*Nineteen hundred and*.............................

as follows:-

....*(Firm's signature)*

Signature.......................................*Full Name* ....

Address ....

....*Occupation* ....

Signature.......................................*Full Name* ....

Address ....

....*Occupation* ....

SIGNED by the above named Purchaser/Tenant at ....

on the ....day of.........................*Nineteen hundred and*.............................

as follows:-

For and on behalf of the Purchaser/the Tenant

....*Director/Authorised Signatory*

....*Director/Authorised Signatory*]

CoWa/P&T 2nd Edition
© BPF, ACE, RIAS, RICS, RIBA 1993

268

Appendix 6
MCWa/F

The JCT Standard Form of Agreement for Collateral Warranty for use where a warranty is to be given to a company providing finance (funder) for the proposed building works by a contractor (2001 Edition)

This document is the copyright of the Joint Contracts Tribunal

Form of Agreement for

Collateral Warranty
for funding institutions
by a main contractor: MCWa/F

The forms in this pad are for use where a Warranty is to be given by a main contractor to a company providing finance ('Funder') for proposed building works let or to be let under:

– a JCT Standard Form of Building Contract 1998 Edition (JCT 98), which may include Performance Specified Work or may have been modified by the Contractor's Designed Portion Supplement;

– or a JCT Intermediate Form of Building Contract for works of simple content 1998 Edition (IFC 98);

– or a JCT Standard Form of Building Contract With Contractor's Design 1998 Edition (WCD 98).

The Forms are **not** for use where the Building Works are situated in Scotland. For such Works the forms issued by the Scottish Building Contract Committee should be used.

GENERAL ADVICE

1 The JCT Warranty MCWa/F is for use with Building Contracts let under JCT 98, IFC 98 and WCD 98 where the Employer has an agreement ('Finance Agreement') with a person ('Funder') who will provide finance for the building works ('Works') required by the Employer.

2 JCT Warranty MCWa/F is collateral to the Building Contract and, generally, should not be entered into before the date of the Building Contract although the enabling clause (provided with each copy of the Warranty in the pads) and which should be included within that contract does envisage the possibility of the earlier execution of the warranty. If the warranty is not entered into prior to or contemporaneously with entry into the Building Contract but the warranty might be required by the Employer during the currency of the Building Contract, then, under the enabling clause, the Contractor can be required to enter into the warranty. If the Employer might so require, it is essential that the enabling clause be included with the tender documents and form part of the conditions of the Building Contract; if the clause is not so included, the Employer can only require the Contractor to enter into the warranty if the Contractor so agrees.

3 Clauses 1(a), 6(3), 6(4), 9 and 13 of JCT Warranty MCWa/F all require completion. Where the warranty is entered into prior to or contemporaneously with the Building Contract being entered into, then such variable information will be known to the Contractor at that time. Where, however, it is entered into after the date of the Building Contract, it is essential that, pre-contract, the Main Contractor is informed:

(a) of the way in which the clauses requiring completion within the JCT warranty will be completed; if subsequently, the warranty is to be completed on different terms then the agreement of the Main Contractor to those different terms would be required; and

(b) if the Employer seeks to make any amendments to the terms of the JCT Warranty MCWa/F which he will require the Contractor to give, then the Employer must set out the amendments as part of the Main Contract tender documentation. **Amendments to the clauses of MCWa/F should however be avoided.**

COMMENTARY ON CLAUSES

Recitals A and B are self-explanatory. The following notes on the clauses are to assist in understanding the use of the document.

Clause 1
This confirms that the Contractor owes the same obligation to the Funder as he owes to the Employer under the Building Contract, but subject to paragraphs (a) and (b).

It is for the Employer and the Funder to agree on the information to be given to the Funder about the content of the Building Contract. It would be difficult for the Funder then to allege, after executing MCWa/F, that he was not aware of the content of the Building Contract and/or of the enforceable agreements arising out of and relating to the Building Contract to which recital B refers.

Paragraphs (a) and (b) qualify and limit in two ways the Contractor's liability in the event of an alleged breach by the Contractor of the Building Contract.

1(a) By this provision the Contractor's potential liability is limited. The Warranty given by the Contractor is based on the assumptions set out in sub-clauses (i), (ii) and (iii) and his liability is limited accordingly. Consequently, this limit on liability applies whether or not the Consultants employed by the Employer have given contractual undertakings and/or Sub-Contractors have given warranties, to the Purchaser/Tenant. Clause 1(a)(i) and 1(a)(ii) should be completed. No such assumptions, as referred to above, are appropriate where the Contractor is designing and building under WCD 98 and clause (a) should be deleted entirely where this contract is being used.

1(b) By this clause (but subject to clause 1(a)) the Funder is in no better and no worse position than the Employer in enforcing the terms of the Building Contract; and the Contractor may raise against the Funder the same defences of liability as he could against the Employer.

1(c) This states that the appointment by the Funder of any consultant (e.g. to survey the building(s) comprising the Works) shall not affect any liability, which the Contractor might otherwise have.

Clause 2
The Contractor warrants that he has not and will not use materials in the Works other than in accordance with the guidelines contained in the publication 'Good Practice in Selection of Construction Materials'. Clause 2 excuses the use by the Contractor of other materials if the use has been required or authorised by the Employer or by the Architect/Contract Administrator on behalf of the Employer. Where (as in WCD 98) the Contractor is the specifier of materials to be used in the Works, he will remain bound by the requirement in clause 2 on the use of materials in accordance with the above publication.

Clause 3
This is included to make it clear that the Funder has no power or authority to instruct the Contractor in his duties to the Employer under the Building Contract. This is however subject to clauses 5 and 6 under which the Funder may take over the rights and duties etc. of the Employer.

Clause 4
This clause makes clear that the Funder has no liability for any amounts due under the Building Contract until he has given a notice under clause 5 following termination of the Finance Agreement or under clause 6 where the Funder takes over as Employer.

Clauses 5, 6 and 7
These entitle the Funder to take over as Employer in which case the Funder also takes over the Employer's rights and duties but on the terms of **clause 7**. The purpose is to permit the Funder to take over as Employer rather than the Contractor determining his employment under the relevant provisions in the Building Contract or exercising any common law right of treating the Building Contract as having been repudiated by the Employer.

Clause 6(1)(a) requires the Contractor to copy to the Funder any preliminary notices which have to be given under the Building Contract before the Contractor can give notice to the Employer that the Contractor's employment is determined; this provides an advance warning so that the Funder can reach a provisional decision on whether he will act under clause 6(4).

When the Contractor

– is in a position to give to the Employer actual notice of determination of the Contractor's employment under the Building Contract,

or

– intends to inform the Employer that he is treating the Building Contract as having been repudiated by the Employer

the Contractor, by clause 6(3), may not give such notice or so inform the Employer without having given the Funder 7 days (or such other period as may have been prescribed in clause 6(3)) notice.

The Funder may within that period act under clause 6(4) and give notice that he will take over the rights and obligations of the Employer but on the terms of clause 7 in regard to the Funder accepting all the liabilities of the Employer at the date of the takeover and agreeing to perform all the obligations of the Employer under the Building Contract. The proviso to clause 6(4) preserves the Employer's right under the Building Contract if he considers that any determination by the Contractor of his employment is wrongful.

Clause 8
This clause will normally only be applicable where the Contractor has carried out a design function in relation to the Works e.g. under WCD 98 or under JCT 98 modified by the Contractor's Designed Portion. Reasonable use by the Funder of drawings and associated documents is necessary in most cases. By this clause the Funder is given the rights that might reasonably be expected but it does not allow the reproduction of the designs for any purpose outside the scope of the Works; such reproduction would require a further agreement with the Contractor.

Clause 9
Attention is called to the side note which states when clause 9 must be deleted. Where clause 9 applies it should be noted:

– the obligation to have and maintain professional indemnity insurance is subject to the proviso that "such insurance is available" and, if available, is "available at commercially reasonable rates"; and that if after the insurance has been taken out the rates upon any renewal cease to be "commercially reasonable" the Contractor must inform the Funder as provided in the clause;

– not all Contractors will have, or can obtain, professional indemnity insurance and Funders must recognise this when seeking to operate the terms of clause 9;

– where this clause is not deleted the amount and period for maintaining cover must be inserted. The words in square brackets must be deleted as necessary depending on whether the amount of cover is to apply "for any occurrence or series of occurrences arising out of one event" or is "in the aggregate". This clause also provides for a limit to the total claims arising out of or in connection with pollution and contamination.

Clause 10
This provision allows the Contractor to comply with obligations imposed upon it under clauses 5 and 6, without being liable to the Employer for a breach of contract.

Clause 11
This clause indicates the right of assignment by the Funder.

Clause 12
This identifies the method of giving notices under the Agreement.

Clause 13
This needs completion. The clause makes clear that any liability that the Contractor has by virtue of this Warranty ceases on the expiry of the stated period of years after Practical Completion of the Works (or, where the Works are completed in sections, after practical completion of the relevant section).

This period should not normally exceed 6 years for Agreements under hand nor 12 years for those executed as a deed.

Clause 14
This clause makes it clear that the Contractor does not, under this Warranty, have any liability for delay in completion of the Works unless and until a notice is served by the Funder under clause 5 or clause 6.

Clause 15
The JCT has agreed that its Forms should contract out of the Contracts (Rights of Third Parties) Act 1999 and this clause so provides.

Clause 16 and Attestation Clause
Clause 16 should not be at variance with the terms of clause 1·10 (JCT 98), clause 1·7 (WCD 98), and clause 1·15 (IFC 98), on the law applicable to the Building Contract.

It will be noted that the agreement is to be governed by English law.

Disputes under the Warranty Agreement are not, unlike disputes under the Building Contract, intended to be referable to arbitration unless the Contractor and the Funder otherwise agree; in which case a provision relating to the reference to arbitration and the conduct of the arbitration should be inserted.

Main Contract: Enabling Clause

for inclusion in the Conditions of Contract where the Building Contract is to be let on the JCT Standard Form of Building Contract 1998 Edition (JCT 98) whether or not the Contract Documents require Performance Specified Work to which clause 42 of JCT 98 applies or is modified by the Contractor's Designed Portion Supplement; on the JCT Intermediate Form of Building Contract for works of simple content 1998 Edition (IFC 98); or on the JCT Standard Form of Building Contract with Contractor's Design 1998 Edition (WCD 98):

Insert as an additional clause 19B (JCT 98), 18B (WCD 98), 3·1B (IFC 98)

*Insert correct clause number where WCD 98 or IFC 98 is used.

*19B Warranty by Contractor – to a person ('Funder') providing finance for the Works

19B·1 ·1 By a notice in writing by actual delivery or by special delivery or recorded delivery to the Contractor at the address stated in the Articles of Agreement the Employer may require the Contractor to enter into a JCT Warranty Agreement ('Warranty MCWa/F') with a Funder identified in the notice. The Warranty must be entered into within _____ days (not to exceed 14) from receipt of the Employer's notice.

19B·1 ·2 A notice under clause 19B·1·1 is only valid if the entries in clauses 1(a), 6(3), 6(4), 13 (and also in clause 9 where that clause is not to be deleted) of the Warranty to be entered into have been notified in writing by the Employer to the Contractor prior to the date of this Contract.

19B·2 Clause 19B·1 shall not apply where the Contractor, on or prior to the date of this Contract, has given a Warranty on the JCT Warranty Agreement (MCWa/F) to the Funder.

Warranty Agreement

This form is for use where a Warranty is to be given by a Contractor to a company providing finance (Funder) for a building project. Where the Funder is acting as an agent for a syndicate of banks and/or other companies providing finance a recital should be added to refer to this as appropriate.

This Agreement

is made the _____ day of _____ 20 _____

Insert name of the Contractor. **(1)** Between _____

_____

whose registered office is at _____

_____ ('the Contractor') and

Insert name of the Employer. **(2)** _____

whose registered office is at _____

_____ ('the Employer') and

Insert name of the Funder. **(3)** _____

whose registered office is at _____

_____ ('the Funder'

which term shall include all permitted assignees under this Agreement).

Whereas

A   The Funder has entered into an agreement ('the Finance Agreement') with the Employer for the provision of finance in connection with the carrying out of building works

Insert description of the building works.

_____

_____

_____

_____ ('the Works'

which term shall include any changes made to the building works in accordance with the Building Contract).

Insert date.

B   By a contract dated _____ ('the Building Contract' which term shall include any enforceable agreements reached between the Employer and the Contractor and which arise out of and relate to the same) the Employer has appointed the Contractor to carry out and complete the Works / the Works by phased Sections.

Delete as appropriate.

274

Now it is hereby agreed

In consideration of the payment of one pound (£1) by the Funder to the Contractor, receipt of which the Contractor acknowledges:

1   The Contractor warrants that it has complied and will continue to comply with the Building Contract. In the event of any breach of this warranty:

Delete the whole of clause 1(a) where using WCD 98.

(a) the Contractor's liability to the Funder for costs under this Agreement shall be limited to the proportion of the Funder's losses which it would be just and equitable to require the Contractor to pay having regard to the extent of the Contractor's responsibility for the same, on the following assumptions, namely that:

Insert the discipline of the Consultant Warrantors.

(i) [_____ the Consultant[s]] engaged by the Employer has/have provided contractual undertakings to the Funder that it/they has/have and will perform their services in connection with the Works in accordance with the terms of his/their respective consultancy agreements and that there are no limitations on liability as between the Consultant and the Employer in the consultancy agreement[s];

Insert the trade or speciality of the Sub-Contractor Warrantors.

(ii) [_____ the Sub-Contractor[s]] has/have provided a warranty to the Funder in respect of design of the Sub-Contract Works that it/they has/have carried out and for which there is no liability of the Contractor to the Employer under the Building Contract;

(iii) the Consultants and the Sub-Contractors have paid to the Funder such proportion of the Funder's losses which it would be just and equitable for them to pay having regard to the extent of their responsibility for the Funder's losses.

(b) The Contractor shall be entitled in any action or proceedings by the Funder to rely on any term in the Building Contract and to raise the equivalent rights in defence of liability as it would have against the Employer under the Building Contract.

(c) The obligations of the Contractor under or pursuant to this clause 1 shall not be released or diminished by the appointment of any person by the Funder to carry out any independent enquiry into any relevant matter.

2   The Contractor further warrants that unless required by the Building Contract or unless otherwise authorised in writing by the Employer or by the Architect/Contract Administrator named in or appointed pursuant to the Building Contract (or, where such authorisation is given orally, confirmed in writing by the Contractor to the Employer and/or the Architect/Contract Administrator), it has not and will not use materials in the Works other than in accordance with the guidelines contained in the edition of the publication 'Good Practice in Selection of Construction Materials' (Ove Arup & Partners) current at the date of the Building Contract. In the event of any breach of this warranty the provisions of clauses 1(a), (b) and (c) shall apply.

3   The Funder has no authority to issue any direction or instruction to the Contractor in relation to the Building Contract unless and until the Funder has given notice under clause 5 or clause 6.

4   The Funder has no liability to the Contractor in respect of amounts due under the Building Contract unless and until the Funder has given notice under clause 5 or clause 6.

5   The Contractor agrees that, in the event of the termination of the Finance Agreement by the Funder, the Contractor shall, if so required by notice in writing given by the Funder and subject to clause 7, accept the instructions of the Funder or its appointee to the exclusion of the Employer in respect of the Works upon the terms and conditions of the Building Contract. The Employer acknowledges that the Contractor shall be entitled to rely on a notice given to the Contractor by the Funder under this clause 5 as conclusive evidence for the purposes of this Agreement of the termination of the Finance Agreement by the Funder; and further acknowledges that such acceptance of the instructions of the Funder to the exclusion of the Employer shall not constitute any breach of the Contractor's obligations to the Employer under the Building Contract.

6   (1) The Contractor shall not exercise any right of determination of his employment under the Building Contract without having first

        (a) copied to the Funder any written notices required by the Building Contract to be sent to the Architect/Contract Administrator or to the Employer prior to the Contractor being entitled to give notice under the Building Contract that his employment under the Building Contract is determined, and

        (b) given to the Funder written notice that he has the right under the Building Contract forthwith to notify the Employer that his employment under the Building Contract is determined.

6   (2) The Contractor shall not treat the Building Contract as having been repudiated by the Employer without having first given to the Funder written notice that he intends to so inform the Employer.

6   (3) The Contractor shall not

        –   issue any notification to the Employer to which clause 6(1)(b) refers

        or

        –   inform the Employer that he is treating the Building Contract as having been repudiated by the Employer as referred to in clause 6(2)

Insert any different number that has been agreed and delete "7".

before the lapse of 7 _____ days from receipt by the Funder of the written notice by the Contractor which the Contractor is required to give under clause 6(1)(b) or clause 6(2).

6   (4) The Funder may, not later than the expiry of the 7 _____ days referred to in clause 6(3) require the Contractor by notice in writing and subject to clause 7 to accept the instructions of the Funder or its appointee to the exclusion of the Employer in respect of the Works upon the terms and conditions of the Building Contract. The Employer acknowledges that the Contractor shall be entitled to rely on a notice given to the Contractor by the Funder under clause 6(4) and that acceptance by the Contractor of the instructions of the Funder to the exclusion of the Employer shall not constitute any breach of the Contractor's obligations to the Employer under the Building Contract. Provided that, subject to clause 7, nothing in clause 6(4) shall relieve the Contractor of any liability he may have to the Employer for any breach by the Contractor of the Building Contract or where the Contractor has wrongfully served notice under the Building Contract that he is entitled to determine his employment under the Building Contract or has wrongfully treated the Building Contract as having been repudiated by the Employer.

7   It shall be a condition of any notice given by the Funder under clause 5 or clause 6(4) that the Funder or its appointee accepts liability for payment of the sums due and payable to the Contractor under the Building Contract and for performance of the Employer's obligations including payment of any sums outstanding at the date of such notice. Upon the issue of any notice by the Funder under clause 5 or clause 6(4), the Building Contract shall continue in full force and effect as if no right of determination of his employment under the Building Contract, nor any right of the Contractor to treat the Building Contract as having been repudiated by the Employer had arisen and the Contractor shall be liable to the Funder and its appointee under the Building Contract in lieu of its liability to the Employer. If any notice given by the Funder under clause 5 or clause 6(4) requires the Contractor to accept the instructions of the Funder's appointee, the Funder shall be liable to the Contractor as guarantor for the payment of all sums from time to time due to the Contractor from the Funder's appointee.

Delete clause 8 if not applicable.

8   The copyright in all drawings, reports, models, specifications, plans, schedules, bills of quantities, calculations and other similar documents prepared by or on behalf of the Contractor in connection with the Works (together referred to as the Documents) shall remain vested in the Contractor but subject to the Employer having paid all monies due and payable under the Building Contract, the Funder and its appointee shall have an irrevocable, royalty-free, non-exclusive licence to copy and use the Documents and to reproduce the designs and content of them for any purpose related to the Works including, but without limitation, the construction, completion, maintenance, letting, sale, promotion, advertisement, reinstatement, refurbishment and repair of the Works. Such licence shall enable the Funder and its appointee to copy and use the Documents for the extension of the Works but such use shall not include a licence to reproduce the designs contained in them for any extension of the Works. The Contractor shall not be liable for any such use by the Funder or its appointee of any of the Documents for any purpose other than that for which the same were prepared by or on behalf of the Contractor.

9   The Contractor has and shall maintain professional indemnity insurance in an amount

Insert amount.

each year of not less than _____ pounds (£_____)

Delete words in square brackets as necessary.

[for any one occurrence or series of occurrences arising out of one event] [in the aggregate]

[but limited to _____ pounds (£ _____) in the aggregate in respect of all claims arising out of or in connection with pollution and contamination]

Insert period.

Delete clause 9 except where the Building Contract is WCD 98 or JCT 98 where modified by the Contractor's Designed Portion Supplement and/or includes Performance Specified Work to which clause 42 of JCT 98 applies.

for a period ending _____ years after the date of Practical Completion of the Works (or of practical completion of a Section of the Works where the Building Contract is modified for completion by phased sections) under the Building Contract, provided such insurance is available at commercially reasonable rates. The Contractor shall immediately inform the Funder if such insurance ceases to be available at commercially reasonable rates in order that the Contractor and the Funder can discuss the means of best protecting the respective positions of the Funder and the Contractor in the absence of such insurance. As and when it is reasonably requested to do so by the Funder or its appointee under clause 5 or clause 6 the Contractor shall produce for inspection documentary evidence that its professional indemnity insurance is being maintained.

10   The Employer has agreed to be a party to this Agreement for the purposes of acknowledging that the Contractor shall not be in breach of the Building Contract by complying with the obligations imposed on it by clauses 5 and 6.

11   This Agreement may be assigned, without the consent of the Contractor by the Funder, by way of absolute legal assignment, to another person (P1) providing finance or re-finance in connection with the carrying out of the Works and by P1, by way of absolute legal assignment, to another person (P2) providing finance or re-finance in connection with the carrying out of the Works. In such cases the assignment shall only be effective upon written notice thereof being given to the Contractor. No further or other assignment of this Agreement will be permitted and in particular P2 shall not be entitled to assign this Agreement.

12   Any notice to be given by the Contractor shall be deemed to be duly given if it is delivered by hand at or sent by special delivery or recorded delivery to the Funder at its registered office; and any notice given by the Funder shall be deemed to be duly given if it is delivered by hand at or sent by special delivery or recorded delivery to the Contractor at its registered office; and in the case of any such notices, the same shall, if sent by special delivery or recorded delivery, be deemed (subject to proof to the contrary) to have been received forty-eight hours after being posted.

Insert number of years.

13   No action or proceedings for any breach of this Agreement shall be commenced against the Contractor after the expiry of _____ years from the date of Practical Completion of the Works under the Building Contract or, where the Works are modified for completion by phased sections, no action or proceedings shall be commenced against the Contractor in respect of any Section after the expiry of _____ years from the date of practical completion of such Section.

14   Notwithstanding any other provisions of this Agreement, the Contractor shall have no liability under this Agreement for delay under the Building Contract unless and until the Funder serves notice pursuant to clause 5 or clause 6. For the avoidance of doubt the Contractor shall not be required to pay liquidated and ascertained damage in respect of the period of delay where the same has been paid to or deducted by the Employer.

15   Notwithstanding any other provision of this Agreement nothing in this Agreement confers or purports to confer any right to enforce any of its terms on any person who is not a party to it.

16   This Agreement shall be governed by English law.

278

Notes

[A1] For Agreement executed under hand and NOT as a deed.

[A1] **AS WITNESS THE HANDS OF THE PARTIES HERETO**

[A1] Signed by or on behalf of the Funder _____

    in the presence of:

[A1] Signed by or on behalf of the Employer _____

    in the presence of:

[A1] Signed by or on behalf of the Contractor _____

    in the presence of:

[A2] For Agreement executed as a deed under the law of England and Wales by a company or other body corporate: insert the name of the party mentioned and identified on page 1 and then use *either* [A3] and [A4] *or* [A5].
If the party is an *individual* see note [A6].

[A3] For use if the party is using its common seal, which should be affixed under the party's name.

[A4] For use of the party's officers authorised to affix its common seal.

[A5] For use if the party is a company registered under the Companies Acts which is not using a common seal: insert the names of the two officers by whom the company is acting *who MUST be either a director and the company secretary or two directors*, and insert their signatures with 'Director' or 'Secretary' as appropriate. *This method of execution is NOT valid for local authorities or certain other bodies incorporated by Act of Parliament or by charter if exempted under s.718(2) of the Companies Act 1985.*

[A6] See page 8.

[A2] **EXECUTED AS A DEED BY THE FUNDER**

[A6]   hereinbefore mentioned namely _____

[A3]   by affixing hereto its common seal

[A4]   in the presence of:

*   OR —

[A5]   acting by a director and its secretary* / two directors* whose signatures are here subscribed:
    namely _____

    [Signature] _____ *DIRECTOR*

    and _____

    [Signature] _____ *SECRETARY* / DIRECTOR**

* *Delete as appropriate*

[A2] For Agreement executed as a deed under the law of England and Wales by a company or other body corporate: insert the name of the party mentioned and identified on page 1 and then use *either* [A3] and [A4] *or* [A5].
If the party is an *individual* see note [A6].

[A3] For use if the party is using its common seal, which should be affixed under the party's name.

[A4] For use of the party's officers authorised to affix its common seal.

[A5] For use if the party is a company registered under the Companies Acts which is not using a common seal: insert the names of the two officers by whom the company is acting *who MUST be either a director and the company secretary or two directors,* and insert their signatures with 'Director' or 'Secretary' as appropriate. *This method of execution is NOT valid for local authorities or certain other bodies incorporated by Act of Parliament or by charter if exempted under s.718(2) of the Companies Act 1985.*

[A6] If executed as a deed by an *individual:* insert the name at [A2], delete the words at [A3], substitute 'whose signature is here subscribed' and insert the individual's signature. The individual MUST sign in the presence of a witness who attests the signature. Insert at [A4] the signature and name of the witness. Sealing by an individual is not required.

[A2] **AND AS A DEED BY THE EMPLOYER**
[A6] hereinbefore mentioned namely _____

[A3] by affixing hereto its common seal

[A4] in the presence of:

* OR

[A5] acting by a director and its secretary* / two directors* whose signatures are here subscribed:
namely _____

[Signature] _____ DIRECTOR

and _____

[Signature] _____ SECRETARY* / DIRECTOR*

[A2] **AND AS A DEED BY THE CONTRACTOR**
[A6] hereinbefore mentioned namely _____

[A3] by affixing hereto its common seal

[A4] in the presence of:

* OR

[A5] acting by a director and its secretary* / two directors* whose signatures are here subscribed:
namely _____

[Signature] _____ DIRECTOR

and _____

[Signature] _____ SECRETARY* / DIRECTOR*

* *Delete as appropriate*

Appendix 7
MCWa/P&T

The JCT Standard Form of Agreement for Collateral Warranty for use where a warranty is to be given to a purchaser or tenant of building works or part thereof by a main contractor (2001 Edition)

This document is the copyright of the Joint Contracts Tribunal

Form of Agreement for

Collateral Warranty
**for purchasers and tenants
by a main contractor: MCWa/P&T**

The forms in this pad are for use where a Warranty is to be given by the main contractor to a purchaser or tenant of the whole or part of the building(s) comprising the Works which have been practically completed under:

– a JCT Standard Form of Building Contract 1998 Edition (JCT 98), which may include Performance Specified Work or may have been modified by the Contractor's Designed Portion Supplement;

– or a JCT Intermediate Form of Building Contract for works of simple content 1998 Edition (IFC 98);

– or a JCT Standard Form of Building Contract With Contractor's Design 1998 Edition (WCD 98).

The Forms are **not** for use where the Building Works are situated in Scotland. For such Works the forms issued by the Scottish Building Contract Committee should be used.

GENERAL ADVICE

1 It is essential that, **pre-contract**, the Main Contractor is provided with a copy of the MCWa/P&T showing the necessary insertions and deletions.

2 It is essential that, **pre-contract**, the Main Contractor is informed:

(a) that the Building Contract includes a provision enabling the Employer after Practical Completion to require the Contractor to enter into a Warranty on MCWa/P&T with the purchaser(s) or tenant(s) whose maximum number has been stated in the tender documents or otherwise notified to the Contractor and confirmed in the Building Contract. **The text of such an enabling clause is included with each copy of the Warranty in the pads;**

(b) if the Employer seeks to make any amendments to the terms of the JCT Warranty MCWa/P&T which he will require the Contractor to give, then the Employer must set out the amendments as part of the Main Contract tender documentation. **Amendments to the clauses of MCWa/P&T should however be avoided.**

COMMENTARY ON CLAUSES

Parties
In (2) the name of the Purchaser/Tenant should correspond with the name given in the Employer's notice under clause 19A·2 in the enabling clause.

Recital A
This needs completion. The building comprising the Works or a part thereof which has been purchased or let should be carefully identified; and should correspond with the identification of the building(s) purchased or let given in the Employer's notice under clause 19A·2 in the enabling clause.

Recital B
This needs completion. Under clause 19A·2 in the enabling clause the Employer cannot require MCWa/P&T to be entered into before the Works have reached practical completion or, where the Works are completed in Phased Sections, before practical completion of the relevant Section. See however clause 19A·5 of the enabling clause to cover the circumstances of a pre-let or a purchase prior to such Practical Completion.

Clause 1
This confirms that the Contractor owes the same obligation to the Purchaser/Tenant as he owes to the Employer under the Building Contract, but subject to paragraphs (a), (b), (c) and (d).

It is for the Employer and the Purchaser/Tenant to agree on the information to be given to the Purchaser/Tenant about the content of the Building Contract. It would be difficult for the Purchaser/Tenant then to allege, after executing MCWa/P&T, that he was not aware of the content of the Building Contract and/or of the enforceable agreements arising out of and relating to the Building Contract to which recital B refers.

Paragraphs (a), (b), (c) and (d) qualify and limit the Contractor's liability to the Purchaser/Tenant in a number of ways in the event of an alleged breach by the Contractor of the Building Contract.

This Warranty provides under 1(a) that the Contactor is liable for the reasonable costs of repair, renewal and/or reinstatement of the building(s) comprising the Works to the extent that the Purchaser/Tenant incurs such costs and/or the Purchaser/Tenant is or becomes liable either directly or by way of financial contribution for such costs. Additionally, the Contractor may be made liable for further losses either under 1(a)(i) where a financial cap *in respect of each breach* is to be inserted or under the alternative 1(a)(i) which provides for a financial cap *in respect of a maximum liability*. Whenever practicable the Purchaser/Tenant, bearing in mind his obligation to mitigate the damages he has suffered by an alleged breach of the Building Contract, should inform the Contractor of the defects etc. prior to having them repaired etc. and consider any offer by the Contractor to carry out the repair etc.

1(b) This clause makes it clear that where 1(a)(i) provisions are both deleted the liability of the Contractor is limited to those costs referred to in clause 1(a).

1(c) By this provision the Contractor's potential liability is limited. The Warranty given by the Contractor is based on the assumptions set out in sub-clauses (i), (ii) and (iii) and his liability is limited accordingly. Consequently, this limit on liability applies whether or not the Consultants employed by the Employer have given contractual understandings and/or the Sub-Contractors have given warranties, to the Purchaser/Tenant.

Clauses (c)(i) and (c)(ii) should be completed. No such assumptions, as referred to above, are appropriate where the Contractor is designing and building under WCD 98 and clause (c) should be entirely deleted where this contract is being used. When the Contractor (see the General Advice above) is invited to tender he should be given a copy of MCWa/P&T with the description of other warrantors to be inserted.

1(d) By this clause (but subject to clause 1(a)) the Purchaser/Tenant is in no better or no worse position than the Employer in enforcing the terms of the Building Contract; and the Contractor may raise against the Purchaser/Tenant the same defences of liability as he could against the Employer.

1(e) This states that the appointment by the Purchaser/Tenant of any consultant (e.g. to survey the building(s) comprising the Works) shall not affect any liability, which the Contractor might otherwise have.

Clause 2

The Contractor warrants that he has not and will not use materials in the Works other than in accordance with the guidelines contained in the publication 'Good Practice in Selection of Construction Materials'. Clause 2 excuses the use by the Contractor of other materials if the use has been required or authorised by the Employer or by the Architect/Contract Administrator on behalf of the Employer. Where (as in WCD 98) the Contractor is the specifier of materials to be used in the Works, he will remain bound by the requirement in clause 2 on the use of materials in accordance with the above publication.

Clause 3

This is included to make it clear that the Purchaser/Tenant has no power or authority to instruct the Contractor in his duties to the Employer under the Building Contract. As the Warranty MCWa/P&T (subject to clause 19A·5 in the enabling clause) is only entered into on or after Practical Completion, clause 3 will relate mainly to the post-Practical Completion obligation of the Contractor under the Building Contract in respect of the Defects Liability Period.

Clause 4

This clause will normally only be applicable where the Contractor has carried out a design function in relation to the Works e.g. under WCD 98 or under JCT 98 modified by the Contractor's Designed Portion. Reasonable use by the Purchaser/Tenant of drawings and associated documents is necessary in most cases. By this clause the Purchaser/Tenant is given the rights that might reasonably be expected but it does not allow the reproduction of the designs for any purpose outside the scope of the Works (or of that part of the Works) purchased or rented by the Purchaser/Tenant): such reproduction would require a further agreement with the Contractor.

Clause 5

Attention is called to the side note stating when clause 5 must be deleted. Where clause 5 applies it should be noted:

- the obligation to have and maintain professional indemnity insurance is subject to the provision that "such insurance is available" and, if available, is "available at commercially reasonable rates"; and that if after the insurance has been taken out the rates upon any renewal cease to be "commercially reasonable" the Contractor must inform the Purchaser/Tenant as provided in the clause;

- not all Contractors will have, or can obtain, professional indemnity insurance and Purchasers/Tenants must recognise this when seeking to operate the terms of clause 5;

- where this clause is not deleted the amount and period for maintaining cover must be inserted. The words in square brackets must be deleted as necessary depending on whether the amount of cover is to apply "for any occurrence or series of occurrences arising out of one event" or is "in the aggregate". This clause also provides for a limit to the total claims arising out of or in connection with pollution and contamination.

Clause 6

This allows the Purchaser/Tenant to assign the benefit of the Contractor's Warranty provided it is done by formal legal assignment and relates to the entire interest of the original Purchaser/Tenant. The assignment by the Purchaser/Tenant can be without the Contractor's consent. Thereafter, only one further assignment of the Purchaser's/Tenant's interest in the Works can be made.

Clause 7

This identifies the method of giving notice under the Agreement.

Clause 8

This needs completion. The clause makes clear that any liability that the Contractor has by virtue of this Warranty ceases on the expiry of the stated period of years after Practical Completion of the Works (or, where the Works are completed in sections, after practical completion of the relevant section).

This period should not normally exceed 6 years for Agreements under hand nor 12 years for those executed as a deed.

The way in which clause 8 is to be completed should be set out in the copy of the Warranty MCWa/P&T which the Contractor should be given as part of the tender documentation.

Clause 9

This clause makes it clear that the Contractor does not, under this Warranty, have any liability for delay in completion of the Works.

Clause 10

The JCT has agreed that its Forms should contract out of the Contracts (Rights of Third Parties) Act 1999 and this clause so provides.

Clause 11 and Attestation Clause

Clause 11 should not be at variance with the terms of clause 1·10 (JCT 98), clause 1·7 (WCD 98), and clause1·15 (IFC 98), on the law applicable to the Building Contract.

It will be noted that the agreement is to be governed by English law.

Disputes under the Warranty Agreement are not, unlike disputes under the Building Contract, intended to be referable to arbitration unless the Contractor and the Purchaser/Tenant otherwise agree; in which case a provision relating to the reference to arbitration and the conduct of the arbitration should be inserted.

Main Contract: Enabling Clause

for inclusion in the Conditions of Contract where the Building Contract is to be let on the JCT Standard Form of Building Contract,1998 Edition (JCT 98) whether or not the Contract Documents require Performance Specified Work to which clause 42 of JCT 98 applies or is modified by the Contractor's Designed Portion Supplement; on the JCT Intermediate Form of Building Contract for works of simple content 1998 Edition (IFC 98); or on the JCT Standard Form of Building Contract with Contractor's Design 1998 Edition (WCD 98):

Insert as an additional clause 19A (JCT 98), 18A (WCD 98), 3·1A (IFC 98)

*Insert correct clause number where WCD 98 or IFC 98 is used.

*19A Warranty by Contractor – purchaser or tenant

19A·1 Clause 19A shall only apply where, before the Contract has been entered into, the Employer has in a statement in writing to the Contractor (receipt of which the Contractor hereby acknowledges) set out the maximum number of Warranties that he may require pursuant to clause 19A.

19A·2 On or after the date of Practical Completion of the Works (or of practical completion of a Section of the Works where the Contract is modified for completion by phased sections) the Employer, by a notice in writing by actual delivery or by special delivery or recorded delivery to the Contractor at the address stated in the Agreement, may require as follows: the Contractor shall enter into a Warranty Agreement on the terms of the JCT Warranty Agreement MCWa/P&T (a copy of which, with clause 1(a) and 1(a)(i) and clause 8 completed and also clause 5 where that clause is not deleted, has been given to the Contractor before the Contract has been entered into, receipt of which the Contractor hereby acknowledges) with the Purchaser/Tenant referred to in clause 19A·4.

(Note: If, after the Building Contract has been entered into, the Employer requires any amendment to the terms of the Warranty Agreement MCWa/P&T given to the Contractor pursuant to this clause, this can only be done by a further agreement with the Contractor.)

Complete as applicable.

19A·3 The Warranty shall be entered into within_____days (not to exceed 14) from receipt of the Employer's notice referred to in clause 19A·2.

Delete as applicable.

19A·4 A notice under clause 19A·2 shall identify the Purchaser / Tenant with whom the Employer has entered into an agreement to purchase / an agreement to lease / a lease and where relevant shall sufficiently identify that part of the building(s) comprising the Works which has been purchased / let.

19A·5 The Employer may require the Contractor to enter into a Warranty Agreement as referred to in clause 19A·2 with a person who has entered into an agreement to purchase / an agreement to lease / a lease as referred to in clause 19A·4 before the date of Practical Completion of the Works (or of practical completion of a Section of the Works where the Contract is modified for completion by phased sections); but on condition that the Warranty Agreement shall not have effect until the date of Practical Completion of the Works (or, as relevant, the date of practical completion of a Section).

Warranty Agreement

This form is for use where a Warranty is to be given by a Contractor to a Purchaser or Tenant of a building project.

This Agreement

is made the _____ day of _____ 20 _____

Insert name of the Contractor.

(1) Between _____

_____

whose registered office is at _____

_____ ('the Contractor') and

Insert name of the Purchaser/ Tenant (delete as appropriate).

(2) _____

whose registered office is at _____

_____ ('the Purchaser / Tenant'

which term shall include all permitted assignees under this Agreement).

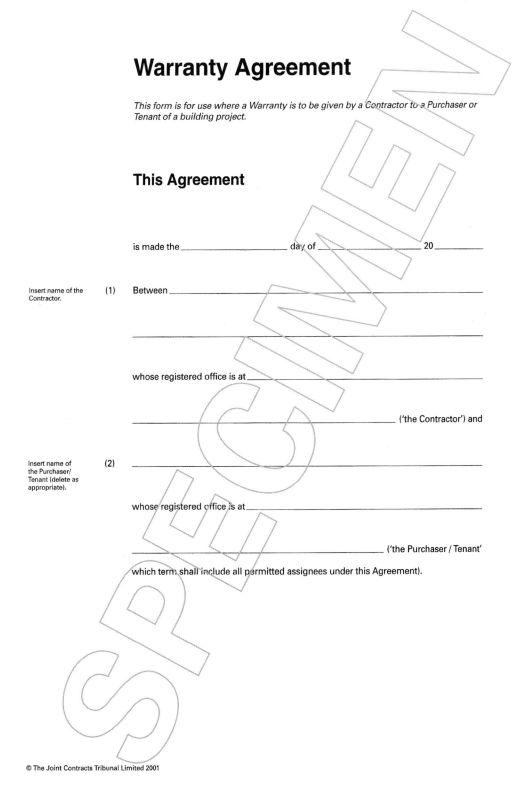

Whereas

Delete as
appropriate.

A The Purchaser / Tenant has entered into an agreement to purchase / an agreement to lease / a lease with

_____ ('the Employer')

Insert description
of the premises
agreed to be
purchased/leased.

relating to _____

_____

Delete as
appropriate.

forming the whole / a part of

Insert description
and address of
the Works.

_____

_____

_____ ('the Works').

Insert date.

B By a contract dated _____ ('the Building Contract' which term shall include any enforceable agreements reached between the Employer and the Contractor and which arise out of and relate to the same) the Employer has appointed the Contractor to carry out and complete the Works.

C This Agreement shall not take effect until the date of Practical Completion of the Works (or, as relevant, the date of practical completion of a Section), or the date of this Agreement, whichever is the later.

Now it is hereby agreed

In consideration of the payment of one pound (£1) by the Purchaser/Tenant to the Contractor, receipt of which the Contractor acknowledges:

1   (a)   The Contractor warrants that it has carried out the Works in accordance with the Building Contract. In the event of any breach of this warranty and subject to clauses 1(b), 1(c) and 1(d):

The Contractor shall be liable for the reasonable costs of repair renewal and/or reinstatement of any part or parts of the Works to the extent that the Purchaser/Tenant incurs such costs and/or the Purchaser/Tenant is or becomes liable either directly or by way of financial contribution for such costs;

*Delete 'AND' and both alternatives if neither is required.

AND*

or

EITHER

Delete the alternative not required.

(a)   (i)   The Contractor shall in addition to the costs in clause 1(a) be liable for any other losses incurred by the Purchaser/Tenant up to a maximum liability of £_____ in respect of each breach.

OR

(a)   (i)   The Contractor shall in addition to the costs in clause 1(a) be liable for all other losses incurred by the Purchaser/Tenant up to a maximum liability of £_____ under this Agreement.

1   (b)   Where clause 1(a)(i) is deleted in its entirety the Contractor shall not be liable for any losses incurred by the Purchaser/Tenant other than those costs referred to in clause 1(a).

Delete the whole of clause 1(c) where using WCD 98.

1   (c)   The Contractor's liability to the Purchaser/Tenant under this Agreement shall be limited to the proportion of the Purchaser's/Tenant's losses which it would be just and equitable to require the Contractor to pay having regard to the extent of the Contractor's responsibility for the same, on the following assumptions, namely that:

Insert the discipline of the Consultant Warrantors.

(i)   [_____ 'the Consultant[s]'] engaged by the Employer has/have provided contractual undertakings to the Purchaser/Tenant as regards the performance of his/their services in connection with the Works in accordance with the terms of his/their respective consultancy agreements and that there are no limitations on liability as between the Consultant and the Employer in the consultancy agreement[s];

Insert the trade or speciality of the Sub-Contractor Warrantors.

(ii)   [_____ 'the Sub-Contractor[s]'] has/have provided a warranty to the Purchaser/Tenant in respect of design of the Sub-Contract Works that it/they has/have carried out and for which there is no liability of the Contractor to the Employer under the Building Contract;

(iii) that the Consultants and the Sub-Contractors have paid to the Purchaser/Tenant such proportion of the Purchaser's/Tenant's losses which it would be just and equitable for them to pay having regard to the extent of their responsibility for the Purchaser's/Tenant's losses.

1   (d)   The Contractor shall be entitled in any action or proceedings by the Purchaser/Tenant to rely on any term in the Building Contract and to raise the equivalent rights in defence of liability as it would have against the Employer under the Building Contract.

1   (e)   The obligations of the Contractor under or pursuant to this clause 1 shall not be released or diminished by the appointment of any person by the Purchaser/Tenant to carry out any independent enquiry into any relevant matter.

2   The Contractor further warrants that unless required by the Building Contract or unless otherwise authorised in writing by the Employer or by the Architect/Contract Administrator named in or appointed pursuant to the Building Contract (or, where such authorisation is given orally, confirmed in writing by the Contractor to the Employer and/or the Architect/Contract Administrator), it has not and will not use materials in the Works other than in accordance with the guidelines contained in the edition of the publication 'Good Practice in Selection of Construction Materials' (Ove Arup & Partners) current at the date of the Building Contract. In the event of any breach of this warranty the provisions of clauses 1(a), (b), (c), (d) and (e) shall apply.

3   The Purchaser/Tenant has no authority to issue any direction or instruction to the Contractor in relation to the Building Contract.

Delete clause 4 if not applicable.

4   The copyright in all drawings, reports, models, specifications, plans, schedules, bills of quantities, calculations and other similar documents prepared by or on behalf of the Contractor in connection with the Works (together referred to as the Documents) shall remain vested in the Contractor but, subject to the Employer having paid all monies due and payable under the Building Contract, the Purchaser/Tenant shall have an irrevocable, royalty-free, non-exclusive licence to copy and use the Documents and to reproduce the designs and content of them for any purpose relating to the Works including, without limitation, the construction, completion, maintenance, letting, sale promotion, advertisement, reinstatement, refurbishment and repair of the Works. Such licence, shall enable the Purchaser/Tenant to copy and use the Documents for the extension of the Works but such use shall not include a licence to reproduce the designs contained in them for any extension of the Works. The Contractor shall not be liable for any use by the Purchaser/Tenant of any of the Documents for any purpose other than that for which the same were prepared by or on behalf of the Contractor.

5   The Contractor has and shall maintain professional indemnity insurance in an amount

Insert amount.

each year of not less than _____ pounds (£ _____)

Delete words in square brackets as necessary.

[for any one occurrence or series of occurrences arising out of one event] [in the aggregate]

[but limited to _____ pounds (£ _____) in the aggregate in respect of all claims arising out of or in connection with pollution and contamination]

Insert period.

for a period ending _____ years after the date of Practical Completion of the Works (or of practical completion of a Section of the Works where the Building Contract is modified for completion by phased sections) under the Building Contract, provided such insurance is available at commercially reasonable rates. The Contractor shall immediately inform the Purchaser/Tenant if such insurance ceases to be available at commercially reasonable rates in order that the Contractor and the Purchaser/Tenant can discuss the means of best protecting the respective positions of the Purchaser/Tenant and the Contractor in the absence of such insurance. As and when it is reasonably requested to do so by the Purchaser/Tenant the Contractor shall produce for inspection documentary evidence that its professional indemnity insurance is being maintained.

Delete clause 5 except where the Building Contract is WCD 98 or JCT 98 where modified by the Contractor's Designed Portion Supplement and/or includes Performance Specified Work to which clause 42 of JCT 98 applies.

6   This Agreement may be assigned without the consent of the Contractor by the Purchaser/Tenant by way of absolute legal assignment, to another person (P1) taking an assignment of the Purchaser's/ Tenant's interest in the Works and by P1, by way of absolute legal assignment, to another person (P2) taking an assignment of P1's interest in the Works. In such cases the assignment shall only be effective upon written notice thereof being given to the Contractor. No further or other assignment of this Agreement will be permitted and in particular P2 shall not be entitled to assign this Agreement.

7   Any notice to be given by the Contractor shall be deemed to be duly given if it is delivered by hand at or sent by special delivery or recorded delivery to the Purchaser/Tenant at its registered office; and any notice given by the Purchaser/Tenant shall be deemed to be duly given if it is delivered by hand at or sent by special delivery or recorded delivery to the Contractor at its registered office; and in the case of any such notices, the same shall, if sent by special delivery or recorded delivery, be deemed (subject to proof to the contrary) to have been received forty-eight hours after being posted.

Insert number of years.

8   No action or proceedings for any breach of this Agreement shall be commenced against the Contractor after the expiry of _____ years from the date of Practical Completion of the Works under the Building Contract, or where the Works are modified for completion by phased sections no action or proceedings shall be commenced against the Contractor in respect of any Section after the expiry of _____ years from the date of practical completion of such Section.

9   For the avoidance of doubt, the Contractor shall have no liability under this Agreement for delay in completion of the Works.

10 Notwithstanding any other provision of this Agreement nothing in this Agreement confers or purports to confer any right to enforce any of its terms on any person who is not a party to it.

11 This Agreement shall be governed by English law.

Notes

[A1] For Agreement executed under hand and NOT as a deed.

[A2] For Agreement executed as a deed under the law of England and Wales by a company or other body corporate: insert the name of the party mentioned and identified on page 1 and then use *either* [A3] and [A4] *or* [A5]. If the party is an *individual* see note [A6].

[A3] For use if the party is using its common seal, which should be affixed under the party's name.

[A4] For use of the party's officers authorised to affix its common seal.

[A5] For use if the party is a company registered under the Companies Acts which is not using a common seal: insert the names of the two officers by whom the company is acting *who MUST be either a director and the company secretary or two directors*, and insert their signatures with 'Director' or 'Secretary' as appropriate. *This method of execution is NOT valid for local authorities or certain other bodies incorporated by Act of Parliament or by charter if exempted under s.718(2) of the Companies Act 1985.*

[A6] If executed as a deed by an *individual*: insert the name at [A2], delete the words at [A3], substitute 'whose signature is here subscribed' and insert the individual's signature. The individual MUST sign in the presence of a witness who attests the signature. Insert at [A4] the signature and name of the witness. Sealing by an individual is not required.

[A1] **AS WITNESS THE HANDS OF THE PARTIES HERETO**

[A1] Signed by or on behalf of the Purchaser / Tenant _____

in the presence of:

[A1] Signed by or on behalf of the Contractor _____

in the presence of:

[A2] **EXECUTED AS A DEED BY THE PURCHASER / TENANT**
[A6]   hereinbefore mentioned namely _____

[A3]   by affixing hereto its common seal

[A4]   in the presence of:

* OR ————————————————————

[A5]   acting by a director and its secretary* / two directors* whose signatures are here subscribed:
  namely _____

  [Signature] _____ *DIRECTOR*

  and _____

  [Signature] _____ *SECRETARY* / *DIRECTOR*

[A2] **AND AS A DEED BY THE CONTRACTOR**
[A6]   hereinbefore mentioned namely _____

[A3]   by affixing hereto its common seal

[A4]   in the presence of:

* OR ————————————————————

[A5]   acting by a director and its secretary* / two directors* whose signatures are here subscribed:
  namely _____

  [Signature] _____ *DIRECTOR*

  and _____

  [Signature] _____ *SECRETARY* / *DIRECTOR*

  * *Delete as appropriate*

Appendix 8
MCWa/F/Scot(Funder)

The Scottish Building Contract Committee Form of Agreement for Collateral Warranty for use where a warranty is to be given by the main contractor to a company providing finance (funder) for proposed building works let or to be let under certain SBCC main contract forms (November 2001)

This document is the copyright of the Scottish Building Contract Committee

Form of Agreement for Collateral Warranty for funding institutions by a Main Contractor: MCWa/F/Scot

This form is for use where a Warranty is to be given by the Main Contractor to a Company providing finance ("Funder") for proposed building works let or to be let under:

a Scottish Building Contract including the version of the Contractor's Designed Portion published by the SBCC incorporating the Conditions of the Standard Form of Building Contract (1998 Edition) published by the Joint Contracts Tribunal which may include Performance Specified Work (SBCC/SF'99)

or a Scottish Building Contract With Contractor's Design published by the SBCC incorporating the Conditions of the Standard Form of Building Contract With Contractor's Design (1998 Edition) published by the Joint Contracts Tribunal (SBCC/WCD'99).

The form is **not** for use where the Building Works are situated in England. For such Works the forms issued by the Joint Contracts Tribunal should be used.

GENERAL ADVICE

1. SBCC Warranty MCWa/F/Scot is collateral to the Building Contract and, generally, should not be entered into before the date of the Building Contract although the enabling clause (provided with each copy of the Warranty) which should be included within that contract does envisage the possibility of the earlier execution of the Warranty. If the Warranty is not entered into prior to or contemporaneously with entry into the Building Contract but the Warranty might be required by the Employer during the currency of the Building Contract, then, under the enabling clause, the Contractor can be required to enter into the Warranty. If the Employer might so require, it is essential that the enabling clause be included with the tender documents and form part of the conditions of the Building Contract; if the clause is not so included, the Employer can only require the Contractor to enter into the Warranty if the Contractor so agrees.

2. Clauses 1(a), 6(3), 6(4), 9 and 13 of SBCC Warranty MCWa/F/Scot all require completion. Where the Warranty is entered into prior to or contemporaneously with the Building Contract being entered into, then such variable information will be known to the Contractor at that time. Where, however, it is entered into after the date of the Building Contract, it is essential that, **pre-contract**, the Main Contractor is informed:

    (a) of the way in which the clauses requiring completion within the SBCC Warranty will be completed; if, subsequently, the Warranty is to be completed on different terms then the agreement of the Contractor to those different terms would be required; and

    (b) if the Employer seeks to make any amendments to the terms of the SBCC Warranty MCWa/F/Scot which it will require the Contractor to give, then the Employer must set out the amendments as part of the Main Contract tender documentation. **Amendments to the clauses of MCWa/F/Scot should however be avoided.**

3. This November 2001 Edition supersedes the May 1999 Edition.

4. The SBCC has decided to remove the whole dispute provisions. The parties accept that the applicable law is Scots Law. There is no provision as to jurisdiction or the Law governing any proceedings unlike the Scottish Building Contract. So far, there is no judicial guidance as to the question of whether or not a collateral warranty is a construction contract relating to construction operations for the purposes of Part II of the Housing Grants, Construction and Regeneration Act 1996.

Copyright of SBCC
ssm pirv\sbcc\sbc472

292

COMMENTARY ON CLAUSES

The Narrative

The name of the Contractor and the Employer should correspond with the names which appear in the Building Contract.

The description of the subjects should correspond with the description in the Building Contract.

The following notes on the clauses are to assist in understanding the use of the document.

Clause 1

This confirms that the Contractor owes the same obligation to the Funder as it owes to the Employer under the Building Contract, but subject to paragraphs (a) and (b).

It is for the Employer and the Funder to agree on the information to be given to the Funder about the content of the Building Contract. It would be difficult for the Funder, given adequate disclosure, then to allege, after executing MCWa/F/Scot, that it was not aware of the content of the Building Contract and/or of the enforceable agreements arising out of and relating to the Building Contract to which the Narrative refers at 'B' on page 6.

Paragraphs (a) and (b) qualify and limit in two ways the Contractor's liability in the event of an alleged breach by the Contractor of the Building Contract

1(a)   By this provision the Contractor's potential liability is limited. The Warranty given by the Contractor is based on the assumptions set out in sub-clauses (i), (ii) and (iii) and its liability is limited accordingly. Consequently, this limit on liability applies whether or not the Consultants employed by the Employer have given contractual undertakings and/or the Sub-Contractors have given warranties to the Purchaser/Tenant. Clause 1(a)(i) and 1(a)(ii) should be completed. No such assumptions, as referred to above, are appropriate where the contractor is designing and building under SBCC/WCD'99 and clause (a) should be deleted entirely where SBCC/WCD'99 is being used.

1(b)   By this clause (but subject to clause 1(a)) the Funder is in no better and no worse position than the Employer in enforcing the terms of the Building Contract and the Contractor may raise against the Funder the same defences on liability as it could against the Employer.

1(c)   This states that the appointment by the Funder of any consultant (e.g. to survey the building(s) comprising the Works) shall not affect any liability, which the Contractor might otherwise have.

Clause 2

The Contractor warrants that it has not used and will not use materials in the Works other than in accordance with the guidelines contained in the publication "Good Practice in Selection of Construction Materials". Clause 2 excuses the use by the Contractor of other materials if the use has been required or authorised by the Employer or by the Architect/Contract Administrator on behalf of the Employer. Where (as in SBCC/WCD'99) the Contractor is the specifier of materials to be used in the Works, it will remain bound by the requirement of clause 2 on the use of such materials in accordance with the above publication.

Clause 3

This is included to make it clear that the Funder has no power or authority to instruct the Contractor in its duties to the Employer under the Building Contract. This is, however, subject to clauses 5 and 6 under which the Funder may take over the rights and duties etc. of the Employer.

Clause 4

This clause makes clear that the Funder has no liability for any amounts due under the Building Contract until it has given a notice under clause 5 following termination of the Finance Agreement or under clause 6 where the Funder takes over as Employer.

Clauses 5, 6 and 7

These entitle the Funder to take over as Employer in which case the Funder also takes over the Employer's rights and duties but on the terms of **clause 7**. The purpose is to permit the Funder to take over as Employer rather than the Contractor determining its employment under the relevant provisions in the Building Contract or exercising any common law rights of treating the Building Contract as having been repudiated by the Employer. Clause 6(1)(a) requires the Contractor to copy to the Funder any preliminary notices which have to be given under the Building Contract before the Contractor can give notice to the Employer that the Contractor's employment is determined; this provides an advance warning so that the Funder can reach a provisional decision on whether it will act under clause 6(4).

When the Contractor

- is in a position to give to the Employer actual notice of determination of the Contractor's employment under the Building Contract,

or

- intends to inform the Employer that it is treating the Building Contract as having been repudiated by the Employer

the Contractor, by clause 6(3), may not give such notice or so inform the Employer without having given the Funder 7 days (or such other period as may have been prescribed in a clause 6(3)) notice.

The Funder may within that period act under clause 6(4) and give notice that it will take over the rights and obligations of the Employer but on the terms of clause 7 in regard to the Funder accepting all the liabilities of the Employer at the date of the take-over and agreeing to perform all the obligations of the Employer under the Building Contract. The proviso to clause 6(4) preserves the Employer's rights under the Building Contract if it considers that any determination by the Contractor of its employment is wrongful.

Clause 8

This clause will normally only be applicable where the Contractor has carried out a design function in relation to the Works e.g. under the Scottish Building Contract Contractors' Designed Portion or the Scottish Building Contract With Contractor's Design. Reasonable use by the Funder of drawings and associated documents is necessary in most cases. By this clause the Funder is given the rights that might reasonably be expected but it does not allow the reproduction of the designs for any purpose outside the scope of the Works; such reproduction would require a further agreement with the Contractor.

Clause 9

The side note states when clause 9 must be deleted. Where clause 9 applies it should be noted:

- the obligation to take out and maintain professional indemnity insurance is subject to the proviso that "such insurance is available" and, if available, is "available at commercially reasonable rates"; and that if after the insurance has been taken out the rates upon any renewal cease to be "commercially reasonable" the Contractor must inform the Funder as provided in the clause;

- not all Contractors will have, or can obtain, professional indemnity insurance and Funders must recognise this when seeking to operate the terms of clause 9;

- where this clause is not deleted the amount and period for maintaining cover must be inserted. The words in square brackets must be deleted as necessary depending on whether the amount of cover is to apply 'for any occurrence or series of occurrences arising out of one event' or is 'in the aggregate'. This clause also provides for a limit to the total claims arising out of or in connection with pollution.

Clause 10

This provision allows the Contractor to comply with obligations imposed upon it under clauses 5 and 6, without being liable to the Employer for a breach of contract.

Clause 11

This clause indicates the right of assignation by the Funder.

Clause 12

This identifies the method of giving notices under the Agreement.

Copyright of SBCC
ssm pirv\sbcc\sbc472

Clause 13

This needs completion. The clause makes clear that any liability that the Contractor has by virtue of this Warranty ceases on the expiry of the stated period of years after Practical Completion of the Works (or, where the Works are completed in Sections, after Practical Completion of the relevant Section).

Clause 14

This clause makes it clear that the Contractor does not, under this Warranty, have any liability for delay in completion of the Works unless and until a notice is served by the Funder under clause 5 or clause 6.

Clause 15

This clause makes it clear that there is no intention to grant any third party rights.

Clause 16

Clause 16 should not be at variance with the proper law clause applicable to the Building Contract. It will be noted that the agreement is to be governed by Scots Law.

Clause 17

The parties consent to registration for preservation and execution. This means that the Collateral Warranty if signed in a manner which makes it self proving can be registered in the Books of Council and Session and an Extract (Official Certified Copy) obtained. The reference to registration for execution means that the Extract is the equivalent of a Court Decree and enforceable as such.

The appropriate method of execution i.e. signature by the Employer, the Contractor and the Funder should be checked carefully. See SBCC: "Note to Users: Attestation" for guidance.

15th November 2001

J M Arnott
Secretary and
Legal Advisor

MAIN CONTRACT: ENABLING CLAUSE

for inclusion in the Conditions of Contract where the Building Contract is to be let on a version of the Scottish Building Contract including the Contractor's Designed Portion published by the SBCC and incorporating the Conditions of Standard Form of Building Contract (1998 Edition) published by the Joint Contracts Tribunal (SBCC/SF'99) whether or not Performance Specified Works to which clause 42 of SF'99 applies are required or on the Scottish Building Contract With Contractor's Design May 1999 Edition incorporating the Conditions of the Standard Form With Contractor's Design published by the Joint Contracts Tribunal (1998 Edition) (SBCC/WCD '99).

Insert as an additional clause 19B (SBCC/SF'99 and SBCC/CDP) or 18B (SBCC/WCD'99).

⁽¹⁾Amend by inserting
appropriate clause
number where
SBCC/WCD '99 is used.

19B⁽¹⁾   **Warranty by Contractor - to a person ("Funder") providing finance for the Works.**

19B.1.1   By a notice in writing by actual delivery or by special delivery or by recorded delivery to the Contractor at the address stated in the narrative of the Agreement ("Agreement") the Employer may require the Contractor to enter into a SBCC Warranty Agreement ("Warranty MCWa/F/Scot") with a Funder identified in the notice. The Warranty must be entered into within ⁽²⁾ ..... days (not to exceed 14) from receipt of the Employer's notice.

⁽²⁾Complete as applicable.

19B.1.2   A notice under clause 19B.1.1 is only valid if the entries in clauses 1(a), 6(3), 6(4) and 13 (and also in clause 9 where that clause is not to be deleted) of the Warranty to be entered into have been notified in writing by the Employer to the Contractor prior to the date of this Contract.

19B.2   Clause 19B.1 shall not apply where the Contractor, on or prior to the date of this Contract, has given a Warranty on the SBCC Warranty Agreement (MCWa/F/Scot) to the Funder.

SCOTTISH COLLATERAL AGREEMENT
(MCWa/F/SCOT – NOVEMBER 2001)

This form is for use where a Warranty is to be given by a Contractor to a company providing finance ("Funder") for a building project. Where the Funder is acting as an agent for the syndicate of banks and/or other companies providing finance a short narrative should be added to refer to this as appropriate.

WARRANTY AGREEMENT

[1] Insert name of the Contractor.

between[1] _____

_____

_____

_____

(hereinafter referred to as "the Contractor")

[2] Insert name of the Employer

and[2] _____

_____

_____

_____ (hereinafter

referred to as "the Employer"),

[3] Insert name of the Funder.

and[3] _____

_____

_____

_____

(hereinafter referred to as "the Funder"),

WHEREAS

A   the Funder has entered into an agreement ("the Finance Agreement") with the Employer for the provision of finance in connection with the carrying out of building works[4]

[4] Insert description of the building works.

_____

_____

at_____

_____("the Works" which term shall include any changes made to the building works in accordance with the Building Contract)

[5] Insert date.

B   By a contract dated _____ [5] ("the Building Contract" which term shall include any enforceable agreements reached between the Employer and the Contractor and which arise out of and relate to the same) the Employer appointed the Contractor to carry out and complete the Works/the Works by phased Sections[6]

[6] Delete as appropriate.

THEREFORE IT HAS BEEN AGREED AND THE PARTIES NOW HEREBY AGREE AS FOLLOWS:

1   The Contractor warrants that it has complied and will continue to comply with the Building Contract. In the event of any breach of this Warranty:

[7] Delete the whole of clause 1(a) where using SBCC/WCD '99

(a)   [7]the Contractor's liability to the Funder for costs under this Agreement shall be limited to that proportion of the Funder's losses which it would be just and equitable to require the Contractor to pay having regard to the extent of the Contractor's responsibility for the same, on the following assumptions, namely that:

[8] Insert discipline of the Consultant Warrantors.

(i) [8]   [_____

_____

"the Consultant[s]"] engaged by the Employer has/have provided contractual undertakings to the Funder that it/they has/have performed and will perform their services in connection with the Works in accordance with the terms of his/her/their respective consultancy agreements and that there are no limitations on liability as between the Consultant and the Employer in the consultancy agreement(s);

[9] Insert the trade or speciality of the Sub-Contractor Warrantors

(ii)   [7][9]["the Sub-Contractor[s]"] has/have provided a warranty to the Funder in respect of design of the Sub-Contract Works that it/they has/have carried out and for which there is no liability of the Contractor to the Employer under the Building Contract

[9](iii)   the Consultants and the Sub-Contractors have paid to the Funder such proportion of the Funder's losses which it would be just and equitable for them to pay having regard to the extent of their responsibility for the Funder's losses.

(b)   the Contractor shall be entitled in any action or proceedings by the Funder to rely on any term in the Building Contract and to raise equivalent rights in defence of liability as it would have against the Employer under the Building Contract.

(c)   The obligations of the Contractor under or pursuant to this clause 1 shall not be increased or diminished by the appointment of any person by the Funder to carry out any independent enquiry into any relevant matter.

2   The Contractor further warrants that unless required by the Building Contract or unless otherwise authorised in writing by the Employer or by the Architect/the Contract Administrator named in or appointed pursuant to the Building Contract (or, where such authorisation is given orally, confirmed in writing by the Contractor to the Employer and/or the Architect/the Contract Administrator), it has not used and will not use materials in the Works other than in accordance with the guidelines contained in the edition of the publication "Good Practice in Selection of Construction Materials" (Ove Arup & Partners) current at the date of the Building Contract.   In the event of any breach of this warranty the provisions of Clause 1(a), (b) and (c) shall apply.

3   The Funder has no authority to issue any direction or instruction to the Contractor in relation to the Building Contract unless and until the Funder has given notice under clauses 5 or 6.

4   The Funder has no liability to the Contractor in relation to amounts due under the Building Contract unless and until the Funder has given notice under clauses 5 or 6.

5   The Contractor agrees that, in the event of the termination of the Finance Agreement by the Funder, the Contractor will, if so required by notice in writing given by the Funder and subject to clause 7, accept the instructions of the Funder or its appointee to the exclusion of the Employer in respect of the Works upon the terms and conditions of the Building Contract. The Employer acknowledges that the Contractor shall be entitled to rely on a notice given to the Contractor by the Funder under this clause 5 as

conclusive evidence for the purposes of this Agreement of the termination of the Finance Agreement by the Funder; and further acknowledges that such acceptance of the instructions of the Funder to the exclusion of the Employer shall not constitute any breach of the Contractor's obligations to the Employer under the Building Contract.

6 (1) The Contractor shall not exercise any right of determination of his employment under the Building Contract without having first

(a) copied to the Funder any written notices required by the Building Contract to be sent to the Architect/the Contract Administrator or to the Employer prior to the Contractor being entitled to give notice under the Building Contract that his employment under the Building Contract is determined, and

(b) given to the Funder written notice that he has the right under the Building Contract forthwith to notify the Employer that his employment under the Building Contract is determined.

6 (2) The Contractor shall not treat the Building Contract as having been repudiated by the Employer without having first given to the Funder written notice that he intends to so inform the Employer.

6 (3) The Contractor shall not

- issue any notification to the Employer to which clause 6(1)(b) refers

or

- inform the Employer that he is treating the Building Contract as having been repudiated by the Employer as referred to in clause 6(2)

(10)Insert the number that has been agreed.

before the lapse of (10) _____ days from receipt by the Funder of the written notice by the Contractor which the Contractor is required to give under clause 6(1)(b) or clause 6(2).

6 (4) The Funder may, not later than the expiry of the (10)_____ days referred to in clause 6(3) require the Contractor by notice in writing and subject to clause 7 to accept the instructions of the Funder or its appointee to the exclusion of the Employer in respect of the Works upon the terms and conditions of the Building Contract. The Employer acknowledges that the Contractor shall be entitled to rely on a notice given to the Contractor by the Funder under clause 6(4) and that acceptance by the Contractor of the instructions of the

Funder to the exclusion of the Employer shall not constitute any breach of the Contractor's obligations to the Employer under the Building Contract. Provided that, subject to clause 7, nothing in clause 6(4) shall relieve the Contractor of any liability he may have to the Employer for any breach by the Contractor of the Building Contract or where the Contractor has wrongfully served notice under the Building Contract that he is entitled to determine his employment under the Building Contract or has wrongfully treated the Building Contract as having been repudiated by the Employer.

7 It shall be a condition of any notice given by the Funder under clause 5 or clause 6(4) that the Funder or its appointee accepts liability for payment of the sums due and payable to the Contractor under the Building Contract and for performance of the Employer's obligations thereunder including payment of any sums outstanding at the date of such notice. Upon the issue of any notice by the Funder under clause 5 or clause 6(4), the Building Contract shall continue in full force and effect as if no right of determination of the Contractor's employment under the Building Contract, nor any right of the Contractor to treat the Building Contract as having been repudiated by the Employer had arisen and the Contractor shall be liable to the Funder and its appointee under the Building Contract in lieu of its liability to the Employer. If any notice given by the Funder under clause 5 or clause 6(4) requires the Contractor to accept the instructions of the Funder's appointee, the Funder shall be liable to the Contractor as guarantor for the payment of all sums from time to time due to the Contractor from the Funder's appointee.

(11)Delete if not applicable - see "Commentary on clauses".

8 (11)The copyright in all drawings, reports, models, specifications, plans, schedules, bills of quantities, calculations and other similar documents prepared by or on behalf of the Contractor in connection with the Works (together referred to as "the Documents") shall remain vested in the Contractor but subject to the Employer having paid all monies due and payable under the Building Contract, the Funder and its appointee shall have an irrevocable, royalty-free, non-exclusive licence to copy and use the Documents and to reproduce the designs and content of them for any purpose related to the Works including, but without limitation, the construction, completion, maintenance, letting, sale, promotion, advertisement, reinstatement, refurbishment and repair of the Works. Such licence shall enable the Funder and its appointee to copy and use the Documents for the extension of the Works but such use shall not include a licence to reproduce the designs contained in them for any extension of the Works. The Contractor shall not be liable for any such use by the Funder or its appointee of any of the Documents for any purpose other than that for which the same were prepared by or on behalf of the Contractor.

(12)Insert amount.

9   The Contractor has and shall maintain professional indemnity insurance in an amount each year of not less than (12)

_____

_____ pounds

(£_____) [for any one occurrence or series of occurrences arising out of one event] [in the aggregate] [but limited to

(13)Insert period. Delete words in square brackets as necessary

_____

pounds

(14) Delete clause 9 except where the Building Contract is on the SBCC Standard Form of Building Contract With Contractor's Design May 1999 Edition (SBCC/WCD '99) or on the SBCC Contractor's Designed Portion (SBCC/CDP) or on SBCC/SF'99 and/or includes Performance Specified Work to which clause 42 of SBCC/SF '99 applies.

(£_____) in the aggregate in respect of all claims arising out of or in connection with pollution and contamination] for a period ending(13) _____ years after the date of Practical Completion of the Works (or of Practical Completion of a Section of the Works where the Building Contract is modified for completion by phased sections) under the Building Contract, provided such insurance is available at commercially reasonable rates. The Contractor shall immediately inform the Funder if such insurance ceases to be available at commercially reasonable rates in order that the Contractor and the Funder can discuss the means of best protecting the respective positions of the Funder and the Contractor in the absence of such insurance. As and when it is reasonably requested to do so by the Funder or its appointee under clause 5 or clause 6 the Contractor shall produce for inspection documentary evidence that its professional indemnity insurance is being maintained.

10   The Employer has agreed to be a party to this Agreement for the purposes of acknowledging that the Contractor shall not be in breach of the Building Contract by complying with the obligations imposed on it by clauses 5 and 6.

11   This Agreement may be assigned, without the consent of the Contractor by the Funder, by way of absolute legal assignation, to another person ("P1") providing finance or re-finance in connection with the carrying out of the Works and by P1 by way of absolute legal assignation, to another person ("P2") providing finance or re-finance in connection with the carrying out of the Works. In such cases the assignation shall only be effective upon written notice thereof being given to the Contractor. No further or other assignation of this Agreement will be permitted and in particular P2 shall not be entitled to assign this Agreement.

12   Any notice to be given by the Contractor shall be deemed to be duly given if it is delivered by hand at or sent by special delivery or recorded delivery to the Funder at its registered office; and any notice given by the Funder hereunder shall be deemed to be duly given if it is delivered by hand at or

sent by special delivery or recorded delivery to the Contractor at its registered office; and, in the case of any such notices, the same shall, if sent by special delivery or recorded delivery, be deemed (subject to proof to the contrary) to have been received forty eight hours after being posted.

13   No action or proceedings for any breach of this Agreement shall be commenced against the Contractor after the expiry of [15]   years from the date of Practical Completion of the Works under the Building Contract or, where the Works are modified for completion by phased sections, no action or proceedings shall be commenced against the Contractor in respect of any Section after the expiry of [15]   years from the date of practical completion of such Section.

[15]Insert number of years

14   Notwithstanding any other provisions of this Agreement, the Contractor shall have no liability under this Agreement for delay under the Building Contract unless and until the Funder serves notice pursuant to clause 5 or clause 6. For the avoidance of doubt the Contractor shall not be required to pay liquidated or ascertained damages in respect of the period of delay where the same has been paid to or deducted by the Employer.

15   Notwithstanding any other provision of this Agreement nothing in this Agreement confers or purports to confer any right to enforce any of its terms on any person who is not a party to it.

16   Whatever the nationality, residence or domicile of the Employer, the Contractor, any Sub-Contractor or supplier and wherever the Works are situated the law of Scotland shall be the law applicable to this Contract.

[16] See "Note to Users: Attestation".

17. The parties consent to registration of these presents for preservation and execution: IN WITNESS WHEREOF[16]

Scottish Standard Form of Agreement

for a Collateral Agreement (MCWa/F/Scot) for use where a Warranty is to be given to a Company providing finance (Funder) for the proposed Building Works by a Main Contractor

(November 2001 Edition)

Collateral Warranty

between _____

_____

_____

_____ (Main Contractor)

and _____

_____

_____

_____ (Funder)

SBCC

The Constituent Bodies of the Scottish Building Contract Committee are:

Royal Incorporation of Architects in Scotland
Scottish Building
Royal Institution of Chartered Surveyors in Scotland
Convention of Scottish Local Authorities
National Specialist Contractors Council – Scottish Committee
Scottish Casec
Association of Consulting Engineers (Scottish Group)

Observers:

Confederation of British Industry
Association of Scottish Chambers of Commerce

Copyright of the SBCC
Excel House,
30 Semple Street
Edinburgh EH3 8BL

Appendix 9
MCWa/P&T/Scot(Purchaser and Tenant)

The Scottish Building Contract Committee Form of Agreement for Collateral Warranty for use where a warranty is to be given by the main contractor to a purchaser or tenant of the whole or part of the building(s) comprising the works which have been practically completed under certain SBCC main contract forms (November 2001)

This document is the copyright of the Scottish Building Contract Committee

Form of Agreement for
Collateral Warranty for
Purchasers and Tenants
by a Main Contractor:
MCWa/P&T/Scot

This form is for use where a Warranty is to be given by the Main Contractor to a Purchaser or Tenant of the whole or part of the subjects comprising the Works which have been practically completed under:

a Scottish Building Contract including the version of the Contractor's Designed Portion published by the SBCC incorporating the Conditions of the Standard Form of Building Contract (1998 Edition) published by the Joint Contracts Tribunal which may include Performance Specified Work (SBCC/SF'99)

or a Scottish Building Contract With Contractor's Design published by the SBCC incorporating the Conditions of the Standard Form of Building Contract With Contractor's Design (1998 Edition) published by the Joint Contracts Tribunal (SBCC/WCD'99).

The form is **not** for use where the Building Works are situated in England. For such Works the forms issued by the Joint Contracts Tribunal should be used.

GENERAL ADVICE

1. It is essential that, **pre-contract**, the Main Contractor is provided with a copy of the MCWa/P&T/Scot showing the necessary insertions and deletions.

2. It is also essential that, **pre-contract**, the Main Contractor is informed.

    (a) that the Building Contract includes a provision enabling the Employer after Practical Completion to require the Contractor to enter into a Warranty on MCWa/P&T/Scot with the purchaser(s) or tenant(s) whose maximum number has been stated in the tender documents or otherwise notified to the Contractor and confirmed in the Building Contract. **The text of such an enabling clause is included with each copy of the Warranty;**

    (b) if the Employer seeks to make any amendments to the terms of the SBCC Warranty MCWa/P&T/Scot which he will require the Contractor to give, then the Employer must set out the amendments as part of the Main Contract tender documentation. **Amendments to the clauses of MCWa/P&T/Scot should however be avoided.**

3. This November 2001 Edition supersedes the May 1999 Edition.

4. The SBCC has decided to remove the whole dispute provisions. The parties accept that the applicable law is Scots Law. There is no provision as to jurisdiction or the Law governing any proceedings unlike the Scottish Building Contract. So far, there is no judicial guidance as to the question of whether or not a collateral warranty is a construction contract relating to construction operations for the purposes of Part II of the Housing Grants, Construction and Regeneration Act 1996.

COMMENTARY ON CLAUSES

Parties

In (2) the name of the Purchaser/Tenant should correspond with the name given in the Employer's notice under clause 19A.2 in the enabling clause.

Narrative A

This needs completion. The subjects comprising the Works or a part thereof which have been purchased or let should be carefully identified; and should correspond with the identification of the subjects purchased or let given in the Employer's notice under clause 19A.2 in the enabling clause.

Narrative B

This needs completion. Under clause 19A.2 in the enabling clause the Employer cannot require MCWa/P&T/Scot to be entered into before the Works have reached practical completion or, where the Works are completed in Phased Sections, before Practical Completion of the relevant Section. See however clause 19A.5 of the enabling clause to cover the circumstances of a pre-let or a purchase prior to such Practical Completion.

Clause 1

This confirms that the Contractor owes the same obligation to the Purchaser/Tenant as it owes to the Employer under the Building Contract, but subject to paragraphs (a), (b), (c) and (d).

It is for the Employer and the Purchaser/Tenant to agree on the information to be given to the Purchaser/Tenant about the content of the Building Contract. It would be difficult for the Purchaser/Tenant, given adequate disclosure, then to allege, after executing MCWa/P&T/Scot, that it was not aware of the content of the Building Contract and/or of the enforceable agreements arising out of and relating to the Building Contract to which Narrative B refers.

Paragraphs (a), (b), (c) and (d) qualify and limit the Contractor's liability to the Purchaser/Tenant in a number of ways in the event of an alleged breach by the Contractor of the Building Contract.

1(a)   This clause provides that the Contractor is liable for the reasonable costs of repair, renewal and/or reinstatement of the subjects comprising the Works to the extent that the Purchaser/Tenant incurs such costs and/or the Purchaser/Tenant is or becomes liable either directly or by way of financial contribution for such costs. Additionally, the Contractor may be made liable for further losses either under 1(a)(i) where a financial cap *in respect of each breach* is to be inserted or under the alternative 1(a)(i) which provides for a financial cap *in respect of a maximum liability.* Whenever practicable the Purchaser/Tenant, bearing in mind its obligation to mitigate the damages it has suffered by an alleged breach of the Building Contract, should inform the Contractor of the defects etc. prior to having them repaired etc. and consider any offer by the Contractor to carry out the repairs etc.

1(b)   This clause makes it clear that where 1(a)(i) provisions are both deleted the liability of the Contractor is limited to those losses referred to in clause 1(a).

1(c)   By this provision the Contractor's potential liability is limited. The Warranty given by the Contractor is based on the assumptions set out in sub-clauses (i), (ii) and (iii) and its liability is limited accordingly. Consequently, this limit on liability applies whether or not the Consultants employed by the Employer have given contractual undertakings and/or the Sub-Contractors have given warranties to the Purchaser/Tenant.

   Clauses c(i) and c(ii) should be completed. No such assumptions, as referred to above, are appropriate where the Contractor is designing and building under the SBCC/WCD '99 and clause (c) should be entirely deleted where this contract is being used. When the Contractor (see the General Advice above) is invited to tender it should be given a copy of MCWa/P&T/Scot with the description of other warrantors to be inserted.

1(d)   By this clause (but subject to clause 1(a)) the Purchaser/Tenant is in no better or no worse position than the Employer in enforcing the terms of the Building Contract; and the Contractor may raise against the Purchaser/Tenant the same defences of liability as it could against the Employer.

1(e)   This states that the appointment by the Purchaser/Tenant of any Consultant (e.g. to survey the building(s) comprising the Works) shall not affect any liability, which the Contractor might otherwise have.

Clause 2

The Contractor warrants that it has not used and will not use materials in the Works other than in accordance with the guidelines contained in the publication "Good Practice in Selection of Construction Materials". Clause 2 excuses the use by the Contractor of other materials if the use has been required or authorised by the Employer or by the Architect/Contract Administrator on behalf of the Employer. Where (as in SBCC/WCD'99) the Contractor is the specifier of materials to be used in the Works, it will remain bound by the requirement in clause 2 on the use of such materials in accordance with the above publication.

Clause 3

This is included to make it clear that the Purchaser/Tenant has no power or authority to instruct the Contractor in its duties to the Employer under the Building Contract. As the Warranty MCWa/P&T/Scot (subject to clause 19A.5 in the enabling clause) is only entered into on or after Practical Completion, clause 3 will relate mainly to the post-Practical Completion obligation of the Contractor under the Building Contract in respect of the Defects Liability Period.

Clause 4

This clause will normally only be applicable where the Contractor has carried out a design function in relation to the Works e.g. under the SBCC/WCD '99 or SBCC/CDP. Reasonable use by the Purchaser/Tenant of drawings and associated documents is necessary in most cases. By this clause the Purchaser/Tenant is given the rights that might reasonably be expected but it does not allow the reproduction of the designs for any purpose outside the scope of the Works (or of that part of the Works) purchased or rented by the Purchaser/Tenant: such reproduction would require a further agreement with the Contractor.

Clause 5

The side note states when Clause 5 must be deleted. Where Clause 5 applies it should be noted:

- the obligation to have and maintain professional indemnity insurance is subject to the provision that "such insurance is available" and, if available, is "available at commercially reasonable rates"; and that if after the insurance has been taken out the rates cease to be "commercially reasonable" the Contractor must inform the Purchaser/Tenant as provided in the clause;

- not all Contractors will have, or can obtain, professional indemnity insurance and Purchasers/Tenants must recognise this when seeking to operate the terms of clause 5.

- where the clause is not deleted the amount and period for maintaining cover must be inserted. The words in square brackets must be deleted as necessary depending on whether the amount of cover is to apply "for any occurrence or series of occurrences arising out of one event" or is "in the aggregate". This clause also provides for a limit to the total claims arising out of or in connection with pollution and contamination.

Clause 6

This allows the Purchaser/Tenant to assign the benefit of the Contractor's Warranty provided it is done by formal legal assignation and relates to the entire interest of the original Purchaser/Tenant without the Contractor's consent. Thereafter, only one further assignation of the Purchaser's/Tenant's interest in the Works can be made.

Clause 7

This identifies the method of giving notice under the Agreement.

Clause 8

This needs completion. The clause makes clear that any liability that the Contractor has by virtue of this Warranty ceases on the expiry of the stated period of years after Practical Completion of the Works (or, where the Works are completed in Sections, after Practical Completion of the relevant Section).

The way in which clause 8 is to be completed should be set out in the copy of the Warranty MCWa/P&T/Scot which the Contractor should be given as part of the tender documentation.

Clause 9

This clause makes it clear that the Contractor does not, under this Warranty, have any liability for delay in completion of the Works.

Copyright of SBCC
smm priv\sbcc\sbc473

309

Clause 10

This clause makes it clear that there is no intention to grant any third party rights.

Clause 11

Clause 11 should not be at variance with the proper law clause applicable to the Building Contract. It will be noted that the agreement is governed by Scots Law.

Clause 12

The parties consent to registration for preservation and execution. This means that the Collateral Warranty can be registered in the Books of Council and Session and an Extract (Official Certified Copy) obtained. The reference to registration for execution means that the Extract is the equivalent of a Court Decree and enforceable as such.

The appropriate method of execution i.e. signature by the Contractor and the Purchaser/Tenant should be checked carefully. See SBCC: "Note to Users: Attestation" for guidance.

15th November 2001

J M Arnott
Secretary and
Legal Adviser

310

MAIN CONTRACT: ENABLING CLAUSE

for inclusion in the Conditions of Contract where the Building Contract is to be let on a version of the Scottish Building Contract including the Contractor's Designed Portion published by the SBCC and incorporating the Conditions of the Standard Form of Building Contract (1998 Edition) published by the Joint Contracts Tribunal (SBCC/SF'99) whether or not Performance Specified Works to which clause 42 of SF'99 applies are required or on the Scottish Building Contract With Contractor's Design May 1999 Edition incorporating the Conditions of the Standard Form With Contractor's Design published by the Joint Contracts Tribunal (1998 Edition) (SBCC/WCD'99)

Insert as an additional clause 19A (SBCC/SF'99 and SBCC/CDP) or 18A (SBCC/WCD'99).

19A[(1)] **Warranty by Contractor - Purchaser or Tenant.**

[(1)]Insert correct clause number where SBCC/WCD '99 is used.

19A.1   Clause 19A shall only apply where, before the Contract has been entered into, the Employer has in a statement in writing to the Contractor (receipt of which the Contractor hereby acknowledges) set out the maximum number of Warranties that the Contractor may require pursuant to clause 19A.

19A.2   On or after the date of Practical Completion of the Works (or of Practical Completion of a Section of the Works where the Contract is modified for completion by phased sections) the Employer, by a notice in writing by actual delivery or by special delivery or recorded delivery to the Contractor at the address stated in the Agreement, may require as follows: the Contractor shall enter into a Warranty Agreement on the terms of the SBCC Warranty Agreement MCWa/P&T/Scot (a copy of which, **with clause 1(a) and 1(a)(i) and clause 8 completed and also clause 5 where that clause is not deleted**, has been given to the Contractor before the Contract has been entered into, receipt of which the Contractor hereby acknowledges) with the Purchaser/Tenant referred to in Clause 19A.4.
(Note: if, after the Building Contract has been entered into, the Employer requires any amendment to the terms of the Warranty Agreement MCWa/P&T/Scot given to the Contractor pursuant to this clause, this can only be done by a further agreement with the Contractor).

[(2)]Complete as applicable

19A.3   The Warranty shall be entered into within[(2)] _____ days (not to exceed 14) from receipt of the Employer's notice referred to in clause 19A.2.

[(3)]Delete as applicable

19A.4   A notice under clause 19A.2 shall identify the Purchaser[(3)]/Tenant with whom the Employer has entered into an agreement to purchase[(3)]/an agreement to lease[(3)]/a lease and where relevant shall sufficiently identify that part of the Works which has been purchased[(3)]/let.

19A.5   The Employer may require the Contractor to enter into a Warranty Agreement as referred to in clause 19A.2 with a person who has entered into an agreement to purchase[(3)]/an agreement to lease[(3)]/a lease as referred to in clause 19A.4 before the date of Practical Completion of the Works (or of Practical Completion of a Section of the Works where the Contract is modified for completion by phased sections); but on condition that the Warranty Agreement shall not have effect until the date of Practical Completion of the Works (or, as relevant, the date of Practical Completion of a Section).

SCOTTISH COLLATERAL WARRANTY
(MCWa/P&T/SCOT – NOVEMBER 2001)

This form is for use where a Warranty is to be given by a Contractor to a purchaser or tenant of a building project.

WARRANTY AGREEMENT

(1) Insert name of the Contractor.

between(1) _____

_____

_____

_____ (hereinafter

referred to as "the Contractor")

(2)Insert name of the Purchaser/Tenant.

and(2) _____

_____

_____

_____ (hereinafter

(3)Delete as appropriate

referred to as "the Purchaser(3)/Tenant"),

312

WHEREAS

A The Purchaser[3]/Tenant has entered into an agreement to purchase[3]/an agreement to lease[3]/a lease with

_____

_____

_____

_____

_____

_____ ("the Employer")

[4]Insert description of the premises agreed to be purchased/leased.

relating to[4] _____

[5]Insert description and address of the Works

_____

_____

forming the whole[3]/a part of at[5]

_____

_____

_____

_____ ("the Works")

[6]Insert date of Building Contract.

B By a contract dated[6] _____("the Building Contract" which term shall include any enforceable agreements reached between the Employer and the Contractor and which arise out of and relate to the same) the Employer has appointed the Contractor to carry out and complete the Works;

C This Agreement shall not take effect until the date of Practical Completion of the Works (or, as relevant, the date of Practical Completion of a Section), or the date of this Agreement, whichever is the later.

THEREFORE IT HAS BEEN AGREED AND THE PARTIES NOW HEREBY AGREE AS FOLLOWS:

1   The Contractor warrants that it has carried out the Works in accordance with the Building Contract. In the event of any breach of this Warranty and subject to clauses 1(b), 1(c) and 1(d):

(a)   The Contractor shall be liable for the reasonable costs of repair, renewal and/or reinstatement of any part or parts of the Works to the extent that the Purchaser/Tenant incurs such costs and/or the Purchaser/the Tenant is or becomes liable either directly or by way of financial contribution for such costs.

AND[7]

[7]Delete 'AND' and both alternatives if neither is required

or

Delete the alternative not required

EITHER

(a)(i)   The Contractor shall in addition to the costs in clause 1(a) be liable for any other losses incurred by the Purchaser/Tenant up to a maximum liability of £_____ in respect of each breach.

OR

(a)(i)   The Contractor shall in addition to the costs in clause 1(a) be liable for all other losses incurred by the Purchaser/Tenant up to a maximum liability of £_____ under this Agreement.

(b)   Where clause 1(a)(i) is deleted in its entirety the Contractor shall be not liable for any losses incurred by the Purchaser/Tenant other than those costs referred to in clause 1(a).

[6]Delete clause where using SBCC/WCD '99

[6](c)   The Contractor's liability to the Purchaser/Tenant under this Agreement shall be limited to the proportion of the Purchaser's/Tenant's losses which it would be just and equitable to require the Contractor to pay having regard to the extent of the Contractor's responsibility for the same, on the following assumptions, namely that:

[9]Insert the discipline of the Consultant Warrantors

[8](i)   [9][

_____

_____

'the Consultant[s]'] engaged by the Employer has/have provided contractual undertakings to the Purchaser/Tenant as regards the performance of its/their services in connection with the Works in accordance with the terms of its/their respective consultancy agreements and that there are no limitations on liability as between the Consultant and the Employer in the consultancy agreement[s];

[8](ii)   ([10][

[10]Insert the trade or speciality of the Sub-Contractor Warrantors

_____

_____

'the Sub-Contractor[s]'] has/have provided a warranty to the Purchaser/Tenant in respect of design of the Sub-Contract Works that it/they has/have carried out and for which there is no liability of the Contractor to the Employer under the Building Contract.

[8](iii)   that the Consultant[s] and the Sub-Contractors have paid to the Purchaser/Tenant such proportion of the Purchaser's/Tenant's losses which it would be just and equitable for them to pay having regard to the extent of their responsibility for the Purchaser's/Tenant's losses.

(d)   The Contractor shall be entitled in any action or proceedings by the Purchaser/Tenant to rely on any term in the Building Contract and to raise the equivalent rights in defence of liability as it would have against the Employer under the Building Contract.

(e)   The obligations of the Contractor under or pursuant to this clause 1 shall not be increased or diminished by the appointment of any person by the Purchaser/Tenant to carry out any independent enquiry into any relevant matter.

2.   The Contractor further warrants that unless required by the Building Contract or unless otherwise authorised in writing by the Employer or by the Architect/Contract Administrator named in or appointed pursuant to the Building Contract (or, where such authorisation is given orally, confirmed in writing by the Contractor to the Employer and/or the Architect/Contract Administrator), it has not used and will not use materials in the Works other than in accordance with the guidelines contained in the edition of the publication "Good Practice in Selection of Construction Materials" (Ove Arup & Partners) current at the date of the Building Contract. In the event of any breach of this warranty the provisions of Clauses 1(a), (b), (c), (d) and (e) shall apply.

3.   The Purchaser[3]/Tenant has no authority to issue any direction or instruction to the Contractor in relation to the Building Contract.

[11]Delete Clause 4 where not applicable

C
s

315

4   [11]The copyright in all drawings, reports, models, specifications, plans, schedules, bills of quantities, calculations and other similar documents prepared by or on behalf of the Contractor in connection with the Works (together referred to as 'the Documents') shall remain vested in the Contractor but, subject to the Employer having paid all monies due and payable under the Building Contract, the Purchaser[3]/Tenant shall have an irrevocable, royalty-free, non-exclusive licence to copy and use the Documents and to reproduce the designs and content of them for any purpose relating to the Works including, without limitation, the construction, completion, maintenance, letting, sale promotion, advertisement, reinstatement, refurbishment and repair of the Works. Such licence, shall enable the Purchaser[3]/Tenant to copy and use the Documents for the extension of the Works but such use shall not include a licence to reproduce the designs contained in them for any extension of the Works. The Contractor shall not be liable for any use by the Purchaser[3]/Tenant of any of the Documents for any purpose other than that for which the same were prepared by or on behalf of the Contractor.

[12] Delete clause 5 except where the Building Contract is on the SBCC Standard Form of Building Contract With Contractor's Design May 1999 Edition (SBCC/WCD '99) or on the SBCC Contractor's Designed Portion (SBCC/CDP) and/or includes Performance Specified Work to which clause 42 of SBCC/SF '99 applies.

[13]Insert amount.

[14]Delete words in square brackets as necessary

[15]Insert period.

5.   [12]The Contractor has and shall maintain professional indemnity insurance in an amount each year of not less than

_____

_____ pounds

(£_____)[14] [for any one occurrence or series of occurrences arising out of one event] [in the aggregate] [but limited to _____ pounds _____

(£_____) in the aggregate in respect of all claims arising out of or in connection with pollution and contamination] for a period ending _____ years[15] after the date of Practical Completion of the Works (or of Practical Completion of a Section of the Works where the Building Contract is modified for completion by phased sections) under the Building Contract, provided such insurance is available at commercially reasonable rates. The Contractor shall immediately inform the Purchaser[3]/Tenant if such insurance ceases to be available at commercially reasonable rates in order that the Contractor and the Purchaser[3]/Tenant can discuss the means of best protecting the respective positions of the Purchaser[3]/Tenant and the Contractor in the absence of such insurance. As and when it is reasonably requested to do so by the Purchaser[3]/Tenant the Contractor shall produce for inspection documentary evidence that its professional indemnity insurance is being maintained.

6   This Agreement maybe assigned without the consent of the Contractor by the Purchaser[3]/Tenant by way of absolute legal assignation, to

another person ("PI") taking an assignation of the Purchaser's[3]/Tenant's interest in the Works and by PI, by way of absolute legal assignation, to another person ("P2") taking an assignation of PI's interest in the Works. In such cases the assignation shall only be effective upon written notice thereof being given to the Contractor. No further or other assignation of this Agreement will be permitted and in particular P2 shall not be entitled to assign this Agreement.

7   Any notice to be given by the Contractor shall be deemed to be duly given if it is delivered by hand at or sent by special delivery or recorded delivery to the Purchaser[3]/ Tenant at its registered office; and any notice given by the Purchaser[3]/Tenant shall be deemed to be duly given if it is delivered by hand at or sent by special delivery or recorded delivery to the Contractor at its registered office; and in the case of any such notices, the same shall, if sent by special delivery or recorded delivery, be deemed (subject to proof to the contrary) to have been received forty eight hours after being posted.

[16]Insert number of years

8   No action or proceedings for any breach of this Agreement shall be commenced against the Contractor after the expiry of [16] _____ years from the date of Practical Completion of the Works under the Building Contract, or where the Works are modified for completion by phased sections no action or proceedings shall be commenced against the Contractor in respect of any Section after the expiry of _____ [16] years from the date of Practical Completion of such Section.

9.   For the avoidance of doubt, the Contractor shall have no liability under this Agreement for delay in completion of the Works.

10.   Notwithstanding any other provision of this Agreement nothing in this Agreement confers or purports to confer any right to enforce any of its terms on any person who is not a party to it.

[17]Where the parties do not wish the law applicable to the Contract to be the law of Scotland appropriate amendments to Clause 11 should be made

11.   [17]Whatever the nationality, residence or domicile of the Employer, the Contractor, any Sub-Contractor or supplier and wherever the Works are situated the law of Scotland shall be the law applicable to this Contract.

12. The parties consent to registration of these presents for preservation and execution: IN WITNESS WHEREOF[18]

[18] See "Note to Users: Attestation".

Scottish Standard Form of Agreement

for a Collateral Warranty (MCWa/P&T/Scot) for use where a Warranty is to be given to a Purchaser/Tenant of Building Works or a part thereof by a Main Contractor

(November 2001 Edition)

Collateral Warranty

between _____

_____

_____ (Main Contractor)

and _____

_____

_____ (Purchaser/Tenant)

SBCC

The constituent bodies of the Scottish Building Contract Committee are:

Royal Incorporation of Architects in Scotland
Scottish Building
Royal Institution of Chartered Surveyors in Scotland
Convention of Scottish Local Authorities
National Specialist Contractors Council – Scottish Committee
Scottish Casec
Association of Consulting Engineers (Scottish Group)

Observers:

Confederation of British Industry
Association of Scottish Chambers of Commerce

Table of Cases

The following abbreviations of Reports are used:

AC	Law Reports, Appeal Cases	Ex.	Law Reports, Exchequer
All ER	All England Law Reports		Division
ALR	Australian Law Reports	FCR	Family Court Reports
Asp MLC	Aspinall's Maritime Law	FTLR	Financial Times Law Reports
	Cases	Gal & Dav	Gale & Davison
Atk	Atkins	HL	House of Lords
BCC	British Company Law Cases	HLR	Housing Law Reports
BCLC	Butterworth Company Law	ICLR	International Construction
	Cases		Law Review
Beav	Beavan	ICR	Industrial Cases Reported
Bing	Bingham	Info TLR	Information Technology Law
BLR	Building Law Reports		Reports
CA	Court of Appeal	IR	Irish Reports
C & P	Carrington & Payne Reports	IRLR	Industrial Relations Law
Ch	Law Reports, Chancery		Reports
	Division	ITCLR	IT and Communications Law
ChD	Chancery Division		Reports
CILL	Construction Industry Law	JP	Justice of the Peace
	Letter	JPL	Journal of Planning and
CL	Current Law		Environmental Law
CLC	Current Law Consolidation	Jur (NS)	Jurist Reports (New Series)
CLR	Commonwealth Law Reports	KB	Law Reports, King's Bench
CLY	Current Law Year Book		Division
Com Cas	Commercial Cases	KIR	Knights Industrial Reports
Com LR	Commercial Law Reports	LGR	Local Government Reports
Con LR	Construction Law Reports	LG Rev	Local Government Review
Const LJ	Construction Law Journal	LJ	Law Journals
Cro Eliz	Croke, Time of Elizabeth I	LJCP	Law Journal, Common Pleas
Deac & Ch	Deacon & Chitty	Lloyds Rep	
De GF&J	De Gex Fisher & Jones	PN	Lloyds Report Professional
DLR	Dominion Law Reports		Negligence
DPC	Davies' Patent Cases	LS Gaz	Law Society Gazette
ECC	European Commercial Cases	LT	Law Times Report
EG	Estates Gazette	LLR	Lloyd's List Law Reports
EGCS	Estates Gazette Case	Lloyds Rep	Lloyds Law Reports
	Summaries	Mont & A	Montague & Ayrton
EWHC		MOO & P	Moore & Payne
Technology	High Court (Technology	My Cr	Myine & Craig
	Court)	New LJ	New Law Journal

NIJB	Northern Ireland Judgment Bulletin	RR	Revised Reports
		Russ	Russell
NPC	New Property Cases	SC	Session Cases
NSW	New South Wales Law Reports	ScotCS	Scottish Court of Session
		SJLB	Solicitors Journal (Lawbrief)
P & CR	Property and Compensation Reports	SLT	Scots Law Times
		Sol J	Solicitors' Journal
PC	Privy Council	SMLC	Smith's Leading Cases
PD	Law Reports, Probate Division	TLR	Times Law Report
		Tr LR	Trading Law Reports
PNLR	Professional Negligence and Liability Reports	WLR	Weekly Law Reports
		WR	Weekly Reporter
QB	Law Reports, Queen's Bench Division	Y & C Ex	Younge & Collyer (Exchequer, Equity)
QBD	Queens Bench Division		

Ailsa Craig Fishing Co Limited *v.* Malvern Fishing Co Limited and Another [1983] 1 WLR 964; [1983] 1 All ER 101; (1983) 80 LS Gaz 2516, HL; (1981) SLT 130, First Division 9.92

Aiolos, The [1983] 2 LLR 25 . . . . 4.22

Albacruz (cargo owners) *v.* Albazero (owners) (The Albazero) [1977] AC 774, HL 3.2, 3.44, 6.24, 7.7

Alfred McAlpine Construction Limited *v.* Panatown Limited [2000] 3 WLR 946, HL; 88 BLR 67; 58 Con LR 46; [1998] CLC 636; (1998) 14 Const LJ 267; [1998] EGCS 19; [1998] NPC 17; *The Times*, 11 February, 1998, CA . . . . 6.21, 6.25, 6.35, 6.38, 6.39, ... 6.40, 7.3, 7.5–7.7, 7.10, 7.14

Allied Maples Group Limited *v.* Simmons & Simmons [1995] 4 All ER 907; [1995] 1 WLR 1602; 46 Con LR 134 . . . . 6.58

A. Monk Construction Limited *v.* Norwich Union Life Insurance Society (1992) 62 BLR 107 . . . . 1.84

Anns *v.* Merton London Borough Council [1978] AC 728; [1977] 2 WLR 1024; 121 Sol J 377; (1977) 75 LGR 555; [1977] JPL 514; (1977) 243 EG 523, 591; (1977) 5 BLR 1; [1977] 2 All ER 492 . . . . 2.10, 2.11, 2.15–219, 2.22–2.24, 2.29, 2.30

Anson *v.* Trump [1998] 3 All ER 331, CA 1.34

Ashcroft *v.* Mersey Regional Health Authority [1985] 2 All ER 96, CA, affirmed [1983] 2 All ER 245 . . . . 5.10

Aswan (M/S) Engineering Establishment Co *v.* Iron Trades Mutual Insurance Co Limited, The Times, 28 July 1988; (1988) CLY 1960 . . . . 8.21

Atlantic Shipping and Trading Company *v.* Louis Dreyfus & Co [1922] 2 AC 250; 91 LJ KB 513; 127 LT 411; 38 TLR 634; 66 Sol J 437; 15 Asp MLC 566; 27 Com Cas 311 . . . . 6.111

Baker *v.* Willoughby [1970] AC 467; [1969] 3 All ER 1528; [1970] 2 WLR 50; 114 Sol J 15 . . . . 6.83

Banco de Portugal *v.* Waterlow & Sons Limited [1932] AC 452; 48 LT 404; 76 Sol J 327 . . . . 6.51

Bank of Nova Scotia *v.* Hellenic Mutual War Risks Association (Bermuda) Limited [1992] 1 AC 233; [1991] 3 All ER 1; [1991] 2 WLR 1279; [1991] 2 Lloyds Rep 191 . . . . 6.7, 6.20, 6.84

Banque Bruxelles Lambert SA *v.* Eagle Star Insurance Co Limited [1997] AC 191; [1996] 3 WLR 87; [1996] 3 All ER 365 . . . . 6.19

Barclays Bank plc *v.* Fairclough Building Limited [1995] QB 214; [1994] 3 WLR 1957;
    [1995] 1 All ER 289; 38 Con LR 86; 68 BLR 1; (1995) 11 Const LJ 2.44
Basildon District Council *v.* J.E. Lesser (Properties) Limited and Others [1985] QB 839;
    [1984] 3 WLR 812; [1985] 1 All ER 20; (1984) 1 Const LJ 57; (1987) 8 Con LR 89;
    (1984) 134 New LJ 330; (1984) 81 LS Gaz 1437 . . . . 6.84
Bateman *v.* Hunt [1904] 2 KB 530; 73 LJ KB 782; 91 LT 331; 20 LT 628; 52 WR 609
    . . . . 4.14
Batty and Another *v.* Metropolitan Property Realisations Limited and Others [1978] QB 554;
    [1978] 2 WLR 500; [1978] 2 All ER 445; (1978) 7 BLR 1 . . . . 2.10
Bawejam Limited *v.* MC Fabrication Limited [1999] 1 All ER (Comm) 377;
    [1999] 1 BCLC 174; [1999] BCC 157, CA 4.47
Beaufort Developments (NI) Limited *v.* Gilbert Ash (NI) Limited [1999] 1 AC 266;
    [1998] CLY 5055, HL 9.109
Beswick *v.* Beswick [1968] AC 58; [1967] 3 WLR 932; 111 Sol J 540; [1967] 2 All ER 1197,
    HL 10.102
Birse Construction Limited *v.* Hastie [1996] 1 WLR 675; [1996] 2 All ER 1; 47 Con LR 162; 76
    BLR 31; [1996] 1 PNLR 8; [1996] CLC 577; (1996) 93 (2) LS Gaz 29; (1996) 140 SJLB 25; The
    Times, 12 December 1995, CA reversing 44 Con LR, QBD 6.85, 6.87, 6.88
Blyth *v.* Birmingham Waterworks Company (1856) 11 Ex 781; 25 LJ Ex 212; 2 Jur NS 333;
    4 WR 294 . . . . 5.2, 5.3
Blyth & Blyth Limited *v.* Carillion Construction Limited [2001] Scot CS 90 . . . . 3.8, 4.54
Bolam *v.* Friern Hospital Management Committee [1957] 1 WLR 582; (1957) 101 Sol J 357;
    [1957] 2 All ER 118 . . . . 5.3, 5.7, 8.27
Bolithio *v.* City & Hackney Health Authority [1997] 3 WLR 1151; [1997] 4 All ER 771;
    (1997) 94 (47) LS Gaz 30; (1997) 141 SJLB 238; The Times, 27 November 1997, HL
    . . . . 5.4
Brett *v.* J.S. (1600) Cro Eliz 755 . . . . 1.48
Brew Bros Limited *v.* Snax (Ross) Limited [1970] 1 All ER 587; [1970] 1 QB 612;
    [1969] 3 WLR 657 . . . . 7.20
Brickfield Properties *v.* Newton [1971] 1 WLR 862; 115 Sol J 307; [1971] 3 All ER 328
    . . . . 6.100
British Steel Corporation *v.* Cleveland Bridge & Engineering Company [1984] 1 All ER 504;
    (1983) BLR 94; [1982] Com LR 54 . . . . 1.84
British Sugar plc *v.* NEI Power Projects Limited and Another (1997) 87 BLR 42; [1998] Info
    TLR 353; [1998] ITCLR 125; (1998) 14 Const LJ 365, CA affirming [1997] CLC 622; The
    Times, 21 February 1997, QBD 9.94
British Westinghouse Electric Company Limited *v.* Underground Electric Railways [1912]
    AC 673; 81 LJ KB 1132; 107 LT 325 . . . . 6.1, 6.21, 6.43
Brocklesby *v.* Armitage and Guest [1990] Lloyds Rep PN 888, CA 6.110
Brown *v.* KMR Services Limited [1995] 4 All ER 598; [1995] 2 Lloyds Rep 513.
    . . . . 6.10, 6.19, 6.20
Business Computers Limited *v.* Anglo African Leasing Limited [1977] 1 WLR 578;
    [1977] 2 All ER 741 . . . . 4.23
Butler Machine Tool Co Limited *v.* Ex-Cell-O Corporation (England) Limited
    [1979] 1 WLR 401; (1977) 121 Sol J 406; [1979] 1 All ER 965, CA 1.37
Canadian Indemnity Co *v.* Andrews & George Co (Canada) [1952] 4 DLR 690 . . . . . 8.23
Caparo Industries plc *v.* Dickman [1990] 2 AC 605; [1990] 2 WLR 358; [1990] 1 All ER 568;
    (1990) 134 Sol J 494; [1990] BCC 164; [1990] BCLC 273; [1990] ECC 313; [1990] LS Gaz 28
    March, 42; (1990) 140 New LJ 248, HL 2.34, 9.10
Cargill International SA *v.* Bangladesh Sugar & Foods Industries Corp [1998] 2 All ER 406;
    [1998] 1 WLR 461 . . . . 1.66

Carlill *v.* The Carbolic Smoke Ball Company [1893] 1 QB 256; 62 LJ QB 257; 67 LT 837;
41 WR 210; 57 JP 325 . . . . 1.34

Cassell & Co *v.* Broome [1972] AC 1027; [1972] 2 WLR 645; 116 Sol J 199;
[1972] 1 All ER 801 . . . . 6.82

Castellain *v.* Preston (1883) 11 QB 380; [1881–85] All ER 493 . . . . 8.15

Chaplin *v.* Hicks [1911] 2 KB 786; 80 LJ KB 1292; 105 LT 285; 27 LT 458; 55 Sol J 580 6.57

Chelmsford District Council *v.* T. J. Evers and Others (1983) CILL 39;
(1983) 25 BLR 99 . . . . 6.100

Conquer *v.* Boot [1928] 2 KB 336; 97 LJ KB 452; 139 LT 18; 44 LT 486 . . . . 6.2

Conway *v.* Crowe, Kelsey & Partner and Another (1994) CILL 927 . . . . 9.11

Cook *v.* Swinfen [1967] 1 WLR 457 . . . . 6.57

Co-operative Retail Services Ltd *v.* Taylor, Young and Others [2000] 2 All ER 865;
(2000) BLR 461; 74 Con LR 12; (2000) 16 Const LJ 347, CA 6.88

Council of the Shire of Sutherland *v.* Heyman (Australia) 157 CLR 424 . . . . 2.23

County & District Properties *v.* Jenner [1976] 2 Lloyds Rep 728; (1974) 3 BLR 41
. . . . 9.102

Crédit Suisse *v.* Beegas Nominees Limited [1994] 4 All ER 803; [1995] 69 P & CR 177;
[1994] 11 EG 151; [1994] 12 EG 189; [1993] EGCS 157; [1993] NPC 123; The Independent, 15
September 1993, ChD 7.22

Croggan ex parte Carbis, Re (1834) 4 Deac & Ch 354; 1 Mont & A 693n 4.20

Cullinane *v.* British 'Rema' Manufacturing Co Limited [1954] 1 QB 292; [1953] 3 WLR 923;
97 Sol J 811; [1953] 2 All ER 1257 . . . . 6.49

Curran *v.* Newpark Cinemas Limited [1951] 1 ALL ER 295 . . . . 4.12

Curran and Another *v.* Northern Ireland Co-Ownership Housing Association and Another
[1987] 2 All ER 13; [1987] 2 WLR 1043; [1987] AC 718; (1987) 38 BLR 1; (1987) 131 Sol J 506;
(1987) 19 HLR 318; (1987) 84 LS Gaz 1574, HL reversing [1986] 8 NIJB 1, CA
. . . . 2.17, 2.18

D. & F. Estates Limited *v.* Church Commissioners for England [1988] 2 All ER 992;
[1989] AC 177; [1988] 3 WLR 368; (1988) 41 BLR 1; [1988] 2 EG 213, 15 Con LR 35, HL
. . . . 2.1, 2.19–2.23, 2.25, 3.10, 7.24, 7.30, 11.36

Darlington Borough Council *v.* Wiltshier Northern Limited [1995] 1 WLR 68;
[1995] 3 All ER 895; (1995) 11 Const LJ 36; (1994) 91 (37) LS Gaz 49; (1994) 138 SJLB 161;
69 BLR 1; The Times, 4 July 1994; The Independent, 29 June 1994, CA reversing 37 Con LR
29, QBD 3.5, 3.44, 6.26, 6.29, 6.33

Dawson *v.* Great Northern and City Railway Company [1905] 1 KB 260; 74 LJ KB 190;
92 LT 137; 21 LT 114 . . . . 4.33, 4.35, 6.63, 6.64, 6.71, 6.72 6.77

Dean *v.* Ainley [1987] 1 WLR 1729; (1987) 131 Sol J 1589; [1987] 3 All ER 784;
(1987) 284 EG 1244; (1987) 84 LS Gaz 2430, CA 6.24, 6.32

Dearle *v.* Hall (1823) 3 Russ 1; 2 LJ Ch 62; 27 RR1 . . . . 4.20

De Lassalle *v.* Guildford [1900–3] All ER 495 . . . . 1.4, 7.37

Dodd Properties Limited *v.* Canterbury [1980] 1 WLR 433; (1979) 124 Sol J 84;
[1980] 1 All ER 928; (1979) 253 EG 1335; (1979) 13 BLR 45 . . . . 6.55

Dominion Bridge Company Limited *v.* Toronto General Insurance Co (Canada) [1964] 1
LLR 194 . . . . 8.23

Donoghue *v.* Stevenson [1932] AC 562; 101 LJ PC 119; [1932] All ER 1; 147 LT 281;
48 LT 494 . . . . 2.4, 2.8, 2.9, 2.11, 2.21, 2.22

Drury *v.* Victor Buckland Limited 85 Sol J 117; [1941] 1 All ER 269 . . . . 1.8

Dunkirk Colliery Co *v.* Lever (1878) 9 Ch 20 . . . . 6.50

Dunlop *v.* Lambert (1839) 6 Cl & F 600 . . . . 6.24, 7.7

Dunlop Limited *v.* New Garage Co Limited [1915] AC 79; 83 LJ KB 1574; 111 LT 862;
30 LT 625 . . . . 6.3

Dunlop Pneumatic Tyre *v.* Selfridge [1915] AC 853 . . . . 3.1
Dutton *v.* Bognor Regis UDC and Another [1972] 1 All ER 462; [1972] 2 WLR 299;
     [1972] 1 QB 373; (1972) 3 BLR 13 . . . . 2.9, 2.10, 2.24
Eames London Estates Limited and Others *v.* North Herts District Council and Others
     (1981) 259 EG 389; (1980) 18 BLR 50 . . . . 6.95
East Ham Corporation *v.* Bernard Sunley & Sons Limited [1966] AC 406; [1965] 3 WLR 1096;
     109 Sol J 874; [1966] 3 All ER 619; 64 LGR 43; [1965] 2 Lloyds Rep 425 . . . . . 6.44, 6.55
Edwards *v.* Skyways Limited [1964] 1 All ER 494; [1964] 1 WLR 349; 108 Sol J 279
     . . . . 1.24, 1.26
Ellis *v.* Torrington [1920] 1 KB 399; 89 LJ KB 369; 122 LT 361; 36 LT 82 . . . . 4.34, 4.35
Entores Limited *v.* Miles Far Eastern Corporation [1955] 2 QB 327; [1955] 2 All ER 493;
     [1955] 3 WLR 48; 99 Sol J 384; [1955] 1 Lloyds Rep 511 . . . . 1.34
Equitable Debenture Assets Corporation *v.* William Moss Group Limited and Others (1984)
     CILL 74; (1984) 1 Const LJ 131; (1984) 2 Con LR 1 . . . . 6.95
Ernst & Whinney *v.* Willard Engineering (Dagenham) Limited and Others (1987) CILL 336;
     (1987) 3 Const LJ 292; (1987) 40 BLR 67 . . . . 7.24
F. & G. Sykes (Wessex) Limited *v.* Fine Fare Limited [1967] 1 Lloyds Rep 53, CA
     . . . . 1.46
First Energy *v.* Hungarian Bank [1993] BCC 533; [1993] 2 Lloyds Rep 194; [1993] BCLC 1409;
     [1993] NPC 34; The Times, 4 March 1993, CA 1.28
Flood, David Charles *v.* Shand Construction and Others 81 BLR 31; 54 Con LR125;
     [1997] CLC 588; [1996] NPC 185; The Times, 8 January 1997, CA 4.46
Foley *v.* Classique Coaches Limited [1934] 2 KB 1; 103 LJ KB 550; 151 LT 242
     . . . . 1.45, 1.46, 1.78
Forsikringsaktieselskapet Vesta *v.* Butcher and Others [1988] 3 WLR 565; (1988) 32 Sol J
     1181; [1988] 2 All ER 43; [1988] 1 FTLR 78; [1988] 1 Lloyds Rep 19; (1988) 4 Const LJ 75;
     [1988] 33 LS Gaz 31 August 6.84
Gable House Estates Limited *v.* The Halpern Partnership and Another (1995) 48 Con LR 1,
     QBD 2.43
Galoo *v.* Bright Grahame Murray [1995] 1 All ER 16; [1994] 2 BCLC 492; [1994] 1 WLR 1360;
     [1994] BCC 319 . . . . 6.5
G. Percy Trentham Limited *v.* Archital Luxfer [1993] 1 Lloyds Rep 25; 63 BLR 44, CA
     . . . . 1.36, 1.51
George Hawkins *v.* Chrysler (UK) Limited and Burne Associates (1986) 38 BLR 36
     . . . . 5.14
George Mitchell (Chesterhall) Limited *v.* Finney Lock Seeds Limited [1983] 3 WLR 163;
     [1983] 2 AC 803 . . . . 9.92
Gibson Lea Retail Interiors *v.* Macro Self Service Wholesalers [2000] BLR 407 . . . . . 1.74
Gilbert *v.* Ruddeard 3 Dyer 272 b,n 1.48
Global Container Lines *v.* State Black Sea Shipping [1999] 1 Lloyds Rep 127, CA . . . 1.46
Gloucestershire County Council *v.* Richardson [1969] 1 AC 480; [1968] 2 All ER 1181; [1968]
     3 WLR 645; 112 Sol J 759; 67 LGR 1; 207 EG 797 . . . . 5.19, 5.23
Good *v.* Parry [1963] 2 QB 418; [1963] 2 WLR 846; 107 Sol J 194; [1963] 2 All ER
     . . . . 6.113
Governors of the Hospital for Sick Children and Another *v.* McLaughlin & Harvey Plc and
     Other (1987) CILL 372 . . . . 6.54
Governors of the Peabody Donation Fund Limited *v.* Sir Lindsay Parkinson & Co Limited
     and Others [1984] 3 WLR 953; [1984] 3 All ER 529; (1984) 28 BLR 1 . . . . 2.14, 2.18
Gray and Others (the Special Trustees of the London Hospital) *v.* T. B. Bennett & Son, Oscar
     Faber and Others and McLaughlin & Harvey Limited (1987) CILL 342; (1987) 43 BLR 63;
     13 Con LR 22 . . . . 6.108

Greater Nottingham Co-operative Society Limited *v.* Cementation Piling & Foundations Limited [1988] 2 All ER 971; (1988) CILL 404; (1988) 41 BLR 43; [1988] 3 WLR 396; [1989] QB 712, CA 9.7

Greaves & Co (Contractors) Limited *v.* Baynham Meikle & Partners [1974] 1 WLR 1261; 118 Sol J 598 . . . . 5.7, 5.13, 5.24

Green & Silley Weir *v.* British Railways Board and Another [1985] 1 All ER 237; (1985) 17 BLR 94; [1985] 1 WLR 570 . . . . 9.102

GUS Property Management Limited *v.* Littlewoods Mail Order Stores Limited (1982) SLT 533 . . . . 2.40, 6.21, 6.63, 6.65, 6.69, 6.70, 6.71, 6.72, 6.73, 6.76

Gwyn *v.* Neath Canal Company CR 3 Ex 209; 37 LJ Ex 122; 18 LT 688; 16 WR 1209 . . . 1.63

Hadley *v.* Baxendale (1854) 9 Ex 341; 23 LJ Ex 179; 2 CLR 517; 18 Jur 358; 2 WR 302; 2 SMLC (13 ed.) 529; 3 LT 69 . . . . 6.13, 6.14, 6.15, 6.16, 6.18

Hall & Tawse South Limited *v.* Ivory Gate Limited (1997) 62 Con LR 117, QBD 1.81

Hancock *v.* B. W. Brazier (Anerley) Limited [1966] 2 All ER 901; [1966] 1 WLR 1317; 110 Sol J 368 . . . . 5.27

Harbutts Plasticine Limited *v.* Wayne Tank & Pump Co Limited [1970] 1 QB 447; [1970] 1 All ER 225, CA; [1970] 2 WLR 198; 114 Sol J 29 . . . . 6.46, 6.53

Harper *v.* Gray and Walker [1985] 1 WLR 1196, (1985) 129 Sol J 776; [1985] 2 All ER 507; [1984] CILL 106; (1983) 1 Const LJ 46; (1985) 82 LS Gaz 3532; [1970] 1 Lloyds Rep 15 . . . . 6.93

Heaven *v.* Pender [1883] 11 QBD 503 . . . . 2.21

Hedley Byrne & Co Limited *v.* Heller & Partners Limited [1964] AC 465; [1963] 3 WLR 101; [1963] 2 All ER 575; [1963] 1 Lloyds Rep 485 . . . . . 2.8, 2.11, 2.25, 2.27–2.30, 2.32, 2.33, ... 2.34, 2.37, 2.42, 9.10

Heilbut Symons & Co *v.* Buckleton [1913] AC 30; 82 LJ KB 245; 107 LT 769 . . . . 1.9

Helstan Securities Limited *v.* Hertfordshire County Council [1978] 3 All ER 262 . . . 4.41

Henderson *v.* Astwood [1894] AC 150 . . . . 1.18

Henderson and Others *v.* Merrett Syndicates Limited and Others [1995] 2 AC 145; [1994] 3 All ER 506; [1994] 3 WLR 761; [1994] 2 Lloyds Rep 468; 69 BLR 26, HL 2.1, 2.32, 2.35, 2.37, 2.39, 2.42, 2.43, 2.44, 9.12

Hendry *v.* Chartsearch Limited [1998] CLC 1382; New Law Digest, 23 July 1998; The Times, 16 September 1998, CA 4.48

Herkules Piling Limited and Another *v.* Tilbury Construction Limited 32 Con LR 112; (1992) 61 BLR 107 . . . . 4.14, 4.22

Hill *v.* Archbold [1968] 1 QB 686; [1967] 3 All ER 110; [1967] 3 WLR 1218; 111 Sol J 543 . . . . 4.35

Hiron *v.* Legal & General Assurance, Pynford South Limited and Others 60 BLR 78; [1992] 28 EG 112; 35 Con LR 71 . . . . 9.9

Holding & Management *v.* Property Holding & Investment Trust [1989] 1 WLR 1313; [1990] 1 All ER 938; (1990) 134 Sol J 262; (1989) 21 HLR 596; [1990] 05 EG 75; (1989) LS Gaz 29 November, 38, CA 7.23

Home Office *v.* Dorset Yacht Co Limited [1970] AC 1004; [1970] 2 WLR 1140; [1970] 2 All ER 294; [1970] 2 Lloyds Rep 453 . . . . 2.11

Hotel Services Ltd *v.* Hilton International Hotels (UK) Ltd (2000) BLR 235, CA 6.18

Hughes *v.* Pump House Hotel Company Limited [1902] 2 KB 190, CA 4.29

Hyde *v.* Wrench (1840) 3 Beav 334; 4 Jur 1106 . . . . 1.33

Independent Broadcasting Authority *v.* EMI Electronics Limited and BICC Construction Limited (1980) 14 BLR 1 . . . . 1.21, 1.23, 5.9, 5.12

Inland Revenue Commissioners *v.* Raphael [1935] AC 96; 51 LT 152 . . . . 1.58

Inntrepeneur Pub Company Limited *v.* East Crown Limited [2000] 2 Lloyds Rep 611; [2000] 3 EG 31; [2000] 41 EG 209; [2000] NPC 93; The Times, 5 September 2000 . . . 1.11

Investors Compensation Scheme Limited *v.* West Bromwich Building Society [1998] 1 WLR 896; [1998] 1 All ER 98; [1998] 1 BCLC 531; [1997] PNLR 541; [1997] CLC 1243; (1997) 147 New LJ 989; The Times, 24 June 1997, HL reversing [1998] 1 BCLC 521; [1997] PNLR 166; [1997] CLC 363; [1997] NPC 104; The Times, 8 November 1998, CA; [1998] 1 BCLC 493; [1997] CLC 348; The Times, 10 October 1996, ChD 1.27, 1.66
J. D. Williams & Co Limited *v.* Michael Hyde and Associates Limited [2000] Lloyds Rep PN 1220; [2000] NPC 78; The Times, 4 August 2000, CA 5.4, 5.7
J. Murphy & Sons Limited *v.* ABB Daimler Benz [1998] EWHC Technology, 278
. . . . 1.51
Jameson and Another *v.* Central Electricity Generating Board (Babcock Energy Ltd) [1998] QB 323; [1997] 4 All ER 38; [1997] 3 WLR 151, CA 6.86, 6.93
John Harris Partnership (a firm) *v.* Groveworld Limited [1999] PNLR 697, QBD . . . 6.33
John Lee & Son (Grantham) Limited *v.* Railway Executive [1949] 2 All ER 581; 65 LT 604; 93 Sol J 587 . . . . 1.69
Johnson *v.* Chief Constable of Surrey, The Times, 23 November 1992, CA 6.105
J. Sainsbury plc *v.* Broadway Malyan (a firm) and Ernest Green Partnership Ltd (1998) 61 Con LR 31 . . . . 6.89, 6.90
Junior Books Limited *v.* Veitchi Co Limited [1982] 3 WLR 477; [1982] 3 All ER 201; [1982] AC 520; (1982) 21 BLR 66 . . . . . . 2.12, 2.25, 2.27–2.30, 2.31, 2.35, 2.36, 2.44, 3.47
Kamouth *v.* Associated Electrical Industries Limited [1980] QB 199; [1979] 2 WLR 795; (1978) 122 Sol J 714 . . . . 6.113
Kemp *v.* Baerselman [1906] 2 KB 604; 75 LJ KB 873 . . . . 4.37
Koufos *v.* Czarnikow Limited (The Heron II) [1966] 2 QB 695; [1966] 2 All ER 593; [1966] 2 WLR 1397; [1966] 1 Lloyds Rep 595 . . . . 6.14, 6.15, 6.16, 6.18
Lancashire and Cheshire Associations of Baptist Churches *v.* Howard and Seddon Partnership [1993] 3 All ER 467 . . . . 9.11
Leigh & Sillivan *v.* Aliakmon Shipping Co Limited [1986] 2 All ER 145; [1986] AC 785; [1986] 2 WLR 902; [1986] 2 Lloyds Rep 1 . . . . 2.16
Letts *v.* Inland Revenue Commissioners [1956] 3 All ER 588, ChD 4.18
Leyland Shipping *v.* Norwich Union [1918] AC 350 . . . . 6.5
Liesbosch Dredger *v.* Edison SS [1933] AC 449 . . . . 6.55
Linden Gardens Trust Limited *v.* Lenesta Sludge Disposals Limited and Others [1994] 1 AC 85; [1993] 3 WLR 408; [1993] 3 All ER 417; 36 Con LR 1; (1993) 9 Const LJ 322, HL reversing in part 57 BLR 57; 30 Con LR 1; (1992) 8 Const LJ 1 . . . . 3.6, 3.44, 4.43, 4.44,
. . . . 4.50, 6.21, 6.22, 6.24, 6.31, 6.63, 6.70, 6.80
Lloyd *v.* Banks (1863) 3 Ch App 488 . . . . 4.20
Lloyd *v.* Lloyd (1887) 2 My Cr App R 192 . . . . 1.59
London Borough of Lewisham *v.* Leslie & Co Limited (1978) 250 EG 1289; (1978) 12 BLR 22, CA 6.107
Lowsley *v.* Forbes [1999] 1 AC 329; [1998] 3 All ER 897; [1998] 3 WLR 501; [1998] 2 Lloyds Rep 577; The Times, 24 August 1998, HL 6.103
M. J. Gleeson (Northern) Limited *v.* Taylor Woodrow Construction (1989) 21 Con LR 71; (1989) 49 BLR 95 . . . . 5.20
McArdle (deceased), McArdle *v.* McArdle, Re [1951] Ch 669; [1951] 1 All ER 905; 95 Sol J 284 . . . . 4.30
Malik *v.* Bank of Credit & Commerce International [1998] AC 20; [1997] 3 All ER 1; [1997] IRLR 462; [1997] 3 WLR 95; [1997] ICR 606 . . . . 6.5
Martin Grant & Co Limited *v.* Sir Lindsay Parkinson & Co Limited (1984) 29 BLR 31
. . . . 1.79
May & Butcher Limited *v.* The King (1929) [1934] 2 KB 17; 103 LJ KB 556; 151 LT 246
. . . . 1.42, 1.46

Midland Bank *v.* Hett Stubbs & Kemp [1979] Ch 384; [1978] 3 WLR 167; (1977) 121 Sol J 830; [1978] 3 All ER 571 . . . . 6.57

Miliangos *v.* George Frank (Textiles) Limited [1976] AC 443; [1975] 3 WLR 758; 199 Sol J 774; [1975] 3 All ER 801 . . . . 6.55

Milroy *v.* Lord (1862) 4 De GF&J 264; 31 LJ Ch. 798; 7 LT 178; 8 Jur NS 806 . . . . 1.56

Mitsui Babock Energy Limited *v.* John Brown Engineering Limited (1996) 51 Con LR 129 QBD 1.51

Monarch Steamship Co Limited *v.* Karlshamms Hjesabrikes (A/B) [1949] AC 196; [1949] LJR 772; 65 LT 217; 93 Sol J 117; [1949] 1 All ER 1; 1949 SC 1; 1949 SLT 51 . . . . 6.11

Moorcock, The (1889) 14 PD 64; 58 LJP 73; 60 LT 654; 37 WR 439 . . . . 1.77

Mourmand *v.* Le Clair [1903] 2 KB 216; 72 LJ KB 496; 88 LT 728; 51 LWR 589 . . . . . 1.64

Muirhead *v.* Industrial Tank Specialists Limited (1985) CILL 216; [1986] QB 507; [1985] 3 All ER 705; [1985] 3 WLR 993; [1985] ECC 225; 129 Sol J 855 . . . . 2.12, 6.1, 6.3

Murphy *v.* Brentwood District Council [1991] AC 398; [1990] 3 WLR 414; [1990] 2 All ER 908; (1990) 22 HLR 502; (1990) 134 Sol J 1076; 21 Con LR 1; 89 LGR 24; (1990) 6 Const LJ 304; (1990) 154 LG Rev 1010; 50 BLR 7; (1991) 3 Admin LR 37, HL reversing [1990] 2 WLR 944; [1990] 2 All ER 269; 88 LGR 333; (1990) 134 Sol J 458; [1990] LS Gaz February 7, 42, CA affirming 13 Con LR 96 . . . . 2.1, 2.8, 2.9, 2.23–2.26, 2.29, 3.10, 9.7, 9.10, 11.36

Nokes *v.* Doncaster Amalgamated Collieries [1940] AC 1014, HL 4.4

Norglen Limited *v.* Reeds Rains Prudential Limited [1996] 1 All ER 945; [1996] 1 WLR 864; [1996] BCC 532 . . . . 4.35

Norta Wallpapers (Ireland) *v.* Sisk & Sons (Dublin), Ireland (1978) IR 114 . . . . 5.25

Northern Regional Health Authority *v.* Derek Crouch Construction Co [1984] 2 All ER 175 . . . . 9.109

Nottingham Community Housing Association Limited *v.* Powerminster Limited (2000) BLR 759 . . . . 1.74

Orion Insurance Co plc *v.* Sphere Drake Insurance plc [1992] 1 Lloyds Rep 239 . . . . 1.25

OTM Limited *v.* Hydranautics [1981] 2 Lloyds Rep 211 . . . . 1.38, 1.80, 1.83

Perry *v.* Sidney Phillips & Son [1983] 1 WLR 1297; 126 Sol J 626; [1982] 3 All ER 705; (1982) 22 BLR 120; 263 EG 888 . . . . 6.43

Perry *v.* Tendring District Council and Others (1985) 30 BLR 118; (1985) 3 Con LR 74; (1985) CILL 145 . . . . 5.8, 6.62

Peter Lind & Co Limited *v.* Mersey Docks & Harbour Board [1972] 2 Lloyds Rep 234 . . . . 1.32

Phoenix Assurance Co Limited *v.* Earls Court Limited (1913) 30 LT 50 . . . . 4.25

Photo Productions *v.* Securicor Transport [1980] AC 827; [1980] 2 WLR 283; (1980) 124 Sol J 147; [1980] 1 All ER 556; [1980] Lloyds Rep 545 . . . . 6.46

Pirelli General Cable Works Limited *v.* Oscar Faber & Partners [1983] 2 AC 1; [1983] 1 All ER 65; (1983) 21 BLR 99; [1983] 2 WLR 6 . . . . 2.29

Plant Construction plc *v.* Clive Adams Associates and JHM Construction Services Limited (2000) BLR 137; (2000) CILL 1598, CA; (1997) 86 BLR 119; 55 Con LR 41, QBD 2.44, 5.5

Post Office *v.* Aquarius Properties Limited [1987] 1 All ER 1055; (1987) 281 EG 798 . . . . 7.21, 7.22

Prenn *v.* Simmonds [1971] 3 All ER 237; [1971] 1 WLR 1381; 115 Sol J 654 . . . . 1.58, 1.66

Prosser *v.* Edmonds (1835) 1 Y & C Ex 481; 41 RR 322 . . . . 4.31

Quick *v.* Taff Ely Borough Council [1985] 3 All ER 321; [1986] QB 809; [1985] 3 WLR 981; (1985) 276 EG 452 . . . . 7.18, 7.21

Quinn *v.* Burch Brothers (Builders) Limited [1966] 2 QB 370; [1966] 2 WLR 1017; 110 Sol J 214; [1966] 2 All ER 283; [1966] KIR 9 . . . . 6.4, 6.6

Radford v. De Froberville & Lange [1977] 1 WLR 1262; sub nom Radford v. De Froberville
　　[1978] 1 All ER 33; (1977) 121 Sol J 319; (1977) 7 BLR 35 . . . . 6.45, 6.46
Raven, The [1980] 2 Lloyds Rep 266 . . . . 4.27
Ravenseft Properties Limited v. Davstone (Holdings) Limited [1979] 1 All ER 929;
　　[1980] QB 12; [1979] 2 WLR 898; (1978) 249 EG 51 . . . . 7.19
Reeves v. Butcher [1981] 2 QB 509; 60 LJ QB 619; 65 LT 329; 39 WR 626 . . . . 6.101
Robinson v. The Post Office [1974] 2 All ER 737; [1974] 1 WLR 1176; 117 Sol J 915 . . . 5.8
Rotherham Metropolitan Borough Council v. Frank Haslam Milan & Co Limited (1996) 78
　　BLR 1; [1996] CLC 1378; (1996) 12 Const LJ 333; [1996] EGCS 59, CA 5.20
Routledge v. Grant (1828) 4 Bing 653; 1 MOO & P 717; 3 C & P 267; 6 LJCP 166
　　. . . . . 1.30, 1.31
Royal Brompton Hospital National Health Trust v. Hammond and Others (No 3) 69 Con LR
　　61; [2000] Lloyds Rep 643, CA 6.92
Ruxley Electronics and Construction Limited v. Forsyth [1996] 1 AC 344;
　　[1995] 3 All ER 268; [1995] 3 WLR 118; 45 Con LR 61; 14 Tr LR 541, 73 BLR 1;
　　(1995) 11 Const LJ 381, HL 6.24, 6.32, 6.41, 6.42, 6.47
Saipem SpA and Conoco (UK) Ltd v. Dredging VO2BV and Geosite Surveys Ltd (the
　　'Volvox Holandia') (No 2) [1993] 2 Lloyds Rep 315 . . . . 6.95
Samuels v. Davis [1943] KB 526; [1943] 2 All ER 3; 112 LJ KB 561; 168 LT 296 . . . . . 5.15
St Martins Property Corporation v. Sir Robert McAlpine Limited, *see also* Linden Gardens
　　Trust Limited v. Lenesta Sludge Disposals Limited and Others
　　. . . . . 6.21, 6.23, 6.24, 6.26, 6.27, 6.30, 6.74, 7.7
Saunderson v. Pier (1839) 7 SC 408; 5 Bing 425; 7 DPC 632; 8 LJCP 227; 50 RR 731
　　. . . . . 1.61
Scarf v. Jardine (1882) 7 AC 345; 51 LJ QB 612; 47 LT 258; 30 WR 893 . . . . 4.51
Shanklin Pier Limited v. Detel Products [1951] 2 KB 854; [1951] 2 All ER 471; 95 Sol J 563;
　　[1951] Lloyds Rep 187 . . . . 1.6, 1.7, 1.8, 1.10, 1.50
Sheldon and Others v. RHM Outhwaite (Underwriting Agents) Ltd and Others [1996] AC
　　102; [1995] 2 WLR 570; [1995] 2 All ER 558; [1995] 2 Lloyds Rep 197;
　　(1995) 145 New LJ 687; (1995) 92 (22) LS Gaz 41; (1998) 139 SJLB 119; The Times, 5 May
　　1995; The Independent, 9 May 1995, HL 6.109
Sidaway v. Governors of Bethlem Royal Hospital (1984) 2 WLR 778; [1985] 2 WLR 480;
　　[1985] 1 All ER 643 . . . . 5.11
Simaan General Contracting Co v. Pilkington Glass [1988] 1 All ER 791; [1988] 2 WLR 761;
　　(1988) CILL 391; [1988] QB 758; (1988) 40 BLR 28 . . . . 2.18, 2.35
Simpson v. Vaughan (1739) 2 Atk 32 . . . . 1.64
Skandia Property (UK) and Another v. Thames Water Utilities Ltd (1999) BLR 338,
　　CA 6.54
Sotiros Shipping Inc and Another v. Sameiet Solholt, The Times, 16 March 1983;
　　(1983) Sol J 305; [1983] 1 Lloyds Rep 605; [1983] Com LR 114, CA; [1981] Com LR 201;
　　[1981] 2 Lloyds Rep 574 . . . . 6.50
Sparham-Souter v. Town & Country Developments (Essex) Limited [1976] 2 All ER 65;
　　[1976] 2 WLR 493; [1976] 1 QB 858; (1976) 3 BLR 72 . . . . 2.10
Spellman v. Spellman [1961] 2 All ER 498; [1961] 1 WLR 921; 105 Sol J 401 . . . . 4.40
Staveley Industries v. Oderbrecht Oil & Gas Services, Unreported 1.74
Sutherland and Sutherland v. C. R. Maton & Sons Limited (1976) 3 BLR 87;
　　(1976) 240 EG 135 . . . . 2.10
Sutro & Co v. Heilbut Symons & Co [1917] 2 KB 348; 86 LJ KB 1226; 116 LT 545; 33 LT 359;
　　14 Asp MLC 34; 23 Com Cas 21 . . . . 1.61
Tai Hing Cotton Mill Limited v. Liu Chong Hing Bank Limited, Privy Council [1986] AC 80;
　　[1985] 2 All ER 947; [1985] 3 WLR 317 . . . . 9.6, 9.7

Tancred *v.* Delagoa Bay Company (1889) 23 QBD 239; 58 LJ QB 459; 61 LT 229
. . . . 4.10, 4.11
Thomas *v.* Thomas (1842) 2 QB 851; 2 Gal & Dav 226; 11 LJ QB 104; 6 Jur 645 . . . . . 1.48
Thomas and Others *v.* T. A. Phillips (Builders) Limited and Taff Ely Borough Council (1985)
CILL 222; (1987) 9 Con LR 72; (1986) 2 Const LJ 64 . . . . 6.49
Thorman and Others *v.* New Hampshire Insurance Co (UK) Limited and the Home
Insurance Co [1988] 1 Lloyds Rep 7; (1987) CILL 364; [1988] 1 FTLR 30 . . . . 8.5
Tito and Others *v.* Waddell and Others No. 2 (The Ocean Island case) Tito *v.* Att-Gen [1977]
Ch 106; [1977] 2 WLR 496; [1977] 3 All ER 129; judgment on damages [1977] 3 WLR
972n 6.48, 6.52, 6.61
Tolhurst *v.* Associated Portland Cement Manufacturers (1900), Associated Portland
Cement Manufacturers (1900) *v.* Tolhurst [1902] 2 KB 660; affirmed sub nom Tolhurst *v.*
Associated Portland Cement Manufacturers (1900), Tolhurst *v.* Associated Portland
Cement Manufacturers (1900) and Imperial Portland Cement Company [1903] AC
414 . . . . 4.36, 4.37, 6.74
Tom Shaw & Co *v.* Moss Empires Limited and Bastow 25 LT 190 . . . . 4.38, 4.42
Torkington *v.* Magee [1903] 1 KB 644; 72 LJ KB 336; 88 LT 443; 18 LT 703 . . . . 4.4
Trendtex Trading Corporation and Another *v.* Credit Suisse [1981] 3 WLR 766;
(1981) 125 Sol J 761; [1981] 3 All ER 520; [1981] Com LR 262 . . . . 4.35
Trolex Products Limited *v.* Merrol Fire Protection Engineers Limited (Unreported) . . . 5.26
Trollope & Colls Limited *v.* North West Metropolitan Regional Hospital Board [1973] 2 All
ER 260; [1973] 1 WLR 601; 117 Sol J 355; 227 EG 653 . . . . 1.60, 1.79
Turriff Construction Limited *v.* Regalia Knitting Mills Limited (1971) 222 EG 1.83
Van Lynn Developments Limited *v.* Pelias Construction Co Limited [1969] 1 QB 607;
[1968] 3 WLR 1141; 112 Sol J 819; [1968] 3 All ER 824 . . . . 4.14
Victoria Laundry (Windsor) Limited *v.* Newman Industries Limited 65 LT 274; 93 Sol J 371;
[1949] 1 All ER 997; [1949] 2 KB 528 . . . . 6.14, 6.15, 6.16, 6.17, 6.43
Viking Grain Storage Limited *v.* T. H. White Installations Limited (1985) 33 BLR 103;
(1985) CILL 206; (1985) 3 Con LR 52 . . . . 5.25
Wagon Mound The (No. 2), [1963] 1 Lloyds Rep 402 (NSW) 6.15
Walford *v.* Miles [1992] 2 AC 128; [1992] 2 WLR 184; [1992] 1 All ER 453; (1992) 64 P & CR
166; [1992] 1 EG 207; [1992] 11 EG 115; [1992] NPC 4; The Times, 27 January 1992; The
Independent, 29 January 1992, HL affirming [1991] 2 EG 185; [1991] 27 EG 114 and [1991]
28 EG 81; (1990) 62 P & CR 410; [1990] EGCS 158; The Independent, 15 January 1991, CA
reversing [1990] EG 212; [1990] 12 EG 107, DC 1.47
Walker *v.* Boyle [1982] 1 All ER 634; [1982] 1 WLR 495; (1982) 261 EG 1089 . . . . 9.17
Walker *v.* Giles (1848) 18 LJCP 323; 13 Jur 588, 753; 77 RR 425 . . . . 1.62
Walter & Sullivan Limited *v.* J. Murphy & Sons Limited [1955] 2 QB 584; [1955] 2 WLR 919;
99 Sol J 290; [1953] 1 All ER 843 . . . . 4.9, 4.10, 4.11, 4.20, 4.21
Warboys *v.* Acme Investments (1969) 210 EG 335; (1969) 4 BLR 133 . . . . 8.32
Weld-Blundell *v.* Stephens [1920] AC 956; 89 LJ KB 705; 123 LT 593; 36 LT 640;
64 Sol J 529 . . . . 6.6
Wells (Merstham) Limited *v.* Buckland Sand & Silica Co Limited [1965] 2 QB 170;
[1964] 1 All ER 41; [1964] 2 WLR 453; 108 Sol J 177 . . . . 1.7, 1.8, 1.10
Wessex Regional Health Authority *v.* HLM Design Limited (1995) 71 BLR 32;
(1994) 10 Const LJ 165, QBD 9.10
White *v.* Jones [1995] 2 AC 207; [1995] 1 All ER 691; [1995] 3 FCR 51; [1995] 2 WLR 187
. . . . 2.37, 2.38, 2.39, 2.40
Whitehouse *v.* Jordan [1981] 1 All ER 267; [1981] 1 WLR 246 . . . . 5.3, 8.27
William Brandts Sons *v.* Dunlop Rubber Co [1905] AC 454; 74 LJ KB 898; 93 LT 495; 21 LT
710; 11 Com Cas 1 . . . . 4.17, 4.19, 4.20

William Hill Organisation Limited *v.* Bernard Sunley & Sons Limited (1982) 22 BLR 1
. . . . 6.106, 6.108
Williams, Williams *v.* Ball, Re [1917] 1 Ch 1; 86 LJ Ch 36; 115 LT 689; 61 Sol J 42 . . . 4.11
Wilson Smithett & Cape (Sugar) Limited *v.* Bangladesh Sugar & Foods Industries
  Corporation [1986] 1 Lloyds Rep 378 . . . . 1.83
Wimpey & Co Limited *v.* BOAC [1955] AC 169; [1954] 3 WLR 932 . . . . 6.93
Wimpey Construction (UK) Limited *v.* D. V. Poole [1984] 2 Lloyds Rep. 499;
  (1984) 27 BLR 58 . . . . 5.10, 8.18
Woodar Investments Development Limited *v.* Wimpey Construction UK Limited [1980] 1
  WLR 277; (1980) 124 Sol J 184; [1980] 1 All ER 571, HL 3.48
Yeandle *v.* Wynn Realisations Limited (in administration) (1995) 47 Con LR 1, CA
. . . . 4.46
Yorkshire Regional Health Authority *v.* Fairclough Building Ltd and The Percy Thomas
  Partnership [1996] 1 WLR 501; [1996] 1 All ER 519; 78 BLR 59; 47 Con LR 120;
  [1996] CLC 366; (1995) 139 SJLB 247, The Times, 16 November 1995, CA 6.104
Young & Marten *v.* McManus Childs [1969] 1 AC 454; [1968] 2 All ER 1169;
  [1968] 3 WLR 630; 11 2 Sol J 744; 67 LGR 1 . . . . 5.18, 5.23, 9.43
Yuen Kun Yeu *v.* Attorney General of Hong Kong, Privy Council [1988] AC 175;
  [1987] 2 All ER 705; (1987) CILL 350; [1987] 3 WLR 776 . . . . 2.18

Table of Statutes & Statutory Instruments

Arbitration Act 1996
  s. 13(1) 6.99
  s. 13(2) 6.99
  s. 13(3) 6.99

Civil Liability (Contribution) Act 1978
  s.1(1) 6.85, 6.87, 6,88, 6.90, 10.88
  s.1(3) 6.88, 6.93
  s.1(4) 6.85, 6.90
  s.1(5) 6.85
  s. 1(6) 6.88
  s.2(1) 6.94
  s.2(3) 6.94
  s.6(1) 9.87

Companies Act 1985
  s.36(a)(4) 1.55
  s.36(a)(5) 1.55

Companies Act 1989
  s. 130 . . . . 1.55

Contracts (Rights of Third Parties) Act 1999   1.3, 3.14, 3.16, 4.1, 7.13, 7.14, 7.24, 7.25, 7.30,
10.65, 10.102, 10.120, 10.145, 11.2–11.4
  s.1(1) 3.20, 3.24 3.27, 3.47, 3.53, 3.54–3.57, 3.59, 3.60, 10.145
  s.1(2) 3.24, 3.26, 3.27, 3.47
  s.1(3) 3.32, 3.59
  s.1(5) 3.30
  s.1(6) 3.31
  s.2 . . . . 3.35, 3.36, 3.59
  s.3 . . . . 3.37, 3.59
  s.5 . . . . 3.38, 3.59
  s.8 . . . . 3.40
  s.9 . . . . 3.18
  s.10(2) 3.17
  s.10(3) 3.17

Copyright, Designs and Patents Act 1988
  s.77(7)(b) 9.63
  s.78 . . . . 9.63

Criminal Law Act 1967
   s.14(2) 4.32

Defective Premises Act 1972 . . . . 5.28, 6.98, 9.97, 11.33–11.38
   s.1(1) 5.27
   s.2 (1) 5.28
   s.6(3) 5.29

House Building Standard (Approved Scheme) Order 1979 SI 1979/381 . . . . 5.28

Housing Grants, Construction and Regeneration Act 1996 . . . . 1.72, 3.41
   s.104 . . . . 1.72, 9.25
   s.105 . . . . 1.72, 9.25
   s.107 . . . . 1.72
   s.108 . . . . 1.75, 9.24, 9.28
   s.109 . . . . 1.75, 9.24, 9.28
   s.110 . . . . 1.75, 9.24, 9.28
   s.111 . . . . 1.75, 9.24, 9.28
   s.112 . . . . 1.75, 9.24
   s.113 . . . . 1.75, 9.28

Judicature Act 1873
   s.25(6) 4.13

Land Clauses Consolidation Act 1845 . . . . 4.33

Latent Damage Act 1986 . . . . 6.96
   s.2 . . . . 6.96
   s.5 . . . . 6.97
   s.8 . . . . 6.97

Law of Property Act 1925
   s.136 . . . . 4.11, 4.13, 4.15, 4.22, 4.28
   s.136(b) 4.46
   s.136(1) 4.7
   s.53(1)(c) 4.28

Law of Property (Miscellaneous Provisions) Act 1989 . . . . 1.53
   s.1(2)(a) 1.54
   s.1(3) 1.54

The Law Reform (Contributory Negligence) Act 1945 . . . . 6.84

Limitation Act 1939 . . . . 6.106

Limitation Act 1980 . . . . 6.96
   s.14A 6.97
   s.14B 6.97
   s.24 . . . . 6.103
   s.29(5) 6.112
   s.32 . . . . 6.105, 6.107, 6.110

s.32(2) 6.106
s.32(1) 6.105, 6.108, 6.109
s.34 . . . . 6.102
s.35 . . . . 6.104

Misrepresentation Act 1967 . . . . 7.39

Sale and Supply of Goods Act 1994 . . . . 5.21

Supply of Goods and Services Act 1982 . . . . 5.21
s.1(3) 5.22
s.4 . . . . 5.21, 5.22
s.4(2) 5.21
s.4(3) 5.21
s.4(4) 5.21
s.4(5) 5.21
s.4(6) 5.21
s.13 . . . . 5.23
s.14 . . . . 1.40
s.15 . . . . 1.40

Unfair Contract Terms Act 1977
s.1(1) 9.14
s.2(1) 9.14
s.2(2) 9.18
s.3(1) 9.16
s.11(1) 9.18
s.11(4) 9.19, 9.92

Index

ACE Conditions of Engagement, 9.2, 9.32, 9.41, 10.83
adjudication, 3.41–3.42, 9.24, 9.27
arbitration, 3.39–3.40, 3.59, 7.6, 9.107–9.110, 10.129
assignment, *see also* collateral warranties
  absolute, 4.9, 4.10
  agent, by, 4.12
  bare rights of action, restriction of, 4.31–4.35
  BPF Form of Warranty, CoWa/F, in, 10.113–10.116
  champerty, 4.31, 4. 32
  charge, by way of, 4.10, 4.21
  chose in action, 4.4
  collateral warranty, of, 9.81–9.86
  comparison of law and equity, 4.6
  conditional assignments, 4.11
  conditions of contract, restrictions imposed by, 4.38–4.49, 9.84–9.86
  consideration, for, 4.18, 4.30, 9.112, 10.81, 10.82, 10.100, 10.122, 10.128
  counterclaims, effect of, 4.25, 4.26
  derivative contractual rights, 4.3
  effect of, 4.15
  equitable assignment, 4.17, 4.18
  equitable chose, 4.4
  form of, 4.12, 4.19
  future chose in action, restriction on, 4.29, 4.30
  intermediate assignees, 4.27
  legal assignment, 4.7, 4.8, 4.12
  legal chose in action, 4.4
  maintenance, 4.31, 4.32
  notice, 4.14, 4.20
  personal contracts, restrictions on, 4.36, 4.37, 9.85
  prior equities, 4.23
  procedural differences, legal and equitable assignment, 4.22
  professional indemnity insurance policy, as to, 8.31, 8.35–8.41
  qualified contractual restriction, 4.48
  set-off, effect of, 4.24, 4.27
  statutory assignments, 4.7, 4.8
  statutory restrictions, 4.28

BPF Agreement for Collateral Warranty, CoWa/F, 7.59, 9.29, 9.51, 9.71, 10.24–10.26, 10.72–10.124
  'appointee', 10.96, 10.102
  clause 1, 10.83–10.89
  clause 2, 10.90–10.92
  clause 3, 10.93
  clause 4, 10.94–10.95
  clauses 5, 6 and 7, 10.96–10.106
  clause 8, 10.107–10.109
  clause 9, 10.110–10.111
  clause 10, 10.112
  clause 11, 10.113–10.116
  clause 12, 10.117–10.118
  clause 13, 10.119
  commentary, 10.72–10.124
  consideration, 10.81–10.82
  general advice of BPF, 10.112–10.124
  notices, 10.117–10.118
  recitals, 10.79–10.80,
BPF Agreement for Collateral Warranty, CoWa/P&T, 9.51, 9.71, 10.2, 10.27–10.28, 10.73,
BUILD, 11.18–11.20

causation of damage
  acts, intervening, 6.6
  causal link, meaning of, 6.4
  events, intervening, 6.11
collateral warranties
  assignment of, *see also* assignment, 9.81–9.86
  consideration, 9.112, 10.81–10.82, 10.100, 10.122, 10.128,
  'construction contract', 9.24–9.28
  contra proferentem, 1.69–1.71, 9.22
  copyright, 7.42, 7.48, 9.61–9.65, 10.107, 10.139

definition of, 1.4
delay, 9.103–9.106
design, 9.30–9.41
disclosure of, to insurer, 8.2–8.3, 8.8–8.14
dispute resolution, 9.107–9.110
entire agreement clause, negating effect
    of, 1.11
examples of, 1.6, 1.7, 1.12
exclusion clauses, 9.14–9.21
execution, 10.12
fitness for purpose, 5.17–5.20, 9.42, 9.43
fund, for, 7.46–7.48
imputation of, 1.6, 1.10
indemnity, 9.102
insurance, professional indemnity,
    9.66–9.72
insurance, use in negotiation, 10.9–10.11
legal costs, 10.7–10.8
limitation of action, 6.96–6.114, 9.14–9.21,
    9.92–9.99
limiting liability, 9.14–9.21, 9.92–9.99
materials, excluded, 9.46–9.60
net contribution, 9.87–9.91
notice, provision for, 9.111–9.112
novation, 9.73–9.80
obligation to enter into, 7.49, 7.55–7.61,
    10.2–10.5
other parties, 9.87–9.91
principal contract, relation with, 9.2–9.4
purchasers, for, 7.41–7.42
standard form, *see* BPF, MCWa/F,
    MCWa/P&T, SCWa/F and SCWa/
    P&T
tenants, for, 7.33–7.35
typical terms, 9.29–9.112
workmanship, 9.44–9.45
commercial lease, *see* lease
Confederation of Construction Specialists
    Employer/Specialist Contractor
    Warranty, 10.67–10.71
consequential loss, 6.3, 6.39–6.41, 6.76–6.81,
    8.29–8.30, 9.93–9.94, 9.97, 11.10, 11.38
consideration, 1.6, 1.17, 1.20, 1.48–1.50, 1.56,
    3.3, 3.13, 4.16, 4.18, 4.30, 4.49, 4.51,
    9.112, 10.81–10.82, 10.100, 10.122, 10.128
construction management, 10.22
construction of contract
    contra proferentem rule, 1.69–1.71, 9.22
    ejusdem generis rule, 1.68
    intention, of parties, 1.58, 1.60, 1.62–1.64
    manuscript, effect of, 1.61

parol evidence rule, 1.65, 1.66
recitals, meaning and effect of, 1.67, 10.79,
    10.80
words, meaning of, 1.59
contract, *see also* construction of contract
    acceptance, of offer, 1.32–1.36
    agreement, 1.26, 1.27
    agreement to agree, 1.47
    capacity, 1.19
    comparison with tort, *see also* tort, 1.14,
        7.25, 8.8, 8.20, 8.21–8.22, 8.25, 8.28,
        9.4–9.13
    consensus ad idem, 1.39
    consideration, 1.6, 1.17, 1.20, 1.48–1.50,
        1.56, 3.3, 3.13, 4.16, 4.18, 4.30, 4.49, 4.51,
        9.112, 10.81–10.82, 10.100, 10.122, 10.128
    contracts under seal, 1.53–1.55, 9.101,
        9.112, 10.12, 10.64, 10.100, 10.103, 10.122
    definition of, 1.13
    implied terms, 1.77–1.79
    incomplete agreements, 1.40–1.47
    invitation to treat, 1.29
    legal relations, intention to create,
        1.20–1.25
    offer, 1.28
    parties, 1.18
    privity, 1.3, 3.1–3.14
    simple contracts, 1.52, 1.56
    tort, relationship with, 1.13, 1.14, 7.25, 8.8,
        8.20, 8.21–8.22, 8.25, 8.28, 9.4–9.13
    withdrawal of offer, 1.30, 1.31
contra proferentem rule, 1.69–1.71, 9.22
contribution, 7.56, 7.59, 9.87–9.91, 10.60,
    10.85, 10.88, 10.127, 10.132
    basis of assessment, 6.85, 6.94, 6.95
    claims under a contract, 6.82
    co-defendants, between, 6.82, 6.84, 6.85
    damage, meaning of, 6.86, 6.87, 6.89–6.92
    liability, meaning of, 6.88, 6.91
copyright, 7.42, 7.48, 9.61–9.65, 10.107,
    10.139
costs, legal, 10.7–10.8
CoWa/F, *see* BPF Agreement for Collateral
    Warranty, CoWa/F
CoWa/P&T, *see* BPF Agreement for
    Collateral Warranty, CoWa/P&T

damages, *see also* measure of damages *and*
    causation of damage
    apportionment of, 6.82–6.84
    ascertainment, difficulties of, 6.56

assessment, date of, 6.55
assignees, entitlement to, 6.60–6.81
betterment, 6.53, 6.54
chance, loss of, 6.57, 6.58
compensation, for loss of amenity, 6.42, 6.47
consequential losses, 6.3, 6.39–6.41, 6.76–6.81, 8.29–8.30, 9.93–9.94, 9.97, 11.10, 11.38
direct costs, 6.61–6.75
economic loss, 6.3, 8.29–8.30
future damages, 6.2
general damages, 6.2
liquidated and ascertained, 6.2, 6.3, 8.37, 8.39, 9.30, 9.103–9.105, 9.108, 10.144
nature of, 6.1
special damage, 6.2
third party loss, recovery of, *see also* third parties, rights of, 6.21–6.24, 6.27–6.34, 6.35–6.39
delay
provision of design, in, 9.103–9.106, 10.56
design, 9.30–9.41
developer, problems of, 7.5–7.14
disputes, resolution of, 7.12, 9.107–9.110
DOM/2, sub-contract, 9.38,
dwellings, 5.27–5.29, 9.97, 11.33–11.34, 11.36

ESA/1, 9.36
estoppel contracts, 1.51
European Community, proposals of, 11.31–11.32
excluded materials, 9.46–9.60
exclusion clauses, 9.14–9.21
execution
under hand, 1.52
under seal, 1.54, 1.55, 9.101, 9.112, 10.12, 10.64, 10.100, 10.103, 10.122
expectation interest
meaning of, 6.42

fitness for purpose, *see also* reasonable skill and care, 5.17–5.26
design and build contractors, 5.24–5.27
professional indemnity insurance and, 8.32–8.34, 9.66–9.72
warranties, in, 9.42–9.43
fund
collateral warranties, from whom, 7.45

contents of collateral warranty, for, 7.46–7.48
problems of, 7.43–7.44

'if' contracts, 1.84
indemnity provisions, 9.102
inspection duties
architects, 9.41
engineers, 9.41
insurance, *see also* professional indemnity insurance
BUILD, 11.18–11.20
building defects, 11.21–11.26
developments in, 11.29–11.30
insurers
changing of, 8.42

JCT 98, 7.55–7.57, 9.39–9.40, 9.44, 9.60, 9.62, 9.97, 9.99, 9.103, 9.109, 10.19–10.20
JCT Contracts
MCWa/F enabling provision, in, 7.55–7.56
MCWa/P&T enabling provision, in, 7.57–7.61,
JCT 'With Contractor's Design' sub-contract DOM/2, 9.38
JCT Works Contract/3, 9.36,

'last shot' doctrine, 1.37, 1.38
law reform, *see* reform of law *and* other solutions
lease
landlord to bear repairing cost, 11.5–11.12
repairing covenant of tenant, 7.15–7.26
letters of intent, 1.80–1.83
liability
limiting, 9.14–9.21, 9.92–9.99
limitation of action
abridgement (of limitation periods), 6.111
acknowledgement and part payment, effect of, 6.112–6.114
cause of action, meaning of, 6.100, 6.101, 6.103
collateral warranties, under, 9.14–9.21, 9.92–9.99
deliberate concealment, 6.105–6.110
meaning of, 6.96
periods of, 6.97, 6.98
liquidated damages, *see* damages, liquidated and ascertained

materials
  see excluded materials *and* collateral
    warranties
Mathurin, Claude, 11.31
MCWa/F-JCT, 7.55, 9.29, 9.71, 9.84, 9.90,
    10.1, 10.29–10.32
  clause 1, 10.126–10.134
  clause 2, 9.53, 10.135
  clauses 3 to 7, 10.136–10.138
  clause 8, 10.139
  clause 9, 10.140–10.142
  clause 11, 10.142
  clause 13, 10.143
  clause 14, 10.144
  clause 15, 10.145
  clause 16, 10.146
  enabling clause, 7.55
MCWa/F/Scot–Scottish Building Contract
    Committee, 10.43–10.45
MCWa/P&T-JCT, 7.55, 9.29, 9.53, 9.71, 9.90,
    10.33–10.35
MCWa/P&T/Scot – Scottish Building
    Contract Committee, 10.46–10.48
measure of damages
  basic rule, 6.13
  expectation interest, meaning of, 6.42
  meaning of, 6.12
  mitigation, general rule, 6.50
  reasonableness, 6.50, 6.51
  reliance expenditure, meaning of, 6.43
  remoteness of damage, 6.12, 6.19, 6.20

negligence
  assumption of responsibility, 2.32–2.44
  neighbourhood test, 2.4
  reliance, 2.27–2.30, 2.43
negligent mis-statement
  reliance, 2.27, 2.28
net contribution, *see* contribution
NHBC
notices, *see also* assignment, 7.52, 7.53, 7.56,
    7.58, 7.60, 9.58, 9.76–9.78, 9.84,
    9.111–9.112, 10.37, 10.97, 10.99, 10.101,
    10.103–10.106, 10.114–10.115,
    10.117–10.118, 10.137–10.138
novation, 4.51, 5.54, 9.73–9.80
NSC/W, 9.36, 9.104–9.105, 10.19

other solutions, *see also* reform of law
  commercial leases, 11.5–11.12
  european proposals, 11.31–11.32

insurance, 11.17–11.26, 11.29–11.30
standard forms, 11.27–11.28
statutory, 11.2–11.4, 11.33–11.38

periods, limitation, *see also* limitation of
  action
  contracts under seal, 6.97
  exclusion of, by statute, 6.98
  principal contract, 6.100, 6.102,
  simple contracts, 6.97
  time limits, calculation of, 6.100–6.102
professional indemnity insurance
  assignment, collateral warranties, of, 8.31,
    9.84, 10.113, 10.142
  BPF Form CoWa/F, in, 10.110
  changing insurer, 8.42
  collateral warranties, provision for, in,
    8.35–8.41
  cover, 8.5
  disclosure, 8.2–8.3, 8.8–8.14
  economic loss, 8.29–8.30
  endorsement for warranties, in, 8.35–8.41
  exclusion provisions, in, 8.26–8.29
  fitness for purpose, 8.32–8.34
  JCT Form of Warranty MCWa/F, in,
    10.140–10.141
  negotiation of warranties, in relation to,
    10.9–10.11
  non-disclosure, 8.4
  operative clause, 8.15–8.25
  proposal form, 8.2
  renewal, 8.5–8.6
  risk, 8.2
  uberrimae fidei, 8.3
purchaser
  collateral warranties, from whom, 7.40
  contents of collateral warranties, for,
    7.41–7.42
  problems of, 7.36–7.39

reasonable skill and care, *see also* fitness for
    purpose, 5.2–5.16, 7.11, 7.24, 7.30, 8.18,
    8.22, 8.25, 9.5, 9.14, 9.15, 9.31, 9.32, 9.34,
    9.36, 9.39, 9.42, 9.43, 9.52, 9.55, 10.60,
    10.68, 10.83, 10.84, 10.91
reform of law, *see also* other solutions
  Defective Premises Act, 11.33–11.38
  repairing covenants, 7.15–7.26, 11.5–11.12
RIBA, *see* Standard Form of Agreement for
    the Appointment of an Architect
RICS Conditions of Engagement, 9.2, 10.83

rights of third parties, *see* third parties,
    rights of

Scottish Building Contract Committee
    Employer/Subcontractor Warranty,
    10.55–10.58
SCWa/F – JCT, 9.53, 9.90, 10.23, 10.36–10.39
SCWa/F/Scot – Scottish Building Contract
    Committee, 10.23, 10.49–10.51
SCWa/P&T – JCT, 9.53, 9.90, 10.23,
    10.40–10.42
SCWa/P&T/Scot – Scottish Building
    Contract Committee, 10.23, 10.52–10.54
specific performance, 7.53, 10.4
Standard Form of Agreement for the
    Appointment of an Architect, 9.41
    Clause 2.1, 9.32
    copyright, 7.42, 7.48, 9.61–9.65, 10.107,
    10.139
sub-contract
    DOM/2, 9.38
    provision for collateral warranties, in,
    9.113–9.114

tenant
    collateral warranties, from whom,
    7.27–7.32
    contents of collateral warranty, for,
    7.33–7.35

problems of, 7.15–7.26
third parties, rights of
    arbitration, and, 3.38–3.40, 3.59
    'construction contracts' and, 3.41–3.42
    defences for promisor, 3.37
    dispute resolution, 3.39–3.40
    double liability, 3.38
    identity, 3.32–3.33
    Law Commission Consultation Paper No.
    121, 3.9–3.11
    Law Commission Report 242, 3.11–3.14
    privity of contract, 1.3, 3.1–3.14
    promisee, definition of, 3.16
    promisor, definition of, 3.16
    Scotland, in, 3.8, 3.19
    statutory rights, 3.20–3.31
    using statutory provisions, 3.51–3.60
    variation or rescission of contract, 3.34–
    3.36
tort
    relationship with contract, 7.25, 8.8, 8.20,
    8.21–8.22, 8.25, 8.28, 9.4–9.13

warranties, *see* collateral warranties
workmanship, 7.24, 7.27, 7.30, 7.46, 9.21,
    9.44–9.45, 9.50, 9.93, 11.11, 11.24
Works Contract/3 – JCT, 9.36
Wren Insurance Association, The, 8.12,
    10.59–10.66